Steel Metallurgy

Properties, Specifications and Applications

Steel Metallurgy

Properties, Specifications and Applications

S. K. Mandal

Formerly, Director (Scientific Services)
Tata Steel Ltd., Jamshedpur
and
Head of Central Metallurgy and Heat Treatment
Tata Motors Ltd., Jamshedpur

Mc
Graw
Hill
Education

New York   Chicago   San Francisco
Athens   London   Madrid
Mexico City   Milan   New Delhi
Singapore   Sydney   Toronto

1 2 3 4 5 6 7 8 9 0   QVS/QVS   19 18 17 16 15

ISBN 978-0-07-184461-1
MHID 0-07-184461-9

The sponsoring editor for this book was Robert L. Argentieri and the production supervisor was Pamela Pelton. The art director for the cover was Jeff Weeks.

This book was previously published as *Steel Metallurgy: Properties, Specifications, and Applications* by McGraw-Hill Education (India) Private Limited, New Delhi, copyright © 2014.

McGraw-Hill Education books are available at special quantity discounts to use as premiums and sales promotions or for use in corporate training programs. To contact a representative, please visit the Contact Us page at www.mhprofessional.com.

This book is printed on acid-free paper.

To

my wife, Chhanda, and my grandson, Nikhil
and
my colleagues, friends and members of metallurgical fraternity at
Tata Motors Ltd. (1969–1989) and Tata Steel Ltd., Jamshedpur (1990–1999)

CONTENTS

PREFACE

Saying goes that *Steel makes the world*; if, at times, steel is not the material that you are using, it is steel made tools and tackles that you need to use to shape or work the other materials. Steel is the most widely used man-made material in this world, second only to cement by volume. Uses of steel will continue to grow with the growth of world economy to support infrastructural growth, manufacturing growth, and growth of social living standards and consumption. Therefore, steel is the concern of all; it touches everybody's life—from a pin that one uses daily to the house in which one lives, the car one drives, domestic appliances one uses, the bridges that one crosses. To cater to such wide and diversified requirements of the society, steelmaking, steel processing and steel-based manufacturing occupies the centre stage of our economy. It evokes the interest of a large cross-section of engineers and engineering students, technicians, entrepreneurs and the society, especially for its uses, utility and applications.

There are numerous books and voluminous literature on steel, covering wide spectrum of steel technology either in general or in a very specialised and comprehensive manner. But, so vast is this literature and so wide is their scholastic coverage that beginners in the field—like the engineering students, non-metallurgical engineers, technicians, entrepreneurs, consumers and others interested in steel—are often clueless or confused to take the call of direction from these books, especially when it comes to the point of selection and application of steels. Engineers and users of steels have traditionally depended on the national and international 'Standards' for selecting steels. There seem to be a few books, at least in the knowledge of the author, that cover the subject of steel metallurgy from its application points of view. This book on *Steel Metallurgy: Properties, Specifications and Applications* attempts to fill that gap—by taking recourse to discussions and illustrations of the relevant parts of applied steel metallurgy in a lucid and simple manner, but keeping intact its focus on application of steels. The book discusses the applied metallurgical parts of steel metallurgy—avoiding discussion on complex thermodynamics based principles of steel-making and steel transformation—for making the book easy to read and understand for the beginners in engineering and metallurgical disciplines. In sum, it is a book of facts on characteristics, properties, factors controlling the properties, specifications, heat treatment and application of steels.

What this book offers is a refreshing change in the way we learn about steel and its applications; it presents a simplified, but highly illustrated narration of 'applied steel metallurgy' with its focus firmly on application of steels. The book is tailored to help students and readers of engineering and metallurgical disciplines to shape-up their knowledge, and build further on it, for appropriate, efficient and effective uses and application of steels.

The book can be conceptually divided into four parts: First, the book introduces and familiarises the readers with some basic metallurgical facts about steel and its characteristics, including the special properties that make steel unique and as most sought-after engineering materials. Then, the book covers the most important aspects of steel metallurgy and its applications, covering the ways microstructures in steel form and can be tailored for developing specific properties for various kinds of uses and applications.

Since steelmaking and steel rolling are at the root of building the quality characteristics in steels for their successful uses and applications—important and fundamental features of steelmaking and rolling processes have been discussed in this part of the book. The book then goes on to highlight the different types of properties of steel and the necessity of its testing and evaluation as an integral part of the process for steel selection and application.

Second, the book discusses the classifications, specifications and properties of steels in a more quantitative manner, based on popular standards and standard-based data, in order to open up the horizon of different steel grades and their merits and properties for selection and applications. Focus of this part of the book is to provide grade specific information and discuss the need-based properties of structural, engineering and stainless steels for efficient selection and effective application.

Third part discusses heat treatment and welding of steels, because heat treatment is at the centre of tailoring the steel properties for exact applications. Various heat treatment methods and their purposes have been highlighted for providing basic understanding of how the steels could be treated, when necessary, for developing the desired properties. This part of the book also highlights some basic aspects of welding and welding precautions in steels, which follow similar metallurgical principles as in the case of heat treatment of steels and has an important bearing on the successful application of steels.

The fourth part of the book dwells on the application of steels; discussing the totality of steel applications from the point of view of reliability and component integrity. This part discusses the importance of cost and quality optimisation in its applications, and emphasises the criticality of design and manufacturing quality for prevention of failures. Various examples of field failures and their prevention have been illustrated and discussed. Finally, this part lays the road map for steel selection and discusses the methods of selection and application of steels through number of illustrations and case studies.

The focus of the book is on the 'application of steels', developed by sequentially discussing the subject matter and using many end-use specific illustrations. The application orientation of the book has been specially adopted for signifying the fact that "the best way to understand and appreciate steel is by understanding its application part, in accordance with the saying—'Begin with the end in mind' ".

The grades of steel that have been discussed in this book include—structural grade steels, engineering and heat treatable grade steels, forming (press forming) grade steels, and some popular grades of stainless steels. In sum, the book has been designed to provide all necessary information and practice based knowledge about steel characteristics, steel properties, steel grades, and steel applications for enabling metallurgical students, metallurgists and non-metallurgical engineers for selecting, processing and using steels with right understanding and for the right purposes. Coverage and mode of presentation in the book is also expected to help, the metallurgical undergraduate students to become 'industry ready'.

To render the book easy to use, an elaborate *index* of contents has been provided for cross-reference as well as a *Bibliography* for further reading or reference. An attempt has been made to make the book user-friendly and maximise its utility by providing *Glossary of important metallurgical terms* at the end. However, the scope of this book is not to act as a substitute of books on 'Steel Treaties' or 'Steel Hand Books'—it is a book that discusses and disseminates the applied metallurgical knowledge required for easy-learning about steels, their properties, specifications, heat treatment and applications. The book has been designed for dual purpose—one to help the students of metallurgy and other engineering disciplines to understand the applied and functional-basics of steels relating to their properties, specification and applications; and the other—to help and support the efforts of engineers and technical personnel in industries for the right selection of steel for the right purposes by providing workable knowledge on steel metallurgy and steel specifications. It is hoped that students of all engineering disciplines, and personnel in industries, businessmen and professionals dealing with steel processing and its uses, will find this book handy and helpful for understanding the functional-basics of steel metallurgy for selection and application of steels.

Disclaimer

It is difficult now-a-days to think, discuss, or write something worthy of knowledge without reference or recourse to the internet world. This book is no exception to this trend. In the course of reference to internet based articles and books, some materials and illustrative figures for this book have been drawn from the internet sources by avoiding any possible infringement of copyrights to the best of my knowledge. In case any material resembles copyrighted publications, it is unintentional and the author should not be held responsible for the same, especially keeping in view the widespread availability of such or similar internet materials and information in the open public domain and their intended use for spreading and enhancing knowledge in the related field.

S. K. MANDAL

Acknowledgements

In the process of writing this book, I have often drawn upon the reference materials from different books that I was familiar with and read during the course of my professional career. These books have been mentioned under the corresponding chapters as 'reference books' along with others that I consulted while writing this book. I am deeply indebted to the authors of these books for their contribution in shaping and refreshing my knowledge on steel metallurgy with their excellent coverage.

During the course of writing this book, I had to frequently refer to various internet based resources, literature and articles for reconfirmation of subject matters, data and reference diagrams. These sources of supply of illustrative information and diagrams, and knowledge, have been both very valuable and indispensable. I wish to thankfully acknowledge the contribution of all the authors and contributors of articles I referred to. I apologise for not individually naming them here, as the list will run into pages.

I also wish to thank all my family members and friends who had been a source of great support in this daunting endeavour. I wish to specially mention the names of my wife, Chhanda, daughter-in-law, Sahila, and grandson, Nikhil, for their love and care that inspired me to go through this project. Finally, I thank my sons—Dhruva and Shuva—for their encouragement and support for my academic pursuit, post retirement from the Tata Steel Ltd., Jamshedpur, India. I owe a lot to all of them.

Dr. S. K. Mandal

CHAPTER 1

Introduction to Steel
Metallurgical Characteristics and Properties

Purpose of this chapter is to lay the foundation for understanding about steel—as structural and engineering material—and to discuss the characteristics of steel, its uniqueness, and its ability to offer different engineering properties based on microstructures (including grain size and inclusions). The chapter highlights different types of properties (including fabrication properties) and their dependence on microstructure, which is the primary source of strength and other mechanical properties in steel. Finally, the character and nature of cast iron group of materials (a close follower of steel for structural and engineering applications) have been outlined and their applications *vis-à-vis* steels have been briefly mentioned.

1.1. Introduction

The world of metallic materials is divided into two classes: Ferrous and Non-ferrous. Steel belongs to the ferrous class. The term *ferrous* is derived from Latin—meaning 'containing iron'. Ferrous material is, therefore, mainly iron based; often containing iron of over 90%, excepting for cases of some high alloy steels where combined alloy can go up to 30% or more, leaving iron content in the material below 75%. A popular example of the latter is the 18Cr-8Ni austenitic stainless steel.

Non-ferrous materials are the other metallic materials, which do not contain any appreciable amount of iron i.e. not iron-based. Iron can be present in non-ferrous material only as incidental metal or alloyed in small amount for improving some property. Examples of non-ferrous materials are copper, zinc, gold, silver or alloys based on these metals. Example of non-ferrous material containing iron as incidental metal is the common brass where iron is present as impurities, coming from the raw-material for melting. Example of iron in non-ferrous material as intentional additive is inconel—an 80Ni-14Cr-6Fe alloy for high temperature applications.

Ferrous materials include pure iron, wrought iron, cast iron and steel. Steel can be plain-carbon steel or alloy steel. All these materials are iron-based, but differ in terms of purity, composition, properties and applications. Because of their compositional differences, their characteristics and properties are also different; steel having the most desirable and sought after mechanical and physical properties amongst them. Cast irons come second from the point of view of industrial applications. Steel and cast irons broadly differ from each other with respect to chemical composition, microstructures and mechanical properties. While further discussions in this chapter will focus on characteristics and speciality of steels, the distinguishing features of cast irons *vis-à-vis* steels will be mentioned at the end of this chapter.

1.2. Introduction to Steel, Steel Types and Grouping of Steels

1.2.1. Crystal Structure and Phase Formation in Steel

Metallurgically, steel is a solid solution of iron (Fe) and carbon (C) where carbon atoms—much smaller than iron atoms—occupy the 'interstitial' position of the iron crystal lattice. The crystal structure consists of atoms that are grouped in an orderly manner, called lattice or lattice structure. In steel, there are two main types of crystal lattice structures which characterises their properties—*α-steel, called ferrite*, which is having body centred cubic (BCC) structure arrangement, and *γ-steel, called austenite* which is having face centred cubic (FCC) structure arrangement. Figure 1-1 shows the schematic arrangement of atomic position in these two crystal structure types.

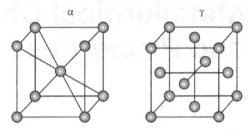

Figure 1-1. The atomic arrangement of BCC (α) and FCC (γ) crystal structures.

When other elements are added or present in the steel, e.g. elements like carbon (C) or manganese (Mn), they take up a preferential position within the respective crystal structure as per their atomic size, thereby forming a 'solid solution' (It is called solid solution, because the exchange or introduction of new atoms take place in solid state). There are strict thermodynamic rules for formation of 'solid solution' but, most importantly it depends on the atomic size and crystal structure of the solvent (parent atom) and solute atoms. When the atomic size between solvent and solute atoms differs by less than 15%, substitutional solid solution can form, as shown in Fig. 1-2A. When the solute atom is even smaller and can fill the interstices of the solvent atoms, an *interstitial solid solution* form, as shown in Fig. 1-2B.

Formation of solid solution by alloying of two or more elements is, however, limited by the solid solubility of the elements concerned, which varies with temperature. For example, carbon, a solute interstitial element in Fe-C alloy system, can dissolve in ferrite iron lattice (α-lattice) upto 0.02% at temperature of about 700°C. When solubility of a solute element in interstitial solid solution exceeds the limit set by thermodynamic conditions, then a separate phase or an intermediate compound forms as per their chemical affinity. For example, in the case of steel, when carbon in ferrite structure exceeds

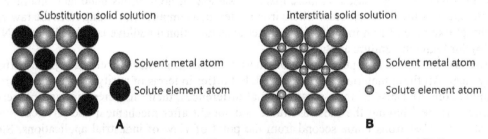

Figure 1-2. A: Illustration of the formation of substitutional solid solution and B: the interstitial solid solution. *Source: Internet source—subtech.com.*

0.02%, excess carbon appears in the microstructure as 'carbide' (Fe_3C)—i.e. an intermediate / intermetallic compound of iron and carbon that contains 6.67%C in it. This gives rise to a microstructure of the steel that contains a mixture of two phases—ferrite and carbide, on cooling.

Solid solution formation in the structure by alloying elements creates local stress field in the lattice, contributing to the strengthening of steel. At the same time, intermediate compound that forms due to excess interstitial element contributes to the strength of steel by precipitation hardening, which is a well-known strengthening mechanism in steel. Thus, strength that is observed in steel is the result of its microstructure, which can be further changed or modified by added processes, like the application of heat and/or deformation.

In steel, alloying elements with smaller atomic diameter, like carbon, nitrogen, titanium, etc. form interstitial solid solution, and most other alloying elements, like the silicon (Si), manganese (Mn), chromium (Cr), nickel (Ni), etc. which have similar or near similar atomic size as the iron, form substitutional solid solution. Solid solution so formed in steel causes increase of mechanical strength, electrical resistivity and decrease in plasticity in the steel. Therefore, by taking note of these changes, steel composition (i.e. alloying addition) can be appropriately designed for maximum gain in strength and other properties. However, solid solubility of alloying elements is not only crystal structure dependent; it is also temperature and composition dependent due to thermodynamic factors controlling the stability of a phase. This situation leads to phase changes/phase transformation in an alloy system—in this case steel—with the change of temperature and composition, which are at the centre of steel metallurgy and are discussed in detail in Chapter 2.

Thus, steel is basically a solid solution based alloy of iron and carbon, where other alloying elements can be added for favourable change in phase formation/structure, leading to different strengthening mechanisms or for change in any specific property, such as oxidation or corrosion resistance of steel.

1.2.2. Definition of Steel and the Steel Types

From the discussions under Section 1.2.1., it can be said that steel is an alloy of iron and carbon to which other *alloying elements*—like Ni, Mn, copper (Cu), Cr, V, etc., can be intentionally added, whenever required, for developing some user-specific properties through microstructural development. When steel is made without any intentional addition of alloying elements, it is called *Plain Carbon Steel* (PC Steel). When intentional alloying is done, it is called *Alloy Steel*.

Generally, %carbon (%C) in steels range from 0.02% to 1.0% with the exception of some wear-resistant steels where %C can go up to or little over 2.0%C in order to produce some amount of *carbide*—a hard and wear resistant phase in the microstructure. If %C far exceeds this higher limits (≥ 2.0%), there would be possibility of excess carbon precipitating as free carbon—called graphite—due to presence of Si or similar graphite forming agent, changing the nomenclature of the material from steel to *cast iron*. This feature of excess carbon in the form of graphite is the critical distinction between the steel and cast iron. *In steels, excess carbon form* carbide *in the microstructure, but in cast irons excess carbon is present as* graphite *in the microstructure, excepting in white cast iron where carbon is present as carbide due to chilling effect.*

Steels are broadly grouped into *plain carbon* (PC) and *alloy steels*. Again, amongst these groups there are a few sub-groups, each of which is meant to serve some user specific purpose at most economical cost and means. Depending on the end use purposes, steels could be made in the grades of either low-carbon, medium carbon and high carbon or low-alloy steel and high-alloy steels. Plain carbon steel made with no intentional alloying can still have few other associated elements like silicon, sulphur, manganese, etc., in limited quantity, which are not intentional addition but incidental to steelmaking process. Steel that is made with intentional alloying (e.g. alloying with Cr, Ni and molybdenum (Mo), etc.) is called *alloy steel*. Here again, if alloy content (total alloying) is low, it is called *low-alloy steel* and if total alloying is high, it is called *high-alloy steel*; general demarcation line between low-alloy and high alloy is variously quoted as 4.0%, 5.0% and 8.0% total metallic alloying. Of these figures, 5.0% seems to be more realistic as it covers majority of low-alloy steel grades under various standards.

Though *plain carbon steels* are made with no intentional addition of alloying elements, some amount of Mn and Si is always present, arising from steelmaking process; though Mn at time is deliberately maintained in the PC steel at or below 0.50% for strength. Similarly, there would be the presence of some small amount of sulphur (S) and phosphorus (P) in PC steel arising from the steelmaking process. If the levels of these elements are below a specified limit, these would not be considered as intentional addition or alloying. Limits of these incidental or residual elements in a grade of steel vary from one standard to the other, but generally Mn above 0.65% and Si above 0.50% are considered intentional alloying; below which they should be considered 'incidental' to steelmaking process. Similarly, the limit for sulphur and phosphorus is below 0.05%. However, respective 'standards of steels'—governing the grade or applications—are the final guide about these permissible limits.

Incidental or residual elements can also be present in alloy steels, but in traces only so that they do not effectively alter the property of the chosen alloy steel. These residuals elements, mostly in traces, come into the steel from the raw-materials and additions used for steelmaking, which contain sulphur and phosphorus as well, along with other residual elements. Hence, the presence of residual elements should be ignored while defining the steel type, such as if the steel is plain carbon or alloy steel. From this standpoint, steel can be said to be—an alloy of iron and carbon with or without any intentional alloy addition and with the possibility of presence of some residuals elements like: silicon, manganese, sulphur, phosphorus etc. that are unavoidable due to economic limitation of steelmaking processes. This situation means that the dividing line between plain carbon and alloy steels is not the presence or absence of certain elements but the level of those elements.

Making *alloy steel* is more difficult than *plain carbon steel*, because of the complexities of recovering and adjusting alloying elements from the ferro-alloy additions. In alloy steels, range of alloy percentage of each element has to be strictly maintained in the finished steel, in order to make the steel composition appropriate for heat treatment. This task becomes difficult because alloying elements are susceptible to rapid oxidation and loss at higher temperature, where steelmaking is undertaken. Hence, making alloy steel requires much more precise control and, at times, additional 'secondary steelmaking facilities'—like use of 'ladle furnace' to control the composition, temperature, and other steelmaking parameters. This makes alloy steel making process more elaborate and expensive. These aspects of steelmaking have been discussed in more details in Chapter 4 on steelmaking and rolling.

1.2.3. Grouping of Steels

For specification and application purpose, steels are grouped and sub-grouped into number of types, as per their composition, applications or properties. For example, *PC steel grades* can be broadly grouped as follows in the order of carbon levels, though the exact limit of carbon for each of these groups may vary a bit in practice.

- *Low-carbon steel*, where carbon content is generally limited to 0.15%, but can go upto 0.25%,
- *Medium carbon steel,* containing carbon ranging from 0.20% to 0.60%, and
- *High-carbon steel*, where carbon content ranges from 0.60% to 1.0%.

Under this grouping, there could be further sub-grouping, namely *extra-low carbon steel, ultra-high carbon steel,* etc.

The other form of grouping of steels is as per shape or applications, irrespective of its carbon content or composition. Examples of shape wise grouping are: sheets, plates, bars, blooms, billets, beams, etc. Examples of application wise grouping are:

1. *Sheet steels,* which generally contain %C below 0.10% for sheet metal applications requiring good cold formability and weldability; the primary requirements of sheet steels. Steels requiring high formability often contain extra-low carbon (≤ 0.05%), which can go down even to below 0.02% carbon. The latter type is made by special steelmaking process and such steels are called *Interstial-free steels* (IF steel), which are used for extra deep drawing sheet metal applications—like the modern automobile body parts.
2. *Wire-rod grade steels* where %C is generally below 0.15% for good cold drawability, but carbon in these steels can be even higher for special applications like spring wire (%C ranges from 0.60% to 0.85%).
3. *Structural steels* with carbon ranging from above 0.10% to 0.25%, which are mostly used for structural applications—like the *reinforced steel bars* for building constructions or *beams* for factory construction or *plate steels* for vessels and container manufacturing that require some tensile strength as well as weldability. At time, %C in structural steel can go above 0.25%, for cases where, increased strength is required in the structural member for specific applications.
4. *Constructional grade plain carbon steels* (also called *engineering steels* for its applications) with higher carbon, ranging from about 0.15% to 0.80%, for uses in engineering constructions—like machine parts, gears, shafts, torsion bars etc. These steels are used mostly after some kind of heat treatment for developing the required properties.

The preceding section outlines different types of steels under plain carbon group; exact composition or carbon range for these steels is covered under different 'Standards' and 'Specifications', which have been discussed in more detail later in the book in Chapters 6 and 7.

Alloy steels grades, where %C generally ranges from 0.10% to upto 2.0% depending on the grade and end uses, contain some specially chosen alloying elements like chromium, manganese (above 0.65%), nickel, molybdenum, niobium, copper etc. for developing some end-use specific properties e.g. high strength in combination with good toughness, corrosion resistance, creep resistance, high temperature strength, wear resistance, etc. Based on alloy contents, alloy steels can be grouped into: (a) *Low-alloy steel*—where total alloy content is limited to less than 5.0%, and (b) *high-alloy steel*—where alloy contents are higher than 5.0%; exact limit is somewhat arbitrary as mentioned earlier.

Examples of popular low-alloy steels are: SAE 5140, SAE 4140, 817 M40 (En-24), 534 A99 (En-31), etc. and the examples of high alloy steels are: 18Cr/8Ni austenitic stainless steel, various martensitic stainless steels, heat resistant steels etc. Alloy steels can also be grouped as per their properties or dominant chemical constituents, for example: (a) Hardening grade steel, (b) Free-machining steel, (c) Stainless steel, (d) Heat resistant steel, etc. for the property related grades and Chromium steels, Manganese steel, Nickel steel, Tungsten steel, etc. for indicating the dominant chemical constituent.

Purpose of expensive alloying addition to steels is to impart specific properties, irrespective of whether the steel is low-alloy or high-alloy steel. Hence, alloy steels must be produced strictly maintaining the specified composition of the steel. For example, stainless steel composition must be appropriate for resistance to corrosion for applications in various corrosive environments; heat resistance grade must be able to withstand the applicable heat without oxidation or creep; and composition of heat treatable grade steels must have sufficient hardenability to ensure martensitic microstructure after heat treatment. More about hardenability and design of alloy steel composition and grades for heat treatment, which require some basic understanding of metallurgical principles for their response to heat treatment and microstructure formation would be discussed separately later in the book. These conditions demand that alloy steels are made with close control of compositions.

In sum, whatever could be the type of steel, composition is the most dominant factor that helps in achieving the end-use properties. Hence, composition of the steel is a primary consideration for classification and standardisation of steels. More about steel classification, standardisation, specification and properties—based on composition and end—uses has been discussed later in this book (Chapters 6 and 7) with focus on steel properties and applications.

1.3. Importance and Uniqueness of Steel

Steel is the second largest man-made material; second only to cement. Steel has been known to the societies for centuries, and has been closely associated with the progress of civilization through the ages. Currently, development of our economy and the resultant high standards of living would not have been possible but for steel. Steel has established itself by far as the most important multi-functional and highly adoptable material. Its production cost is low compared to other materials, it is less energy consuming and most competitive, when its *high strength to weight ratio* is considered *vis-à-vis* cost. Steel is eco-friendly material; steel scraps are 100% recyclable and steel plant by-products are largely re-usable for gainful purpose like in the manufacturing of cement and fertilisers. Steel is said to be the material whose applications are not limited by choice, but by imaginative ability of our mind. Principal reasons for dominant position of steel as structural and engineering construction material are:

- Properties of steels—like strength, toughness, machinability, weldability, formability, corrosion resistance, fatigue resistance, creep resistance etc., and can be precisely developed by suitable choice of composition, alloying, thermal-processing and heat treatment;
- Excellent cold formability and hot deformation characteristics for ease of shaping, forming and fabrication—which can be developed by controlling carbon and impurities in the steels along with control of microstructure, temper and texture;
- Excellent heat treatment characteristics to respond to wide varieties of microstructural requirements for wide ranging engineering applications,
- Good weldability and surface engineering possibility, which can be gainfully used for complex fabrication and difficult to form component manufacturing, requiring special surface properties;
- Good machinability, which can be further improved by appropriate chemical addition of Sulphur, Tellurium etc.;
- Can be made to withstand extreme corrosive and hot environment by special alloying, coating, surface treatment etc.; and
- Very high familiarity, availability (to exact quality) and recyclability of steels.

The foregoing characteristics of steels combined with superior physical properties like: *Elastic modulus* (E), *Shear modulus* (G) make the steels most designer friendly material. The popularity and uniqueness of steels has led to extensive research for further improvement of steel technology, quality and steel usages. Steel also offers distinct advantage with regard to *Life-cycle cost* (LCC) compared to other materials for similar usage. Life-cycle cost comprises of: (a) purchasing cost, (b) cost of shaping, forming, fabricating, heat treatment, assembly etc., (c) cost of maintenance and repair during usage, (d) cost of disposal, scrapping and re-cycling, and (e) cost of deleterious influence and effect on environment.

Life-cycle cost is a helpful tool to decide which material is the most suitable for a given application over its span of life. The calculation of LCC relies upon the concept of the 'time value of money'. It uses the economic evaluation method of discounted cash flow, to reduce all the costs involved in LCC calculation to the present value. The present day value represents the amount of money which would have to be invested today in order to meet all the future operating costs—including running costs, maintenance, replacement and production loss through downtime and environmental penalty. When these are added to the initial costs to arrive at the total LCC, a real comparison can be made of the options available and the potential long-term benefits of using steels *vis-à-vis* other materials. Steel stands out over other materials in this respect. Because of LCC advantage and high strength to weight ratio, steel is by far the most preferred material for mass production of automobile parts and engines, car body and frames, white goods manufacturing, containers, ship-building, railways, machine-tool manufacturing etc. The list of steel usages can be endless, depending on the human skills and creativity.

To add to this competitive advantage of steels, new steels are emerging every year with improved strength, cleanliness and excellent cold formability and fabrication properties. Allied technologies like the heat treatments, cold-rolling, surface coating, surface engineering, in one hand, and promotion of value-added application of steels, on the other, are helping the growth of steel *vis-à-vis* other materials. Combined together, steels are working miracles in facilitating developments in automobile technology, ultra-lightweight car-body manufacturing, rocket engineering, chemical processing, and in many other trying and difficult engineering applications. Steel is, therefore, not only a dominant and important material at present, but also most indispensable material. Steel will certainly continue to remain popular and versatile material in the future. No other material offers the combined advantages of steel with regard to properties, manoeuvrability, availability, disposability, and LCC. Hence, steel occupies the first place in the choice of materials for wide varieties of applications—especially in structural, engineering and construction applications.

1.4. Introduction to Properties of Steels: Property Types and Influencing Factors

1.4.1. Types of Properties of Steels

Properties of steels can be grouped into: (a) Chemical properties, (b) Physical properties and (c) Mechanical properties. Very often, the term physical property is also used to describe mechanical property, but there are some differences between them; basic physical property of materials like the 'Young's modulus', 'Shear modulus' etc. is characteristic of the physics of the material (such as crystal structure, density, etc.) and do not vary appreciably with the composition or microstructure, whereas mechanical properties—like strength, toughness, etc.—are strongly composition and microstructure dependent. Amongst the properties of steels, physical properties are nearly constant for most applications, but other two properties, namely chemical and mechanical properties can be tailored to suit any specific applications: chemical property by appropriate alloy addition; and mechanical property by changing the composition, heat treatment or the state of mechanical working (cold-rolling, drawing, etc.) of steels. These different types of properties are all important for the application of steels, but mechanical properties get the most attention due to their wide implications on the behaviour of steels in forming, manufacturing and applications.

Chemical properties are related to the chemical composition of the steel. There are some special alloying elements that impart the required chemical properties—like the corrosion and oxidation resistance; degree of which depends on the chemical potential of the element. Such elements are copper, chromium, nickel, etc. For instance, there are number of stainless steels that find applications for corrosion resistance, but their corrosion resistance potential is dependent on the exact chemical composition of the steel e.g. 18-8 Chrome-Nickel stainless steel has much superior corrosion resistance than 12% Chrome ferritic stainless steel. Similarly, copper adds to oxidation resistance to the steel—even when added in small amount. These properties—like corrosion resistance, oxidation resistance, weather resistance etc.—are related to the chemical composition and not on the microstructure of the steel, *per se*; hence they are called *chemical properties*.

There are some basic *physical properties*—like Young's modulus or modulus of elasticity (E), Shear modulus (G), and Poisson's ratio (v), which are related to physics of the metal and their crystal structures; they are more or less constant for all steels. They do not vary appreciably with composition or microstructure in steel. Because of their independent character, there are formulae to calculate or correlate them for application purpose. Some examples are as follows:

Young's modulus, $E = \sigma/e$, where σ is stress and e is strain within the limit of proportionality between stress and strain. Commonly, 210 GPa is taken as the E value of structural steels for most design purposes (where 1 GPa = 10^9 N/m^2) and 130 GPa is taken for cast iron. Shear modulus, G, and Bulk modulus (K) can be calculated by using the Young's modulus, E and Poisson's ratio, v, which are taken as 210 GPA and 0.29, respectively, for steels. Formula for calculating Shear modulus, G and Bulk modulus, K in steel using E and v are:

$$G = \frac{1}{2}\frac{E}{1+v} \quad \text{and} \quad K = \frac{1}{2}\frac{E}{1-v}$$

Mechanical properties, on the other hand, are not constant like the physical properties; they vary with the microstructures, which, in turn, can vary due to the chemical composition of the steel, the state of mechanical work done on the steel or the heat treatment meted out to the steel. All three factors exert their influence in changing or producing a characteristic microstructure in steel and, thereby, influencing the mechanical properties. Important mechanical properties of steels include: hardness, tensile strength, toughness, elongation, impact strength, fatigue strengthand wear resistance strength, etc. These properties are greatly influenced by the nature of microstructures in the steel matrix, which, in turn, depend on the chemistry and heat treatment that the steel is subjected to.

1.4.2. Factors Influencing the Properties of Steels

Properties of steels, as discussed in the preceding section, are primarily dependent on two factors: composition and microstructure. However, in steel, they are interrelated; composition influences the formation of microstructure. Chemical composition has pronounced influence on the microstructure due to its influence on phase formation and microstructure generation, which has been elaborated in Chapter 2. Due to the influence of composition on microstructure and in turn microstructure's influence on properties of steel, chemical composition is of considerable importance.

The influence of composition on the properties of steels can arise due to different strengthening mechanisms in steel such as: solid solution strengthening, precipitation hardening, heat treatment or

Table 1-1. Summary of steel strengthening processes and their general metallurgical features.

Steel Strengthening Process	Dependence on Chemistry	Degree of Influence*
Solid solution strengthening	Strong to moderate; depends on the size of solute atom diameter	Moderate
Precipitation/dispersion strengthening	Strong; must have elements promoting precipitation/ dispersion after treatment	High
Work hardening	Indirect	High; increases with degree of working and temperature; lower temperature of working and higher deformation add more strength
Grain refinement	Indirect (e.g. Al-killing of steel for grain size control) Exception: micro-alloying by Ti, V or Nb for grain refinement	Moderate; strengthening increases rapidly with grain refinement above ASTM size 10 and above
Heat treatment (Bulk)	Strong	Very high; widely used for achieving wide ranging properties
Heat treatment (Surface)	Strong	Very high; used for wear and fatigue resistance

*Relative to martensitic strengthening.

work-hardening. Some of these methods are listed in Table 1-1 for illustration. All these methods directly or indirectly depend on the composition of the steel in order to bring about specific changes in the microstructure, and, thereby, influence the properties of steels.

While all mechanisms mentioned in Table 1-1 can influence the properties of steels, heat treatment has by far the strongest influence on altering or tailoring the microstructure of steels for strength. Heat treatment is the process involving uses of controlled heating and cooling, for effecting change in microstructure and properties in steel. Some examples of such heat treatments are: annealing for softening; normalising for grain refinement and contribution to strength; hardening and tempering to increase strength and toughness by producing martensitic microstructure in steels. The latter is also used for various surface hardening processes for localised hardening of steel surfaces for fatigue resistance, wear resistance etc. The uniqueness of steel lies in its ability to respond to various strengthening mechanisms, especially to different heat treatment processes, for developing required properties.

Properties of steels are, in fact, products of different microstructures or microstructural combinations in steel. Mechanical properties—such as hardness, tensile strength, elongation, impact strength, ductility, toughness, etc.—in steel can be developed by choosing the right combination of composition, hot/cold working, and/or heat treatment. Therefore, most important criteria in the selection and applications of steels for mechanical properties are the choice of chemical composition, which can be then suitably treated/heat treated for the development and control of appropriate microstructures. Because of microstructure dependence of steel properties, steel is called a *structure sensitive material*. This means that to get the right type of properties, focus of steel selection has to be—the basic composition of the steel and the process of developing right microstructure. Right chemistry with right type of heat

Table 1-2. Different properties of steels and their controlling factors.

Property Types	Properties	Controlling Factors	Remarks
Physical property (Intrinsic physical property)	Young's modulus, E Shear modulus, G Bulk modulus, K Poisson's ratio, v	Crystal structure	Generally taken as constant in steels for all practical purposes
Chemical property	Corrosion resistance Oxidation resistance Pitting resistance	Composition Alloying elements having favourable electro-chemical behaviour Residual stresses in the steel body	Elements like Cu, Cr and Ni are known to increase corrosion and oxidation resistance
Mechanical property	Hardness, Tensile strength, UTS Yield strength, YS, Elongation, Ductility, Toughness, Impact strength, etc.	Composition and Micro-structure of the steel Also, influenced by Grain size and Inclusion	Changes with heat-treatment, cold working etc.

treatment (including cooling/heating cycle) is the key to the development of desired microstructures for a given application, and this is at the core of applied steel metallurgy (refer Chapter 2).

Though mechanical properties of steels are of prime importance for engineering application of steels, there are other types of properties in steels, each with their characteristic controlling factors. Table 1-2 lists out such properties and their controlling factors.

Other than foregoing types of classical properties, there are some application related properties of steels as well—for example: fabrication or forming property, welding property, wear resistance, heat resistance, etc. Out of these application related properties, fabrication property is very strongly microstructure dependent (e.g. volume fraction of ferrite and pearlite, grain size, inclusions, etc.) and welding property is dependent on the chemical composition of the steel namely carbon equivalent (CE); a factor that is calculated based on carbon and other elements present in the steel. Wear and heat resistance properties of steel are also microstructure dependent, but those microstructure formations calls for presence of certain special alloying elements (e.g. Cr and Mo for wear resistance and W and Mo for heat resistance) to give the characteristic attributes to the structure.

Therefore, major factors that contribute or control the strength and other properties of steels are:

• The *chemical composition of the steel*, which can influence the chemical property by virtue of the chemical characteristics of elements present, as well as the mechanical properties by way of facilitating development of different microstructures upon cooling or heat treatment e.g. ferrite and pearlite or martensite and bainite etc., and

• *Microstructure of the steel*, which could be changed or developed as per requirement by appropriate hot working, cold working and heat treatment for a given composition of the steel.

In effect, any process that changes the microstructural characteristics of steel could influence the steel properties. In this regard, the strengthening modes and mechanism of steel listed in Table 1-1 are important sources of strength and properties. These strengthening processes influence the strength of steel by bringing about changes in microstructure, which may involve change in the crystal lattice structure or precipitation or change in the shape and size of grains and other microstructural fibers. Example of strengthening mechanism involving change in the crystal lattice structure, is the solid solution strengthening of steel.

Solid solution strengthening of steel is particularly relevant for structural steels, where the process is an important source for imparting strength in the steel. Structural steels are seldom heat treated for strengthening. Hence, structural steels generally depend on strengthening by interstitial and substitutional solid solution plus strength arising from microstructural phases. *Solid-solution-strengthening* and associated modification of crystal lattice structures are physical phenomena, but dependent on the chemistry. An important example of this is the strengthening effect observed in low-carbon steels with increasing carbon and manganese. This effect is due to solid solution strengthening effect of carbon and manganese atoms, where carbon results in interstitial strengthening and manganese results in substitutional strengthening.

Any element that has some solid solubility (i.e. atoms that can take position in the solid crystal lattice of iron) contribute to solid-solution strengthening. There can be two types of solid solution strengthening in steel—one by the atoms having similar or near similar sizes as iron atoms e.g. Mn, Si, Ni, etc. and the other, by smaller atoms than iron atoms e.g. C, N, Ti, etc. However, strengthening by interstitial atoms (e.g. C, N, Ti, etc.) is more effective in increasing the yield strength of steel than the substitutional atoms, like Mn, Si, Cr, etc.

Composition of the steel and mechanism of strengthening or heat treating the steel, should be carefully planned to meet the processing and/or end-use requirements (e.g. forming of steel parts and/or end application). For example, for processing by cold forming, steel requires good formability, which can be achieved by micro-alloying of low-carbon steel, using V, Ti or Nb, for interstitial strengthening as well as precipitation hardening. Such micro-alloyed steels can be thermo-mechanically control-rolled and cooled during hot rolling, causing grain size refinement and precipitation hardening. Due to fine grain size and precipitation, such steels are known to give high yield strength with high elongation. This type of steels are known as high strength low-alloy steels (HSLA steel) and are used widely for high strength structural applications. Thus, by appropriate choice of composition and thermo-mechanical controlled rolling, steel can be made good for both critical forming as well as high strength structural applications. But, if steel is for load bearing component, such as automobile shaft which has to carry some dynamic load in application, the steel composition should be so selected, as to correctly respond to heat treatment, for development of strong and tough microstructure. Strength and toughness required for such dynamic load bearing applications are higher than that HSLA steel can offer. Hence, for the latter application, composition of the steel must have sufficient carbon and other alloying elements (e.g. Cr, Mn, Ni, Mo, etc.) that improve hardenability of the steel and produces a combination of microstructures in the steel that is optimum in strength and toughness.

1.4.3. Factors Influencing the Choice of Steels

Thus, at the core of steel application is the understanding of structure-property relationship in steels. The preceding two examples of steel selection may illustrate this point. In these two cases, it may seem that the means of achieving the end results are choosing the right chemistry, but the end results actually get accomplished through change of microstructures. The *HSLA steel* gets its properties from precipitation hardening and grain size effect, and the *shaft steel* gets its strength from hardening to martensitic microstructure. What these situations imply is that mechanical properties of steels are highly structure sensitive. Therefore, unlike for chemical property (e.g. oxidation or corrosion resistance), mechanical properties of steel will depend on the microstructure it can develop by appropriate treatment, be it controlled-rolling or heat treatment.

Hence, for mechanical properties, the selection of steel has to be based on the consideration of what microstructure can be economically developed in the steel, so that the required properties—either forming property or the strength related property—can be obtained. And, this is the key to the selection of steel for a given application. The choice between low-carbon and HSLA steel for forming properties or choice between plain carbon and alloy steel for load-bearing applications is, by and large, guided by such structure–property relationship considerations. As such, criterion for selection of steel, either from processing or application point of view, should be examined from this structure-property angle.

For forming and bending applications, steel should have a softer microstructure with lower strength and higher elongation, and for that matter, choice of low-carbon steels is essential. However, at times, forming and stretching property obtainable from low carbon steel by standard rolling practices may not be adequate. Hence, planning could be made to roll this type of steel under controlled manner for inducing higher forming and stretching property by alteration of grain structure and texture, e.g. cold-rolled and temper passed for improved deep drawability during forming.

Similarly, for load bearing applications or applications requiring minimum strength, what is important is not the chemistry but the microstructure, that has sufficient strength and ductility; where chemistry of the steel is a means to the strength. Such steel should have appropriate microstructure that imparts the required strength. If the microstructural requirement is (ferrite + pearlite) for moderately load bearing applications, selected steel could be simply a medium carbon steel, like En-8 (080M45) or its equivalent grades from other standard e.g. SAE 1045 or 45C8, and used after normalising for uniform properties. But, for higher strength, the choice could be between right quality plain-carbon steel and an economic grade of low-alloy steel which can respond to suitable heat treatment for developing martensite or (martensite + bainite) structure, as per the chemistry of the steel. Chemistry of chosen steel should be adequate for the required structure with economy in alloy addition.

Martensite provides the highest strength to steel and next comes bainite; though bainite is often associated with a martensitic structure in alloy steels, due to slack cooling at some spots in the steel body during quenching or fast cooling (see Chapter 2 for more information). And, when martensite structure is suitably tempered, toughness of the steel increases while strength gets moderately dropped. More about the influence of different microstructures on the mechanical properties of steels has been discussed in the next chapter.

Sometimes, steels are also required to have good *wear resistance property*, *heat resistance property* etc. These are again influenced by microstructures, but the nature of chemistry of steel also plays a dominant role. Presence of higher carbon and some hard carbide forming elements—like Cr, W, V, etc.—in the chemistry of the steel results in high wear and heat resistant characteristics. This phenomenon is an example of how combining *physical* and *chemical state* of the steel, certain properties can be developed in the steel. Adjusting chemical composition for high temperature oxidation and scaling resistance, as well as to form stable and hard complex carbides, is said to be the solution to heat resisting steels. For wear resistance, chemistry should be so controlled as to produce a very hard structure with uniformly dispersed fine and stable hard carbide particles. Microstructural improvements in these steels are generally brought about by careful hot-working (e.g. hot forging) and heat treatment in order to make the carbides finer in size and physically more uniform in distribution. Thus, composition and heat treatment of steels can help in developing different application specific properties for industrial applications.

Heat treatment is a versatile tool for imparting desirable properties to steels. Purpose of heat treatment in steel is to alter the initial microstructures (including grain size) to a desirable form and state that can impart the required properties for end-uses and applications. This, however, requires careful choice of *alloy addition to steel*. With appropriate carbon and alloying combination in steel, the steel can be tailor-made to respond to particular sets of mechanical properties through heat treatment and, thereby, making the steel fit for critical applications, which few other materials can do successfully and economically. Hence, *heat treatment* takes the centre stage in making the steels most useful material in the world. Since heat-treatment is so critical to the steels, and the process is very widely used for applications in the manufacturing of engineering components, heat treatment of steels has been discussed in more details in Chapter 8.

To sum up, steels like any other metals can have three types of basic properties: physical, chemical and the mechanical property. All of them have some role in the application and uses of steel, but mechanical properties are the most decisive one for most applications, excepting applications involving corrosion and oxidation in service. For corrosion and oxidation resistance, especially alloyed steels containing Cu, Cr, Ni, etc. are used, and these properties are not appreciably microstructure sensitive (see Table 1-3). Other two properties which are chemistry dependent are heat-resistance and welding. Heat resistance property of steel is dependent on special chemistry for high temperature oxidation resistance and formation of alloy carbides that are stable at higher temperatures. Welding of steel is dependent on the CE of the steel, which controls the tendency for formation of martensite in the heat affected zone (HAZ). But mechanical properties—like strength, ductility, toughness, fatigue, creep, low-temperature properties, wear etc.—are strongly microstructure sensitive, including the presence of inclusions and grain sizes. Formation of microstructures and their influence on mechanical properties has been discussed in Chapter 2, and influence of grain size and inclusion on the properties of steels has been discussed in Chapter 3.

1.5. Fabrication Properties of Steels

For applications, steels are to be formed into parts and components, and used under a specific situation, e.g. for load-bearing members, protection from chemical environment, working under temperature etc. End-use application of steel calls for judicious selection of steel based on required physical, chemical, or mechanical properties. But, to start with, steel has to be formed, shaped, or fabricated to parts and components as per design. This demands certain fabrication properties, for example:

- Formability for forming under dies and presses;
- Bendability for bending and shaping to some contour;
- Forgability for forging and shaping the parts;
- Machinability for machining the component to design; and
- Weldability for jointing different steel parts in the process of fabricating a constructional or engineering part etc.

In general, fabrication properties of steels are also microstructure dependent, as in the case of mechanical properties. Chemical composition, which is fundamental to steel, influences the fabrication property through the resultant microstructures that we get, after the hot-working, cold-working or

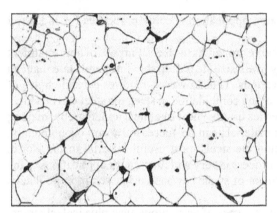

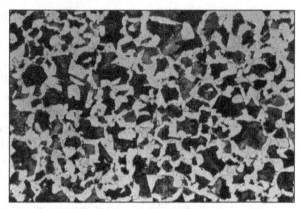

Figure 1-3. Essentially ferritic structure of low-C sheet steel—with patches of pearlite (dark phase).

Figure 1-4. A mixture of ferrite and pearlite microstructure in 0.4%C steel.

heat-treating. For example, very low carbon steel gives softer microstructure after rolling, which is essentially ferrite or with very low pearlite content (see Fig. 1-3). Such steels will exhibit good bending and drawing properties, facilitating higher draw and stretch under die-tool formation. The cold formability by drawing and stretching can be further improved by cold-rolling and annealing of such steels. If the chemical composition is enriched with higher carbon or alloying, then the steel will be harder with more pearlite volume fraction, making the steel less formable. But, such harder steels are preferred for machining (see Fig. 1-4). *Machinability* of low carbon steel improves with pearlite content, because in harder steels chips that form during machining break easily and avoid tool tip wear. More about machinability of steels will be discussed in a subsequent chapter.

All mechanical properties of steel, including fabrication property, are strongly structure dependent. Variation of strength in low and higher carbon steel arises from its microstructural difference. Figures 1-3 and 1-4 illustrate microstructure of a very low carbon steel (Fig. 1-3) and a medium carbon steel (Fig. 1-4), showing distinct difference in their microstructures, especially with respect to pearlite content. While ferrite (light phase) is very soft, pearlite (darker phase) is harder; much harder than ferrite. Hence, for cold forming and bending, low carbon ferritic steel with lower strength and higher elongation is preferred. But for machining, softer steel will pose problem; hence, a medium carbon ferrite-pearlite structure, as shown in Fig. 1-4, is preferred.

Steels with higher volume of pearlite content might have higher strength but much lower elongation/ductility than the lower carbon steel. As a consequence, steels with higher pearlite contents are difficult to cold form where ductility/elongation of steel is a critical criterion. Therefore, for cold forming and bending, low-carbon steel with very low pearlite content is preferred, provided the strength requirement of the component permits use of such lower strength steel. If strength requirement for a cold formed part is higher, then the options are: (a) to use steel with appropriate carbon content with lower ductility, if the application permits or, (b) use controlled rolled HSLA steel where a combination of higher strength and ductility can be obtained. However, the former type steels with lower ductility for cold forming can be given an additional annealing/normalising treatment for improving the ductility to some extent.

Pearlitic steels cause difficulties in cold forming due to crack formation while bending or forming from the interface of hard pearlite colony and soft ferrite in the matrix; cracks can also originate from the brittle carbide phase in the pearlite colony. Hence, for *sheet-metal forming* using dies and tools or for *low-carbon wire rod* drawing, steel should have microstructure with low pearlite. With progressive increase of pearlite with increasing carbon, cold forming of steel becomes difficult, making the steel unsuitable for bending and stretching. This is the reason why low-carbon steel is recommended for sheet-metal forming. If steel of higher strength (i.e. of higher carbon) has to be used for the end-application, steel can be HSLA type or can be given a separate softening heat treatment (e.g. annealing or normalising) for improved ductility than as cold-rolled steel.

Now-a-days, advancement of steel making and rolling technology has made possible controlled rolling of steel in the rolling mill itself, in order to get suitable strength as well as ductility for forming applications. Example of this is the production of HSLA steel, which is suitably micro-alloyed and control rolled within a narrow range of intermediate temperature, to produce fine grained ferritic structure with higher strength and ductility. Such steels are stronger with higher yield strength and better elongation. Forming by bending and stretching is an important part of the application of steels—under the category of sheet-metal forming—where mostly low-carbon cold-rolled steels are used. With developing steelmaking technology, steels can be made with extra-low carbon and in a state of high cleanliness (i.e. freedom from inclusions), which can be appropriately cold-rolled to give higher forming textures that help in severe stretching and forming. Demands for such steels are coming from automobile industry, where stronger and thinner steel sheets for manufacturing of complex shapes are increasing.

The other area of steel uses is in the manufacturing of engineering components. Most engineering components—like gears, cams, shafts, levers, etc.—are required to carry fluctuating or dynamic load where higher strength, hardness and toughness are demanded. Such parts are generally formed by *forging* and *machining,* and thereafter, subjecting these to a suitable heat treatment. Forging is a process of hot shaping steels under hammer or press using appropriate dies at higher temperatures. Currently, warm or cold forging is becoming more favoured due to energy saving and cleaner surface of the jobs. Forging or forgabilty requires hot-ductility of steel for good flow of material inside the die cavity. This can be ensured in steels by controlling the chemistry—to make it free from harmful tramp elements like tin, zinc, lead, etc. which are low-melting metal and can cause embrittlement of the steel—and the inclusion level in the steel. Machining, on the other hand, is a cold cutting process using suitable machine tools, where steel part is cut to finish shape as per design with appropriate surface finish. Machining requires not only easiness to cut steel, but also easiness in chip removal from the cutting interface of the job, for better surface finish and tool life. It is in this context, carbon content in the steel or pearlite content in the microstructure becomes important for machining. Harder microstructure, such as the pearlitic steel or hardened and tempered steel, is known to give better machining property—with some limits of higher hardness—in terms of surface finish due to easy cheap breaking characteristic. Thus, in general, either by composition adjustment and/or heat treatment, most steels can be rendered suitable for good forming, machining and fabrication.

The other features in steel microstructure that considerably influences the mechanical and forming properties of steels are the *grain size* and *inclusion*. Grains are a physically distinguishable microstructural feature in steels that get revealed when a steel surface is polished and chemically etched. What we see on the polished surface under magnification is dark outlines of two-dimensional areas, but actually grains are a three dimensional feature, with two dimensions seen on the flat surface and the other dimension is inside the body. Examination of Fig. 1-3 will reveal such two-dimensional ferrite

Table 1-3. Different microstructural features that favourably influence the corresponding fabrication process.

Forming Types →	Cold Forming by Bending	Cold Forming by Drawing and Stretching	Machining	Welding
Microstructural features Favourable for forming	Softer micro Finer grains Lower inclusions	Extra soft micro Fine to extra fine grain sizes Very low inclusions	Mixed micro Moderate to higher strength Not so fine grain sizes Lower inclusions or special type of tool lubricating inclusions e.g. sulphides	Lower carbon or lower carbon equivalent Fine to medium grain size Low inclusions

grain boundaries in a microstructure. Grain boundaries generally act as obstacle to plastic flow of metal while under deformation, meaning that presence of more number of grain boundaries will give higher deformation resistance, i.e. higher strength.

Mathematically, finer the grain size, higher would be the number of grain boundaries that will be observed in a microstructure. Hence, fine grains are a source of higher strength and such grains would naturally pose a problem in machining in terms of cutting force. Therefore, for machining, coarser grained steels are preferred. However, finer grain sizes exhibit higher elongation during deformation and, as such, they favour better cold forming and bending properties. Inclusions, on the other hand, are generally deleterious for most forming and machining operations, except for softer sulphide inclusion which is beneficial for machining. Harder inclusions are bad for machining, because these inclusions can cause faster wear of tool tips; and it is also bad for forming because of the possibility of fracture starting from the inclusion to metal interface. These are some general aspects of grain size and inclusion; more details about their formation and influence will be discussed later in Chapter 3.

Thus, to sum up, fabrication properties of steels are influenced by the microstructure, which includes micro-phases, grain size and inclusion content (size and distribution). Table 1-3 provide general guidelines about favourable factors that influence different fabrication properties, including welding, which is dependent on the CE of the steel; lower the CE better is the weldability, in general.

1.6. Influence of Microstructures on the Properties of Steels

It would now be useful to introduce how various steel microstructural constituents influence different applications and fabrication properties of steels. And, in this context, it should be appreciated that versatility of steel comes from the fact that different microstructural features can be introduced in steels by adopting different kinds of heat treatments. *Heat treatment refers to the process involving heating to and/or cooling from a set temperature at a pre-determined rate; the rate could be slow cooling for softening, fast cooling for hardening or controlled cooling at an intermediate rate for developing some special microstructure for an end application.* The latter can be accomplished even during rolling of the steel in a rolling mill—called the *controlled rolling and cooling*. A number of heat treatment processes are available to steel users, and they will be discussed in greater detail in the relevant chapter on 'heat treatment'. In general, heat treatment operations are directed towards producing a desirable set of microstructures that can fit the needs of intended applications. Thus, at the centre of steel application is the fact that steels can be appropriately rolled or heat treated to develop a set of desirable microstructure that gives it the unique forming or end-use related properties, which no other material can offer at a comparative economic rate.

However, getting appropriate properties, either for forming or service specific end uses, would necessitate careful balancing of few features in the microstructure, which can exhibit any one or more of following microstructural phases in some combination and proportion. These microstructural phases are: *Austenite, Ferrite, Pearlite, Martensite* (generally tempered martensite), *Bainite* and *Carbide*. Formation of these phases has been discussed in Chapter 2. In general, all of these are not thermodynamic phases, but can be generally described as microstructural phases with distinct individual characteristics that form during heating/cooling of steels; such as either during rolling, forging, or during heat treatment. The qualitative effects of each of these microstructural constituents (including grain size and inclusions) on various mechanical and chemical properties of steel are indicated in Table 1-4.

The table indicates that *inclusion* is an important parameter for determining certain properties of steels such as—*fatigue and impact strengths, cold forming*, and *ductility and toughness*. Inclusion content in steel plays vital role in the successful forming or machining of the steel as well. All good quality cold forming steels that are used for severe cold bending or drawing e.g. automobile frame parts

Table 1-4. Influence of microstructural constituents on properties of steels.

Microstructural constituents	Srength Tensile	Strength Impact	Strength Fatigue	Ductility/ Toughness (D)	(T)	Cold Forming	Weldability	Machining	Corrosion
Austenite	↓	↑ ↑	↓	↑	↑	↑	↓	↓	↑
Ferrite	↓ ↓	↑	↓ ↓	↑	—	↑ ↑	↑ ↑	—	—
Pearlite	↑	↓	↑	↓	—	↓	↓	↑	—
Bainite	↑ ↑	↓	↑		↑	↓	↓	↓ ↓	—
Martensite*	↑ ↑ ↑	↑	↑ ↑		↑ ↑	↓	↓	†	—
Carbide	↑ ↑	↓ ↓	↓	↓ ↓	↓ ↓	↓ ↓	↓ ↓ ↓	↓ ↓ ↓	—
Grain size	↑	↑	↑	↑	↑	↑	↑	↓	↑
Inclusions		↓	↓ ↓	↓	↓	↓		↓	—

* Martensite referred here relates to tempered martensite

† On tempering

Note:

1. Note adverse effect of fine grain size on machining.

2. Toughness is not same as ductility. Toughness is indicated by the energy absorbed on fracture, whereas ductility is represented by % elongation of the steel at fracture. Toughness is generally referred with respect to heat treated steels where toughness is developed in the structure by appropriate tempering. Ductility is considered for cold forming operations.

3. Austenite as a phase in steel is predominant in stainless steel grades and in some high-alloy steels where austenite is stable at room temperature.

and body part manufacturing, call for very low inclusion content; and that too in finer and globular size and in uniform dispersion. Because, inclusions as such are harmful to steel in many respects, but more so when present in coarser and irregular sizes. In general, inclusions in cold forming or drawing steels give rise to cracks or tears during forming or pressing. Similarly, inclusions in steels for engineering applications that are subjected to sudden or fluctuating load in services may give rise to poor *impact* and *fatigue strength*, respectively. Hence, inclusion content in steel is an important feature of steel quality specification and should be limited to as low a level as possible. More on the influence of inclusions and effect of grain size on steel properties has been discussed in Chapter 3 and in discussions on steelmaking and steel specifications.

Table 1-4 indicates broad influence of the microstructural phase/constituents as per their individual characteristics, which may differ when these microstructural phases/constituents are present in combination with others in different shape, sizes, proportion and distribution. Actual effects will, thus, depend on:

- Chemical composition and the volume fraction of different phases (microstructural phases) present in the steel;
- Location and distribution of these phases;
- Nature of the bainite and quality of tempering of martensite;
- Size and uniformity of grains;
- Chemistry of the inclusions (e.g. oxides, silicates, sulphides, etc.), inclusion size and shape, and distribution.

Thus, controlling steel properties by manipulating microstructure and microstructural constituents is a tricky task. For getting the optimum combination of forming and application of related properties, steel composition, steelmaking route and steel rolling process, as well as heat treatment process, if involved, have to be carefully chosen and executed. In general, *mechanical properties of steels*—like the strength, toughness, elongation, ductility etc. which are related to forming and final behaviour of steel—are directly dependent on the microstructure, grain size and cleanliness (freedom from harmful inclusions). *Chemical properties*—like the corrosion and oxidation resistance (the latter property also influences the heat resistance quality)—are dependent on the presence of certain alloying elements, like Cu, Cr, Ni, Mo etc. in the steel. *Welding,* another important fabrication technique in steels is dependent on the carbon equivalent of the steel and the inclusion content. By controlling these parameters and basic internal quality of steels as regards their internal soundness, all application related properties of steels can be very well controlled and optimised for most applications. More about the influence of microstructure and microstructure types will be discussed Chapter 2.

1.7. Difference between Steel and Cast Iron: Structure, Properties and Applications

Introduction about steel would not be complete without discussing about the cast irons and their characteristics *vis-à-vis* steel; because cast irons are close competitors of steels for many applications and it is also a structure sensitive material. Structurally, there are many similarities between steel and cast iron, but there are differences too. Hence, examination of these two materials for their metallurgical character and properties are necessary.

1.7.1. Microstructure and Composition of Cast Irons

Like steel, microstructures of cast irons could be either ferritic or pearlitic or bainitic or martensitic and a combination of them. The only difference is the presence of free carbon as 'graphite' in cast irons (except in white cast iron where it is carbide), whereas any extra carbon in steel, other than required to form different crystal phases, is present in combined form, called 'carbide' e.g. Fe_3C or alloy carbide. Like steel, cast iron is also an alloy of Fe and C, but carbon content in cast irons is much higher than that in steels—generally ranging from 2.0%C to 4.3%C in cast irons; see Fe-C diagram in Fig. 1-5, which indicates at the bottom of the diagram about respective carbon range of steel and cast iron.

Figure 1-5 indicates that iron-carbon diagram has two distinct parts—steel and cast iron. Cast iron part starts at 2.0% carbon and can go over 4.30% as per this diagram. Generally, cast iron composition extends upto 4.3% carbon, the 'eutectic' carbon composition—but in special cases carbon in cast iron can exceed 4.3% level (see extreme right hand part of the diagram) with cementite and ledeburite, instead of graphite, in the microstructure. These are special wear resistant grade alloy cast irons under the category 'white irons'.

The Fig. 1-5 also indicates that in the pure Fe-C system—where composition is not having any other element or alloying—cast iron, even with carbon below or upto 4.3% (eutectic composition) will

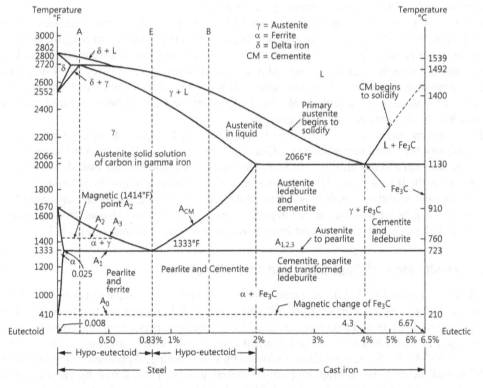

Figure 1-5. Fe-C diagram showing the phase details of steel and iron part with change in carbon content and temperature.

decompose to 'cementite, pearlite and ladeburite' under slow equilibrium cooling; see the microstructure indicated in the cast iron portion of the diagram. *Ladeburite* is a mottled like eutectic structure in cast iron containing cementite and pearlite. If the cast iron composition contains sufficient silicon—which is known to promote graphite—then graphite, instead of cementite or ladeburite, will form upon equilibrium cooling or slow cooling. Thus, for the production of graphitic cast iron, addition of some Si is essential. Therefore, cast iron types can be of *grey cast iron*—containing 'graphite' as free carbon in the microstructure or *white cast iron*—containing cementite, pearlite and ladeburite, depending on the composition. White cast iron is brittle and cannot be used in industry without further heat treatment, like *malleabilising,* excepting for wear plate and others under compressive forces only.

Silicon, is an essential element in grey cast iron that ranges from 1.0% to 5.0% in grey iron compared to 0.45% in common steels. Silicon, being a strong graphite former,

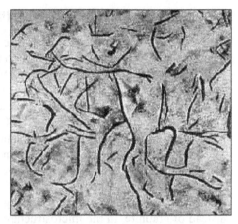

Figure 1-6. Typical grey cast iron microstructure containing flake shaped graphite (dark) in grey matrix of pearlite (magnification: ×100, light acid etch).

promotes precipitation of carbon as 'free carbon' i.e. *graphite*—a free form of carbon, which is porous and very light. Free carbon or graphite occupies large volume in the matrix, making the occupied space in the matrix virtually void. As a result, grey cast irons are weak in tension, because these voids can act like notches inside. The difference in mechanical properties between steel and grey cast iron is primarily due to this difference of character and form of carbon. Shape, size and distribution of graphite in the matrix of grey cast iron structures play a critical role about its property and applicability. Figure 1-6 shows a typical grey cast iron microstructure with flake graphite.

Properties of cast iron family are largely influenced by the form of carbon present and determine their suitability for different applications. *Graphite,* which is virtually a void in the matrix, drastically lowers the strength and ductility of cast iron. On fracture, a graphitic cast iron shows grey to dull grey appearance on the fracture face. Hence, cast iron with carbon as graphite is called *grey cast iron.* Cast iron with cementite and no graphite is called *white iron,* based on fracture face appearance. Cementite in cast iron can be in the form of carbide in pearlite (Fe_3C + ferrite) and also as 'free primary cementite'. A traditional classification chart of cast irons based on structure is shown in Fig. 1-7.

1.7.2. Classification of Cast Irons and their Structures

Classification chart in Fig. 1-7 divides cast irons into two primary groups—grey machinable irons with graphite and white, un-machinable irons with no graphite. White irons are brittle and hard; hence their uses are limited. If white cast iron has to be used for engineering parts involving some tensile strength, it has to be given a *malleablising* treatment. Malleablising involves long annealing cycle in the temperature range of 900°C–950°C for several days, followed by slow cooling. Approach of the process is to cause breakdown of the carbides to graphite (temper carbon graphite) and ferrite by carefully balancing the composition and heat treatment (see Fig. 1-8D). Malleable cast iron calls for presence of silicon and sulphur in the composition for nucleation of graphite (temper carbon) in the structure during annealing. However, due to long annealing cycle and energy cost required for production of malleable castings, the

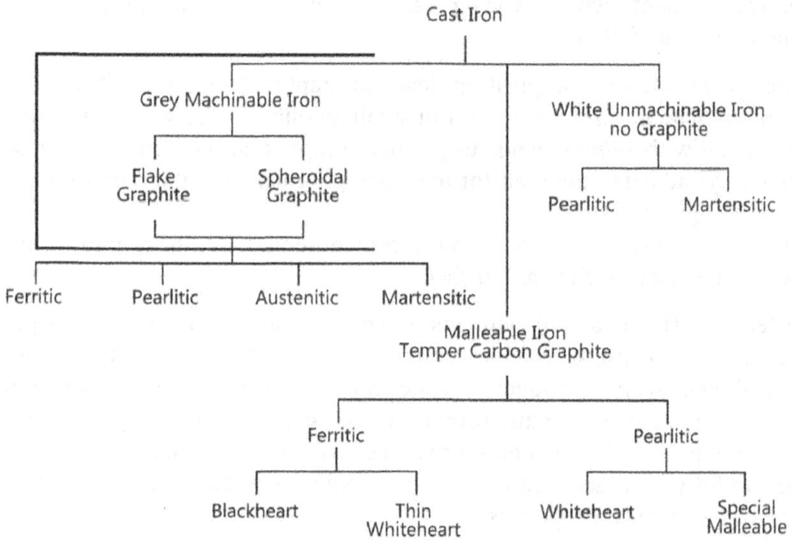

Figure 1-7. A traditional classification of cast iron family based on structure.

process is gradually going out of favour from automobile industries—where malleable castings were earlier used for many housing and casing applications (e.g. rear axle housing, differential housing, gear box housing, etc.). They are being replaced by spheroidal graphite iron castings.

Major usages of cast irons now-a-days are due to the machinable grey iron branch of the classification chart, namely 'flake graphite' and 'spheroidal graphite iron' (SG Iron). SG iron form of cast iron is very popular, because it allows modification of graphite forms and manipulation of matrix microstructures as per need; either by specially treating the liquid metal with some modifier or inoculants, alternatively by following an economical heat treatment process. This part is shown in a box on the left side in Fig. 1-7.

Thus steel and cast iron—which are extensively used in engineering and machine manufacturing industries—differ in their composition, microstructure and properties. Primarily, it is due to the form, shape, size and distribution of carbon-containing phases like graphite and cementite. Steel does not contain any graphite in the matrix, and most part of carbon in steel (other than small %C in the ferrite) are in the combined form of cementite (Fe_3C), which may occur in the form of lamellar, globular or fine precipitates, unless the steel is austenitic at room temperature—like the 18-8 stainless steel. Combined carbon (Fe_3C) in cast irons can, however, be changed by controlling the cooling rate from casting temperature or by separate heat treatment to form lamellar structure (pearlite) or martensite like fine precipitates on ferrite or completely ferritic structure by—expulsion of all carbon from combined carbon to form spheroidal graphite in the matrix of ferrite. Properties of cast irons will, therefore, change accordingly i.e. as per the final structure in the matrix. Major factors influencing the character of the carbon in cast irons include:

- The chemical composition, especially the amount of carbon and silicon;
- The rate of cooling of the casting; this implies that carbon character may vary with section thickness as the cooling rate will vary with that; and
- Presence of 'nuclei of graphite' and other micro-substances or addition of 'graphite modifier'.

Prominent amongst elements which when present in certain quantity promote formation of either graphite or cementite are as follows:

Silicon, when present in the composition, leads to graphite formation. This gets further supported by aluminium, which is mostly present in small amounts in the scraps used for melting.

Manganese, along with *sulphur* tends to promote graphite in the structure. Phosphorus content—which is always added to the melt for increasing fluidity of liquid metal for casting—also helps in formation of graphite.

Boron, chromium and molybdenum promote cementite and tend to promote chilling (a very hard white cementite layer) of the cast surface.

Ultimate effect of different added elements in the cast iron composition will depend on the section thickness of the casting. Casting with wide variation in section thickness may produce variation of the carbon character; thinner section producing white iron containing cementite, while heavier section of the same casting is giving rise to graphite formation, the shape and size of which may again differ with the variation of cooling rate with changing cross section. Hence, casting with widely varying section will have to be carefully moulded and cooled for avoiding structural variations that adversely affect either machining or mechanical properties.

In order to determine what would be the structure of cast iron of a given composition, CE value is worked out by using the formula:

$$CE = \text{Total \%C} + \frac{1}{3}(\%Si + \%P)$$

The CE value is used to determine how close the composition of cast iron is to 'Eutectic carbon' (4.3%) indicated by arrow on composition line in Fig. 1-5. CE value gives indication of how much free graphite would be present—more the carbon towards and upto the eutectic point more will be free graphite, provided the composition contains some silicon.

1.7.3. Properties of Cast Irons

Traditional grey irons are planned with higher CE value so that the casting gives more free graphite in the matrix of ferrite and pearlite for ease of machining. Figure 1-6 shows a grey iron microstructure with the presence of flake shaped graphite in the matrix of lightly etched pearlite. However, such structures are associated with lower strength and virtually no ductility, because flaky graphite acts as stress raisers and facilitate early crack formation from the tips at lower load. However, grey cast iron is generally chosen not for application under tensile strength but application under compressive load or very low tensile load, e.g. machine tool base plate.

Presence of free graphite also improves *damping capacity,* i.e. vibration absorption capacity of the material, which is successfully used for applications relating to foundations and base plates of heavy machineries. In this context, it should be noted that grey iron may be limited for using in tension, but in compression there is no limitation for using grey cast irons. Rather, robust cast iron structures are preferred in foundation and other uses involving compressive forces due to high compressive strength along with high damping capacity of grey iron. Because of these and other advantages of 'free carbon' structure in cast irons and lower cost due to easier manufacturing possibility, methods have been developed now to specially inoculate and heat-treat cast iron to develop steel like microstructures—i.e. structure containing ferrite, pearlite, bainite or martensite, but with free carbon in the form of 'nodular' or rosette like temper carbon structure; see microstructures under Fig. 1-8.

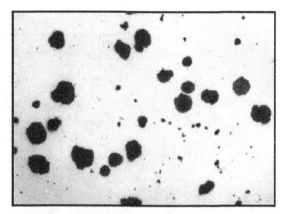

Figure 1-8A. Ferritic ductile iron.

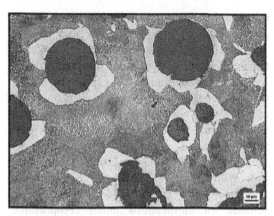

Figure 1-8B. Pearlitic ductile iron, showing bulls-eye type spheroidal graphite.

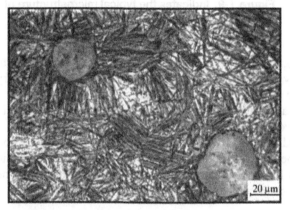

Figure 1-8C. Austempered ductile iron with bainitic structure.

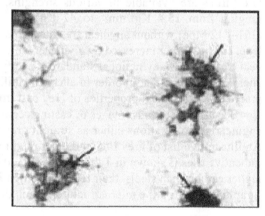

Figure 1-8D. Malleable cast iron with temper carbon in ferrite matrix.

Microstructures in Figs. 1-8A and B are obtained by special inoculation treatment during casting, producing spheroidal shape of the graphite. The structure could vary between 100% ferrite to 100% pearlite, with nodular graphite or with ferrite–pearlite combination in between, depending upon the carbon content. But, for producing structures in Figs. 1-8C and D, the casting requires special heat treatment. While austempered ductile iron (ADI) is produced by austempering alloy cast iron of suitable composition, malleable iron is produced by prolonged malleablising heat-treatment, which converts all graphite into rosette structure, leaving the matrix ferritic.

Castings with flake graphite have good machining property, much better than the steel; because discontinuity of matrix at graphite flake sites act as chip-breakers. Also, the graphite present in the structure lubricates the cutting tool tip, as and when the tip comes in contact with the graphite. Thus, another effect of graphite is to reduce frictional wear as it would act like lubricant to reduce friction. But, flakes also act as notches inside the metal and lead to poor toughness—as would be evident from the fact that flaky grey iron fractures without any perceptible elongation. For this reason, grey cast iron is not suitable for application in tensile mode. If cast iron has to be considered for some applications involving tensile stress, the recommended permissible limit is only ¼th of actual ultimate tensile strength of the cast iron. Figure 1-9 illustrates a pearlitic grey iron at higher magnification.

However, there is no limitation for grey iron with graphite flakes for using under compression. In fact, grey cast iron is most suitable for applications under compressive stresses e.g. machine foundations, housings, etc. where tensile component of stress is marginal. Grey irons are available in various strengths, ranging from 15.4 Kgf/mm^2 to 42 Kgf/mm^2 (151–412 Mpa) without any heat treatment; the level can be further increased by giving suitable

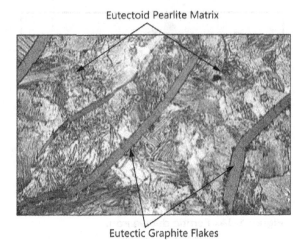

Figure 1-9. Indicates the typical microstructure of pearlitic grey iron showing flake graphite at high magnification (×1000).

heat treatment. Many national standards specify grey irons with strength only, without binding it with chemical composition, in order to allow flexibility in the choice of composition as per specific foundry practice. Some typical properties of grey cast iron as per ASTM A48 standard is shown in Table 1-5.

Because of cost advantages of castings compared to steels, cast iron group of materials are used in engineering applications either as straight grey iron castings or modified or in heat treated condition. An illustrative list of grey iron and other modified cast irons along with their mechanical property and indicative uses is shown in Table 1-6. It should be noted from this table that though cast irons contain higher carbon than steels, their tensile strength is lower than comparable steels. This is because of the 'graphite effect' in the microstructure, which does not contribute to strength.

Table 1-5. Properties of ASTM A48 classes of gray iron.

Grade	Tensile Strength (ksi)	Compressive Strength (ksi)	Tensile Modulus (10^6 psi)
20	22 (151 Mpa)	33 (227 Mpa)	10
30	31 (213 Mpa)	109 (748 Mpa)	14
40	57 (391 Mpa)	140 (961 Mpa)	18
60	62.5 (429 Mpa)	187.5 (1287 Mpa)	21

* 1 ksi = 0.70 kgf/mm^2; 14,223 Psi = 98.1 Mpa.

Table 1-6. Some popular grades of cast irons and their properties, typical composition, treatment and applications.*

Name/ Grade	Nominal Composition	Form/ Condition	Y.S. (0.2% PS) (Kgf/mm²)	Tensile Strength (Kgf/mm²)	Elongation (%)	Hardness (BHN)	Uses
Cast grey iron (ASTM A48)	C 3.4, Si 1.8, Mn 0.5	Cast	—	18	0.5	180	Engine blocks, fly-wheels, gears, machine-tool bases
White Iron	C 3.4, Si 0.7, and Mn 0.6	Cast (as-cast)		—	0	450	Bearing surfaces
Malleable iron (ASTM A47)	C 2.5, Si 1.0, Mn 0.55	Cast (annealed) Special H&T	23	37	12	130	Axle bearings, track wheels, automotive crankshafts
Ductile or Nodular iron (SG Iron)	C 3.4, P 0.1, Mn 0.4, Ni 1.0, Mg 0.06	Cast	24	37	18	170	Gears, cams, crankshafts
Ductile or Nodular iron (ASTM A339) Heat treated	—	Cast (quench and tempered)	72	95	5	310	Special Applications like high duty gears.

* For Metric-English stress conversion, see Appendix D: Physical data and conversion tables.

Table 1-6 lists the tensile strength and elongation of cast irons, which can be compared with each other. The table shows that white iron, which is known for its brittleness, has very high hardness (450 BHN), making it suitable for wear plate and related applications but with zero elongation. The list also gives some examples of uses of different cast irons. Grey iron castings are preferred for uses in compression or lightly loaded parts like fly-wheels, engine block, etc. that involve heavy and accurate machining and no tensile load in applications. If similar structure is required for machining but the component is subjected to some tensile force or fatigue load, cast grades like malleable iron castings or ductile iron castings (with or without alloying) should be considered. Because of high cost of steel forgings, alloy cast irons that can withstand higher tensile or fatigue load have been developed. These castings can be made with special alloying and heat treated like steel—to impart reasonable high tensile strength and toughness for applications involving dynamic stresses (life stress). Suitably alloyed and heat treated cast iron imparts properties—like wear resistance, heat resistance, corrosion resistance,—etc. similar to steels, but at lower cost. These castings are becoming quite popular now-a-days.

To sum up, limitation of grey cast irons comes from its flaky graphite structure, which acts not only as void but also as stress raiser under tensile load. Hence, even though some cast irons can have comparable tensile strength e.g. 35 kgf/mm² or above, they will have very little elongation or toughness. To introduce elongation in the cast irons, the microstructure has to be modified by special heat treatment. White cast irons could be given a special malleablising treatment, which is a long heat treatment cycle that causes graphite to precipitate as *temper carbon* in the matrix. It is an expensive energy consuming operation; hence not much in use now-a-days. Preferred route to induce ductility in cast iron is to go for SG iron, which produces *nodular* graphite when small percentage of magnesium (Mg) is added for inoculation. Nodular graphite structure of suitable composition can be further heat treated by *austempering* process to induce acicular ferritic matrix with structure like lower bainite or tempered martensite, which is very strong. Thus, cast iron group of materials can be tailored to suit many ordinary applications, but the combination of high strength and toughness of steels for safety critical applications cannot be reproduced in cast irons.

The major limitations of cast irons are their lack of adequate toughness and ductility for cold forming, cold working or applications involving high dynamic fatigue load and low temperature applications. However, at time, very high nickel (%Ni ≥ 22) SG iron casting can be considered for cryogenic applications due to fabrication difficulties in steel by forging big parts while casting of such parts is easier and more convenient. High nickel SG iron shows much improved low-temperature toughness than standard SG irons. Some typical microstructures and properties of SG iron have been presented in Fig. 1-8 and Table 1-6.

Because of the importance of SG iron, some more data about SG iron is indicated in Table 1-7 indicates minimum properties that can be expected as per popular British Standard, BS 2789 of 1961. Grades with higher elongation and ferritic matrix will also have some impact properties like steel but not to the same level. Because of the ductility of SG iron, the material is finding increasing applications in low and medium duty crankshafts, which are subjected to fatigue loading in applications. The choice of SG iron casting for crankshaft over steel forging is due to lower cost of casting. Even where higher strength with some degree of toughness is required in service, SG irons are being increasingly used after heat treatment in the same way as hardening of steels. Obviously, strength and elongation of these structures vary considerably; following the Fig. 1-8, structure A will have highest elongation with relatively lower strength, and structure type C showing the maximum strength and minimum elongation. Malleable structure shown in D has similar structure as A, but with rosette type carbon. This structure has similar property like A.

Thus, cast iron family of materials follows similar metallurgical principles for phase transformation as steel, and some of them can be heat treated as steel. But, because of the form of carbon, namely graphite, cast irons have limited strength, ductility and toughness as compared to steels. However, presence of graphite offers some positive advantage for machining, which is an expensive manufacturing

Table 1-7. List of some typical grades and properties of SG iron along with their structure.

SG Iron Grades	Tensile Strength (Kgf/mm^2)	0.5% Permanent Set (Kgf/mm^2)	%Elongation	Matrix
SNG 24/17	37	24	17	Ferrite Impact resistant
SNG 27/12	42	27	12	Mostly ferrite + pearlite
SNG 32/7	50	32	7	Mostly pearlite
SNG 42/2	65	42	2	Pearlitic Can be hardened and tempered
SNG 47/2	75	47	2	Hardened and tempered at about 350°C

operation. With the development of metallurgical understanding and casting technology, these materials can also offer reasonable varieties of properties for different end-uses. It is, therefore, no wonder that the combination of steel and cast iron family of materials, offer the most versatile and economic choice of materials for engineers. There are cast irons of suitable compositions that can be given special heat treatment (e.g. austempering) for developing steel-like properties—higher strength and toughness—extending the scope of application of cast irons, especially the SG irons. Such applications are plenty in industries, especially for low to medium duty gears, crankshafts and similar load bearing applications, where a combination of wear and fatigue strength is required.

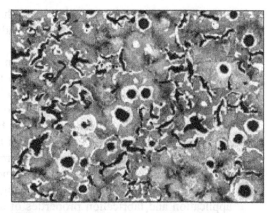

Figure 1-10. An illustrative micrograph of CG iron (at ×100) showing thicker and shorter graphite. The structure is mixed with few 'bulls-eye' type graphite.

There is another variety of grey cast iron that is gaining popularity for applications, requiring either higher strength or lower weight than standard grey iron castings. This type is called 'compacted graphite' cast iron (CG cast iron). Figure 1-10 illustrates the microstructural difference between normal grey iron and CG iron.

The CG iron is also known as 'vermicular graphite iron' due to shape of the graphite, which is shorter, curly and thicker. This results in higher strength due to better adhesion between graphite and iron matrix. CG iron castings are finding increasing uses in automotive brake drums, crankshaft housings/ engine block, turbo housing and engine exhaust gas manifolds.

In a cost competitive market, continuous attempts are being made to upgrade the quality and properties of both cast-irons and steels, which is prompting many new developments in the areas of their industrial applications. Surface corrosion, a big issue in steel applications, is being successfully addressed now by developing cleaner steels with superior surface coating technology. Strength and toughness combination, which has been the traditional limitation of cast irons, is being addressed by developing new heat treatment technology, such as austempering, in order to produce steel-like structures, but retaining some goodness of cast iron technology. Thus, these two parts of ferrous technology—steels and cast irons—are complementing each other for many challenging and cost-effective applications in industries at large.

SUMMARY

1. The chapter defines the nature and character of steel and then introduces the common methods of grouping the steels as per chemistry and uses. Compositional characteristics of plain carbon and alloy steels have been highlighted and the role of composition in achieving the ultimate properties of steels has been emphasised.
2. The uniqueness of steel as an engineering material has been discussed and the principal reasons for its superiority over other materials have been mentioned. Steel is most competitive when it

comes to using material with high strength to weight ratio for industrial applications. In this regard, components of 'life-cycle' cost and the advantage of favourable life cycle cost of steel have been pointed out.

3. Different properties of types of steel, namely physical, chemical and mechanical properties have been discussed and also the importance of mechanical properties and its structure sensitivity has been pointed out. Different properties, their controlling factors and uses have been listed.

4. Since strength of steel is a dominant property for applications, sources of strength in steel and different strengthening mechanisms have been briefly discussed in this chapter.

5. The chapter points out the importance of fabrication properties of steels and separately discusses the aspects of forming and fabrication properties along with illustrations of their dependence on microstructure, including grain size and inclusions.

6. Finally, the chapter deals at greater length, the subject of microstructure dependence of both application and fabrication properties of steels, and lists out the influence of microstructures on such properties. It has been pointed out that uniqueness of steel comes from the fact that properties of steel can be exactly tailored by choice of appropriate chemistry followed by different treatments like rolling, drawing or heat-treating which no other material offers.

7. The chapter also describes and discusses the compositional and microstructural difference between cast-iron group of materials and steel, in order to compare these two materials which often compete with each other for some industrial applications. In this context, development of higher strength SG iron castings and ADI has been pointed out.

FURTHER READINGS

1. Rollason, E.C. *Metallurgy for Engineers*. London: The English Language Book Society and Edward Arnold Ltd., 1973.
2. Reed-Hill, Robert E. *Physical Metallurgy Principles*. New York: Van Nostrand Reinhold Co., 1973.
3. Dieter, George E. *Mechanical Metallurgy*. New York: McGraw-Hill, 1961.
4. Higgins, R.A. *Engineering Metallurgy (Applied Physical Metallurgy)*. New Delhi: Viva Books, 2004.
5. Bhadeshia, H. and Honeycombe, R. *Steels: Microstructures and Properties*. Oxford: Butterworth-Heinemann, New York, 2006.
6. Smallman, R.E. *Physical Metallurgy and Advance Materials*. Oxford: Butterworth-Heinemann, 2007.
7. Raghavan, V. *Physical Metallurgy: Principles and Practice*. New Delhi: PHI Learning, 2006.
8. Boyer, H.E. *Practical Heat Treating*. American Society for Metals. Ohio: ASM International, 1984.

CHAPTER2

Formation of Microstructures in Steel

Structure–Property Relationship

The Purpose of this chapter is to describe and discuss the mechanisms of formation of different kinds of microstructures in steel and their contribution to mechanical properties. Discussions in the chapter centre on the process of decomposition of austenite and its control—for obtaining the right type of structure for required mechanical properties. The chapter discusses with illustrations, the formation of different microstructures and highlights their distinguished features contributing to mechanical properties of steels. Finally, the chapter discusses the details of character and morphology of different structures such as: ferrite, pearlite, bainite and martensite, as part of understanding the structure–property relationships in steel (i.e. variation of properties with structures in steel), which is at the heart of all steel processing and heat treatment mechanisms.

2.1. Introduction

Different microstructures in steels and their qualitative influences have been discussed in the previous chapter. It is evident from the Table 1-4 of Chapter 1 and the subsequent discussions that steel could have a number of microstructural constituents namely: *austenite, ferrite, pearlite, bainite,* and *martensite*. In addition to these products, microstructure of steel also contains *carbides, grains* and *inclusions*. Carbides and inclusions are chemical compounds and their properties are based on their chemistry. Grains are three-dimensional physical volumetric enities (crystals) surrounded by boundaries. They are an integral part of steel body, arising from 'nucleation and growth' process during solidification or transformation of steel. More about grains boundaries and inclusions have been discussed in Chapter 3 and the formation of carbides will be discussed along with the austenite decomposition in this chapter.

Austenite decomposition may produce a number of products—like ferrite, pearlite, bainite and martensite—depending on the cooling rate and the composition of the steel. Of these structures, ferrite and carbide are thermodynamic phases of steel (other than austenite from which they form). This is because true thermodynamic phase is a portion of an alloy system which is physically, chemically and crystallographically homogeneous throughout, which is a condition fulfilled by the austenite, ferrite and carbide only. Other products like pearlite, bainite and martensite (tempered) are, in effect, mixture of ferrite and carbide, and their structural characteristics are determined by cooling condition and composition of the steel. The pattern of their formation and existence determines their influence on the mechanical properties of steels. For example, ferrite and carbide in pearlite exist in lamellar form and inter-lamellar spacing between these two phases determines the strength of the steel. Similarly, bainite is, in effect, a mixture of ferrite and carbide where carbide precipitates on the ferrite plates and boundaries. Strength and toughness of bainite structure also depend on the location and intimacy of carbide precipitates. Similarly, martensite, which is primarily a distorted ferrite structure with super-saturated carbon in it, forms carbide on tempering, making the structure an intimate mixture of acicular ferrite and carbide. Properties of martensitic structure are also influenced by the nature of ferrite and carbides in the tempered structure. Thus, steel can give rise to number of structural combinations with differing structural morphology from its limited number of phases—namely austenite, ferrite and carbide. This is a very unique feature in steel and leads to the possibility of developing different microstructures with different properties in steels.

Different structures in steel can be formed by varying the cooling rate during austenite decomposition and changing the steel composition. By producing different structures and by controlling austenite decomposition, properties of steels can be changed and, thereby, provide the opportunity for tailoring the properties of steels as per requirements of different applications. Aim of this chapter is, therefore, to discuss how different structures can form in steel from austenite decomposition, and highlight their structural and morphological differences leading to different mechanical properties. Structures that get produced from austenite decomposition are ferrite, pearlite, bainite and martensite.

Other micro-constituents in steel microstructure are carbide, grains and inclusions; they will come up for reference and discussions of austenite decomposition. Hence, a very brief introduction of these micro-constituents is necessary here. Carbides form due to chemical reaction between carbon and iron, or some alloying elements having affinity for carbon. But, this requires the steel to have carbon in excess of what ferrite can hold in solid solution, which is 0.025%C at 723°C as per Figure 1-5 of Chapter 1. Since all commercial steels contain carbon above 0.05%, carbides in some quantity are always present in all engineering and constructional grade steels, excepting austenitic stainless steel. For example, carbides are present as iron-carbide (Fe_3C) in all plain carbon steels (with the exception of interstitial-free steel) or as alloy carbide in alloy steels. Iron carbide contains 6.67%C. Carbide can be present as separate phase or associated with another structure; for example, lamellae of carbides in *pearlitic* microstructure or carbide precipitates on ferrite plates and plate boundaries in *bainitic* structure or precipitated carbide (either Fe_3C or alloy carbide in case of alloy steel) in *tempered martensitic* structure.

Grains form due to very basic nature of solidification process by 'nucleation and growth' (N&G) and the *inclusions* form due to unavoidable chemical reactions at the time of high temperature steelmaking, by reactions between gases and the metallic present in the molten steel or due to trapped slag particles. The aspects of grain boundary formation and inclusion formation, and their effects on the properties of steels have been discussed in Chapter 3. This chapter will focus on the microstructure formation, by decomposition of austenite under different cooling conditions from higher temperatures. However, before proceeding with the discussion of austenite decomposition under different cooling conditions, it is necessary to introduce the 'Iron-Carbon' diagram and basic principles of phase separation/microstructure formation.

2.2. Introduction to Iron-Carbon Equilibrium Diagram

A full Fe-C diagram has been already illustrated in Fig. 1-5 of Chapter 1. The diagram represented the full range of iron-carbon system covering both steel and cast iron. This type of diagram facilitates follow-up of solidification as well as phase transformation processes, for a given composition of steel

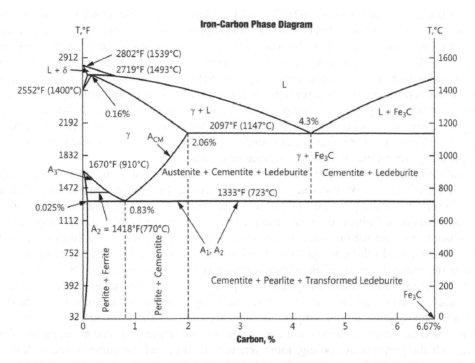

Figure 2-1. Fe-C phase diagram, depicting the equilibrium products at different temperatures. Steel portion of the diagram has been demarcated by thin lines on the left. *Courtesy: Subtech.com.*

and the resultant microstructure. Figure 2-1 depicts a similar diagram as Fig. 1-5, but highlights the steel portion of the diagram on the left and marking up different microstructures that can co-exist at various temperatures and carbon level. The diagram is called 'Iron-carbon phase diagram'.

This type of temperature-composition diagram is called *phase diagram* or *equilibrium diagram,* because the diagram depicts the stages in phase changes and development of microstructure under equilibrium cooling condition. Equilibrium cooling implies slower rate of cooling, where the corresponding phases can react with each other for establishing a 'reversible equilibrium' condition such as:

$$\text{Austenite} \underset{\text{Heating}}{\overset{\text{Cooling}}{\rightleftharpoons}} \text{Ferrite} + \text{Cementite (Fe}_3\text{C)}$$

Though this equilibrium reaction is reversible in cooling and heating, there is a hysteresis in heating and cooling, which is again influenced by the rate of heating or cooling. In fact, this hysteresis effect influences the 'critical temperatures' in steels, namely 'upper critical temperature' (A_3) and 'lower critical temperature' (A_1). As such, these critical temperatures in steel are often symbolised differently, for indicating whether it refers to heating or cooling. For example, A_{C3} and A_{C1} are used to denote critical temperatures on heating and A_{r3} and A_{r1} are used to denote critical temperatures on cooling. Faster the heating and cooling rate, greater is the gap between the A_C and A_r points. This effect is most pronounced on lower critical temperature (A_1). Consequently, the A_1 temperature line for heating and cooling is variously quoted between 695°C to 723°C, depending on the exact cooling rate within the equilibrium cooling conditions—faster cooling rate suppresses the A_1 temperature and *vice-versa*.

The Fe-C phase diagram in Fig. 2-1 shows the presence of four critical temperatures in steel—namely A_1, A_2, A_3 and A_{Cem}, in order of increasing temperature—which are important points for phase changes. Amongst these critical temperatures: A_1—called the lower critical temperature—is independent of composition and nearly constant (except for hysteresis in cooling and heating); all others critical temperatures strongly vary with the composition; A_2 and A_3 sharply decrease with increasing carbon, but A_{Cem} sharply increases with increasing carbon in the steel. For discussions in this book, these critical temperatures referring to phase changes or transformation have been simply and uniformly referred to as A_1, A_2 and A_3 without the notation for heating or cooling.

Crystallisation of liquid steel to solid 'austenite' (γ, face centred cubic crystal structure) starts at or around the peritectic point, marked by arrow at 0.16%C (peritectic composition of steel) in the diagram. Considering a composition of 0.50%C in the steel for illustration (as shown in Fig. 2-1), the entire liquid steel gets solidified to 100% austenite on cooling below 1400°C; see the top part of Fig. 2-1 with reference to the vertical line drawn on 0.5%C composition. This austenitic structure will remain more or less unchanged till the temperature reaches the A_3 line, see Fig. 2-1. A_3 temperature denotes the point in the phase diagram below which austenite is not stable and ferrite (α, body centred cubic crystal structure) formation is thermodynamically preferred. Therefore, upon reaching A_3 temperature line, austenite will start separating out ferrite whose composition (carbon content) will be determined by the corresponding point on the temperature line A_2. Along with the separation of ferrite, austenite composition will be richer in carbon and composition of austenite will change along the sloping line A_3. Thus, with the progress of cooling, more ferrite will form and the composition of ferrite and the remaining austenite will change along A_2 and A_3 lines following the principles of 'lever rule' (see definition of lever rule in Glossary at the end of the book). Finally, on reaching A_1 temperature, the remaining austenite, which is richer in carbon by now due to ferrite separation, will transform in 'eutectoid fashion' to 'pearlite'—a lamellar combination of ferrite and carbide (Fe_3C) stringers. Once the ferrite-pearlite structure has formed from the mother austenite, the structure will remain the same down to room-temperature, i.e. there will not be any further change in the resultant structure. This is because, below A_1 temperature, there is virtually insignificant change in the carbon content of ferrite.

Figure 2-2 shows a ferrite-pearlite micrograph of medium carbon steel after solidification casting, where ferrite is the light area and pearlite is the darker area. Solidified cast structure produced from high temperature austenite is rather coarse and with irregular grains. Depending on the solidification conditions, the structure can have columnar and dendritic pattern of the structure. Because of coarseness and inhomogeneity of the structure, *as-cast structure* is seldom used for industrial applications due to lack

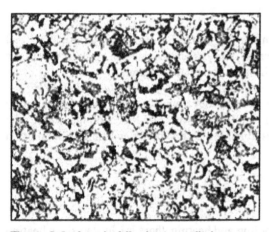

Figure 2-2. A typical (ferrite 1 pearlite) structure obtained after solidification of steel (medium carbon) shown at 3100. The structure pertains to transverse section and indicates presence of feathery ferrite, indicating relatively faster cooling—a phenomenon common during solidification casting.

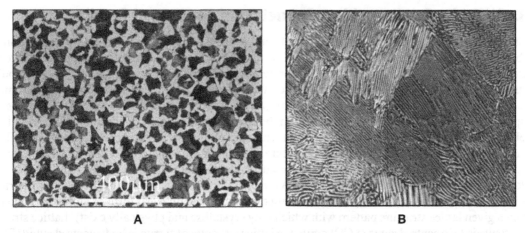

A   B

Figures 2-3A&B. Figure **A** indicates general appearance of normalised (ferrite + pearlite) structure in steel at X200 magnification, and **B** illustrates the dense lamellar structure of pearlite of a near eutectoid steel at high magnification (X2000).

of adequate mechanical properties. The cast structure, therefore, needs to be homogenised and refined by hot-working (e.g. rolling and forging) and heat treatment (e.g. annealing, normalising, etc.) before put to final uses. This is where the importance of phase transformation and structure formation begins in steel technology, and understanding of Fe-C diagram becomes important.

The refinement of as-cast structure not only depends on rolling and forging, but also on the type of rolling/forging and the degree of deformation given to the steel during rolling/forging. The refined structure, having finer and regular grain sizes and more uniform and denser ferrite-pearlite structure, is the starting point for further industrial uses and applications.

Figures 2-3A and B represent forged and normalised structure of medium carbon and high carbon steel, indicating refinement of structure compared to Fig. 2-2. Figure 2-3B shows the magnified view of pearlite colony showing closely spaced lamellae of ferrite and carbide. The figure illustrates that pearlitic structure consists of light ferrite and darker carbide lamellae, closely interspaced with each other, and further demonstrates that pearlite is not a phase—it is a mixture of ferrite and carbide phases in the form of lamellae. Structures shown in Figs. 2-3A and B are the outcome of how steel of a given composition cools from higher temperature.

Therefore, to understand as to what type of microstructure can be expected from steel under a given cooling condition, understanding and interpretation of Fe-C diagram forms the starting point. Figure 2-1 depicts the iron-carbon equilibrium diagram, illustrating the possibility of formation of different phases and structures, while the steel cools under equilibrium condition. This is the basic phase diagram of Fe-C alloy system, providing information about conditions of phase separation and structure formation with respect to composition and cooling. Though the phase diagram refers to equilibrium cooling condition, it can help to understand the obtainable structure and structural modification, that can be expected when cooling departs from exact equilibrium cooling. Thus, the diagram is widely referred for processing of steels, such as by rolling, forging and non-hardening heat treatments (e.g. annealing and normalising). Study of Fe-C diagram, therefore, forms an indispensible part of understanding *austenite decomposition*, especially for ferrite and pearlite formation.

2.3. Austenite Decomposition: Formation of Ferrite and Pearlite

Austenite decomposition for formation of ferrite and pearlite under equilibrium cooling condition can be best studied by reference to Fe-C diagram as shown in Fig. 2-1. When a solid piece of steel is heated from room temperature to austenitisation temperature, the reverse of phase changes on cooling occurs, as discussed in Section 2.2. For example, with reference to Fig. 2-1, austenite starts forming on reheating upon reaching the critical temperature (A_1). Austenite forms on reaching and crossing this temperature at the expense of pearlite present in the steel structure. This signifies that all pearlite in the steel will get converted to austenite, if held for a while just above this A_1 temperature, producing (ferrite + austenite) microstructure. When the temperature is further raised and crosses the A_3 temperature all ferrite present in the steel also gets converted to austenite (see vertical line at 0.5%C in Fig. 2-1). This makes the steel 100% austenite (γ) above the A_3 temperature. Change in steel structure from ferrite to austenite is associated with change in crystal lattice structure (the physical form of atomic arrangement within a given lattice structure pattern with which they crystallise and physically exist). Lattice structure of austenite is face centred cubic (FCC) while the lattice structure of ferrite is body centred cubic (BCC). Solubility of carbon in this FCC structure is much higher than that in BCC structure e.g. austenite can hold up to 2.06%C at higher temperature (see Fig. 2-1) in solution compared to 0.025%C for ferrite at the A_1 temperature. Thus, on heating the steel above A_3 temperature, austenite of corresponding composition (called parent austenite) is obtained, which is 0.50%C in this case, as indicated by vertical line in Fig. 2-1. This parent austenite is then cooled for transformation. Depending on cooling condition and composition, the transformed product could be ferrite, pearlite, bainite, martensite or a combination of them. The process of transformation of austenite to ferrite, pearlite, bainite or martensite is popularly known as 'austenitic decomposition'.

Formation of these phases from austenite depends on the cooling condition. Austenite decomposition can take place under equilibrium cooling (slow cooling) or at a faster cooling rate (accelerated cooling). Slow cooling produces structures like ferrite and pearlite, and accelerated cooling produces bainite and martensite. There can be some variations within these cooling rate bands—slow cooling can be by cooling in closed chamber or in open air, and accelerated cooling can be by isothermal cooling or continuous cooling. Isothermal cooling refers to, initial fast cooling to an intermediate temperature and then holding at that temperature and maintaining the constant temperature. Based on the cooling condition, microstructure or the mix of microstructure produced by the austenitic decomposition also changes. Isothermal cooling gives uniform structure over a cross section, but continuous cooling produces mixed structure under continuously changing cooling conditions. Slow furnace cooling produces coarser structure of ferrite and pearlite (as in annealing) and open air cooling produces finer structure of ferrite and pearlite (as in normalising).

With reference to equilibrium cooling, decomposition of austenite starts just after crossing A_3 temperature (see Fig. 2-1) on cooling. This is because solubility of carbon in austenite increases with lowering of temperature below 910°C, which necessitates formation of another phase which has lower carbon solubility in that phase, in order to provide extra carbon to austenite necessary for its stability with the lowered temperature. Thus, formation of ferrite starts with temperature coming below A_3 temperature of the steel. This decomposition of austenite, partly to ferrite and partly to carbon rich austenite, will continue with lowering of temperature till A_1 temperature has been reached. On reaching A_1 and below, carbon rich austenite in the structure will decompose to *pearlite*—a lamellar structure of alternate layers of ferrite (α) and carbide (Fe_3C); see micrographs in Fig. 2-3B. This decomposition of austenite to pearlite can take place only on crossing A_1 temperature on cooling and is not temperature dependent thereafter, under equilibrium cooling conditions. This is called *eutectoid decomposition* of austenite. On further cooling below A_1, there would not be any significant change in ferrite and carbide distribution in the pearlite, due to very limited solid-solubility of carbon in iron below A_1.

However, faster cooling of the steel than the equilibrium cooling from the austenitisation temperature will produce finer ferrite and pearlite or other products like bainite and martensite, depending on the exact cooling rate and chemistry. These changes are due to change in kinetic factors—like nucleation and growth of phases—that control the decomposition process.

The process of decomposition of austenite is driven by thermodynamical considerations as well as the kinetics of the process. Change of 'free energy' associated with a reaction or phase change at a given temperature is a thermodynamical factor (such as change of austenite to ferrite below A_3), but the growth would be controlled by the kinetic factors like available nucleation sites and nucleation rate. Presence of prior grain boundaries or defects in the structure can influence the nucleation process during phase changes and growth. In fact, this condition that nucleation rate at a given temperature is influenced by the crystallographic defects in the structure is often exploited in thermo-mechanical working and heat treatment of steel for producing fine and strong microstructure.

Generally, thermodynamical condition, such as temperature for free energy change, is a must for any equilibrium phase change to take place. A phase change can only occur, if the free-energy changes associated with that change in a given temperature are favourable for thermodynamic stability of the phase, i.e. to lower the overall energy of the system. But the kinetic factors, such as grain boundaries or dislocations from structural defects or under-cooling—which can facilitate nucleation of the new phase—help or deter the process of phase nucleation and growth by their presence or absence. Thus, finer prior austenite grains or faster cooling of austenite will produce more nuclei and thereby make the transformed structure finer. This situation is utilised in industrial heat treatment practices, such as normalising, for change in the nature of ferrite and pearlite microstructure.

Thus, decomposition of austenite can produce a different morphology of microstructure when cooled differently. This essentially happens because actual decomposition process of austenite is controlled by 'nucleation and growth' (N&G) processes, which are influenced by the cooling rate, parent austenite grain size and various impurities and tramp elements present in steel (not considering now the effect of prior deformation). A generalised picture of N&G process in austenite decomposition is shown in Fig. 2-4, where the influence of prior austenite grain size has been also depicted.

From Fig. 2-4, it can be further concluded that:

- Nuclei for ferrite are ferrite, but nuclei for pearlite are 'carbide' particles that separate out during decomposition;
- Nuclei for bainite are 'ferrite' upon which carbide precipitates;
- At higher temperature nucleation (N) is small and growth (G) is fast;
- At lower temperature, N is large and G is slow; and
- Faster cooling will produce more nucleation (N), requiring lower growth, G.

However, the figure is silent on the nuclei of martensite, which is now believed to be the 'dislocation groups' and 'strained grain boundary plates' that exist in the steel due to prior working or due to strain produced during rapid cooling. But, growth rate of the nuclei for martensite (which forms only at very fast cooling rate) is very high and corresponds nearly to the velocity of sound in the matrix.

Thus, decomposition or transformation products of austenite (such as the morphology of ferrite, pearlite and bainite) are the result of N&G process, which is initiated by the 'free-energy' changes,

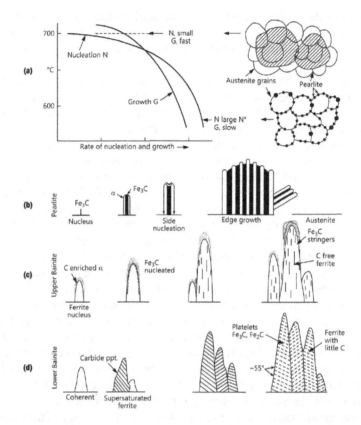

Figure 2-4. Schematic diagram of the nucleation and growth phenomenon in steel. Note the interrelationship between nucleation and growth phenomenon; lesser nucleation means larger growth and more nucleation means less growth. *Source: Rollason, E.C. Metallurgy for Engineers, 1973.*

but the nature and features of those products are controlled by the kinetics of phase changes. It is now believed that martensitic transformation is also controlled by the nucleation, which could be crystal defect sites like dislocation tangles or strain-induced embryo of highly strained ferrite with super-saturated carbon. Nuclei of martensite, once formed, grow very fast due to high strain in the matrix—thus making the growth process 'athermal' (without the assistance of further change in temperature). This means transformation of martensite is nucleation dependent and not growth dependent; once it gets nucleated, it can grow very fast without assistance from temperature change. This condition implies that, for all practical purposes, once nuclei for martensite have formed at a temperature, the process of martensite transformation would be complete instantly and would not depend on further cooling or time. But, the very fact of formation of nuclei being temperature dependent, 100% transformation of austenite to martensite will be dependent on the temperature to which the steel has been cooled; else some amount of austenite may remain untransformed along with the transformed martensite. This situation necessitates precautions to avoid untransformed austenite called 'retained austenite' during heat treatment of steels. More about formation of martensite will be discussed under austenite decomposition under continuous cooling.

Transformation and growth rate of other phases, like ferrite, pearlite and bainite, is fully nucleation and growth dependent. Figure 2-4 suggests that if the cooling rate is faster, more nucleation will be possible and, thereby, the phases or grains will be finer due to lower growth, necessary with higher nucleation sites available for growth in a given volume. This is the reason why, faster cooling always produces finer structure and this opportunity is vastly used in industries for cost-effective heat treatment and steel applications. For instance, when steel is cooled in air for normalising, the faster air cooling than equilibrium cooling will cause instantaneous under-cooling effect at the start of transformation. This under-cooling effect will lead to higher number of nuclei formation and higher number of nuclei will, in turn, produce more number of centres for growth and finer structure; vide the logic of Fig. 2.4. However, if the steel is slow cooled like in annealing, there will be very little under-cooling effect because faster the cooling rate higher is the amount of under-cooling (refer Glossary for under-cooling). As a result, there will be lesser nuclei and less growth centre, leading to relatively coarser structure produced by annealing than normalising.

Thus, sensitivity of austenite decomposition to N&G and cooling rate can be gainfully exploited in industrial heat treatment for producing the right type of microstructure. This situation is exploited in the heat treatment of steels by making use of *isothermal cooling* or *continuous cooling* for producing non-equilibrium products like bainite and martensite, instead of equilibrium cooling which is largely used for transformation of austenite to ferrite and pearlite. However, there could be some departure from exact equilibrium cooling in industrial heat treatment practices, such as normalising which involves open air cooling. Whatever could be the cooling condition, N&G processes as per the prevailing temperature and cooling condition play the critical role in the transformation of austenite.

Figure 2-5 sequentially analyses the process of austenite decomposition with reference to N&G process under normal cooling rate. It could be observed from this figure that austenite on crossing the temperature line A_3 on cooling will first precipitate *ferrite* nuclei at the prior austenite grain boundaries and then those ferrite nuclei will start growing with cooling. If the cooling is somewhat faster than equilibrium cooling (e.g. air cooling in normalising), there would be larger number of stable ferrite nuclei due to under-cooling effect, which will produce more growth centres and finer structures. Ferrite formation by following the N&G process will continue till temperature reaches lower critical temperature line A_1 marked as $723°C$ in Fig. 2-5.

On reaching and crossing A_1, pearlite will start forming from remaining austenite in areas free from prior ferrite formation. Pearlite formation is also controlled by 'nucleation and growth' process, but nuclei for pearlite growth are 'carbide' (Fe_3C). On reaching and crossing A_1 (marked as $723°C$ line in Fig. 2.5), austenite, which is by then richer in carbon than the original steel composition (0.4%C) due to ferrite separation, will momentarily precipitate *carbides* from carbon-rich austenite and this will act as nuclei for pearlite formation (see Fig. 2-4). Carbide precipitation, in turn, will momentarily reduce carbon from the surrounding austenite and this will induce ferrite precipitation adjacent to carbides, giving rise to 'lamellar formation' as schematically indicated in Fig. 2-5. Thus, pearlite that will form on decomposition of parent austenite will be alternate lamellae of carbide and ferrite, as schematically shown in the Fig. 2-5.

While Fig. 2-5 depicts a schematic presentation of pearlite formation, Fig. 2-3B illustrates the actual micrographs of pearlite lamellae, showing alternate precipitation of ferrite and carbide stringers. If the cooling rate is faster than equilibrium cooling, like normalising, the lamellar spacing in the pearlite colony will be closer due to larger number of stable carbide nuclei under faster cooling, which leads to higher under-cooling effect. Micrographs shown in Fig. 2-3 are, in fact, products of faster air cooling (normalising) than slow equilibrium cooling (annealing).

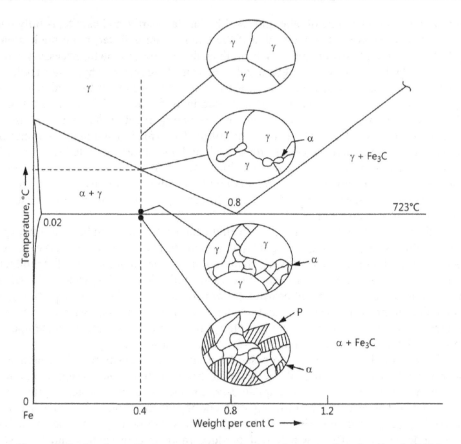

Figure 2-5. A schematic representation of microstructure changes in a 0.40%C steel from austenitising temperature (dotted vertical line) at normal cooling rate. Symbols: γ = Austenite; α = Ferrite, P = Pearlite and Fe₃C is Cementite.

The foregoing elaboration points to the fact that the nature of the ferrite and pearlite formed during austenite decomposition will be characteristic of the N&G process under a specific cooling condition. The general trend of N&G under different temperatures of formation has been shown in Fig. 2-4 (a), which indicates that at higher temperature of formation, nucleation rate will be low and growth will be higher for completion of transformation. At lower temperature of formation, nucleation rate will be higher and need for growth would be lower. What this implies is that when phases form at higher temperature, they will tend to be coarser than when they form at relatively lower temperature.

The N&G process is also influenced by the cooling rate. When cooling rate of austenite is higher than normal, then the steel will momentarily experience sudden under-cooling effect—similar to liquid to solid crystallisation process, where formation of solid nuclei requires 'nucleation under-cooling'. Faster cooling rate causes more under-cooling (also called super-cooling effect), momentarily bringing down the temperature and making possible greater number of nuclei to form in correspondence to under-cooled temperature. Thus, more stable nuclei are available for growth with faster cooling rate, which can facilitate early completion of transformation. This leads to finer microstructure. For pearlite formation, this means finer inter-lamellar spacing in the pearlite colony of the microstructure. This is the reason why we get finer structures by normalising the steel as against coarser structures by annealing.

However, if cooling rate is increased faster than air cooling, there will be a retardation effect on N&G process for standard thermodynamical phases like ferrite and pearlite. This is because there is need for some 'incubation time' for nucleation and growth to start. Hence, faster cooling than air cooling may overshadow the ferrite/pearlite separation due to lack of sufficient incubation time. This can be also understood from the hysteresis effect on cooling. If the cooling rate is very rapid (as in quenching of steels), it can suppress the transformation for several hundreds of degrees due to sharp decrease in reaction rate with decreasing temperature. Hence, cooling rate (time to cool) is an important factor in the decomposition of austenite, especially under fast cooling condition.

If cooling rate is faster than air cooling and reaches the level of quenching, ferrite/pearlite separation from the austenite may be totally or partially suppressed. Under such circumstances, new phases of non-equilibrium structure, such as bainite and martensite, will start forming, depending on the cooling rate and composition. Very fast cooling rate will tend to produce martensite and relatively lower rate will tend to produce bainite, if the composition of the steel is amenable to bainite formation.

Bainite and martensite are non-thermodynamical products with distorted structure of ferrite, super-saturated with carbon. Due to faster cooling, normal N&G process for ferrite or carbide nucleation gets suppressed and the steel matrix at that stage becomes strained with super-saturation of carbon. With further cooling below the ferrite/pearlite separation temperature zone, this matrix transforms to martensite or bainite, depending on the composition and cooling rate.

Therefore, depending on the cooling rate and composition of the steel, the parent austenite can form a combination of any of the following structures on decomposition:

Ferrite: A granular crystallographic structure in BCC lattice. Grain size in ferrite will depend on the rate of cooling during decomposition; if cooled faster, as in air cooling (normalising), ferrite grain size will be finer. If cooled slowly (annealed), grain size will be coarser and steel will be softer.

Pearlite: A mixture of ferrite and carbide in lamellar form. Pearlitic lamellar spacing would be coarser if cooled slowly (e.g. furnace cooling) or finer if cooled faster. Faster cooling tends to produce more pearlite and finer pearlite due to N&G effect.

Bainite: A feathery ferrite structure with fine carbide precipitates over the ferrite plates, which can be predominantly on the ferrite plate boundaries or all over the ferrite plate at random. The former pattern is called 'upper bainite' and the latter pattern is called 'lower bainite'. Bainite will form when cooling rate is faster to suppress austenitic decomposition to ferrite and pearlite, but not fast enough to form martensite, provided the steel composition is amenable to bainite formation.

Martensite: A heavily strained and distorted ferrite with super-saturated carbon, taking the shape of platelets or needles in the microstructure. It is a highly distorted structure with distorted tetragonal ferrite lattice structure supersaturated with carbon. Martensite forms when the steel is cooled from austenitic temperature at very high speed, like water or oil quenching, suppressing the possibility of nucleation of any other phase.

Formation of these phases by austenite decomposition holds the key to all heat treatment processes of steels. Tailoring the austenite decomposition process by designing appropriate heating and cooling methods, properties of steels can be suitably modified for any given application. For example, if a softer structure is required for the given steel, austenite can be slow cooled inside a furnace, called *annealing* in heat treatment, for producing coarser ferrite grain size and pearlitic structure. If a stronger structure is required of the same steel, the steel can be subjected to air cooling to produce finer ferrite grains and denser pearlite, called *normalising* in heat treatment. If the strength of the steel has to be further improved, the same can be quenched (*hardening*) to form martensite or (martensite + bainite), as the case may be, by appropriate quenching. Quenching means fast cooling of the steel in a manner that suppresses the formation of higher temperature phases like ferrite or pearlite in the structure.

Other than cooling conditions, other factors that may influence the nature of pearlite or the grain sizes of ferrite in the structure are: the carbon and alloying elements in the steel and the history of prior working of the steel. Higher carbon and/or alloying elements promote easier and more stable carbide nuclei due to heterogeneity in parent austenite, producing more pearlite in the steel and, as a result, ferrite content in the structure becomes lower and grain size becomes finer. Prior working of the steel piece before heat treatment causes the original structure to be broken down and fragmented, providing more nucleation sites during heating or recrystallisation and influencing the austenite decomposition process. The other factor is the grain size. Grain size also influences the decomposition of austenite process by influencing the nucleation sites. Grain boundaries can act as preferred nucleation sites due to higher surface energy available with them and thereby influencing the process (see Fig. 2-4). This explains why coarser grain sizes of austenite are preferred for easy hardening to martensite—because coarser grains with lower grain boundary areas allow less opportunity of transformation to other phases (like ferrite, pearlite or bainite) by grain boundary nucleation. If the parent austenite grain size is fine, providing plenty of areas for easy nucleation of high temperature phases, it may tend to produce ferrite and pearlite in the structure despite faster cooling, but the morphology of the structures will be finer with improved mechanical properties.

2.4. Austenitic Decomposition under Isothermal and Continuous Cooling Conditions: Formation of Bainite and Martensite

Discussions under Section 2.3 with reference to Figs. 2-1 and 2-5 relate to 'equilibrium cooling' conditions. Equilibrium cooling is slow cooling (e.g. air or furnace cooling) and has limited applications in the heat-treatment of steels. Since austenite decomposition exhibits strong sensitivity of N&G dependence, the process can be gainfully used for widening the scope of steel applications by creating a condition where conventional higher temperature nucleation process could be suppressed, facilitating the formation of non-equilibrium products like martensite and bainite. It is in this respect, study of austenite decomposition under *isothermal* or *continuous cooling* conditions assumes great significance for the formation of non-equilibrium products, like bainite and martensite in steels, through heat treatments. *Isothermal cooling* implies cooling at a predetermined rate to a specific temperature and then holding the steel at that temperature in isothermal condition. *Continuous cooling* implies cooling the steel piece continuously under a given cooling condition till the temperature reaches the room temperature or below.

A typical isothermal transformation diagram of 0.6%C steel is shown in Fig. 2-6. The time-temperature-transformation (TTT) curve diagram refers to isothermal cooling conditions, where the steel piece is cooled from austenitising temperature and held at an intermediate constant temperature for a given period of time, to examine how much of the austenite has been transformed to what phases. This type of study unearths important information about designing the thermal processing or heat treatment of steels, for tailoring the microstructures in order to obtain desired properties for various uses and end-applications.

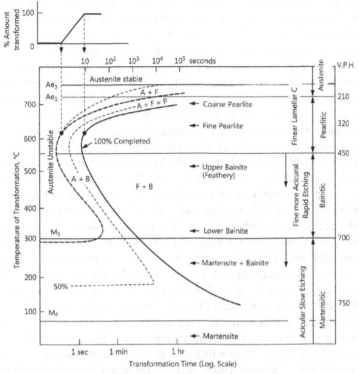

Figure 2-6. An ideal TTT curve for 0.6%C steel. Diagram depicts time interval required for 50% and 100% transformation of austenite—produced by heating above Ae_3 temperature—at a constant temperature. Notations: A = Austenite, F = Ferrite, P = Pearlite, B = Bainite.

An examination of Fig. 2-6 will show that the TTT diagram not only reveals the nature of transformation products like bainite or martensite, it can also reveal the nature and character of ferrite or pearlite under a given isothermal cooling condition. In other words, a TTT diagram lends itself to the study of the entire span of austenite decomposition, starting from ferrite precipitation at the higher holding temperature to martensite transformation when sharply cooled to lower temperature and held.

Isothermal cooling is used not only for TTT study, but also can be used for special heat-treatment process, to reduce thermal stresses or to produce a desirable microstructure. Some such isothermal cooling curves have been superimposed in the figure (see top left part of the diagram), indicating how the steel under study can be sharply cooled initially and then held at constant temperature for isothermal transformation. These cooling curves correspond to the study of ferrite–pearlite separation at higher temperature, ferrite–bainite separation at the intermediate temperature and martensite transformation at the lower temperature of holding. All these cases require sharp cooling to a predetermined temperature avoiding touching the nose of corresponding 'C'-curve (as shown in Fig. 2-6) and then holding at that temperature over a time period to observe the phase transformation. The gap of time between the start of austenite transformation and the temperature axis is termed as *incubation period* for starting the transformation of austenite to the phase corresponding to the temperature of holding. It would be evident from Fig. 2-6 that at higher temperature of holding (e.g. at around 630°C), it takes about 10–15 seconds (after the incubation period) for 100% austenite transformation to (ferrite + pearlite).

The bold curves in the Fig. 2-6 indicate the start and finish of austenite transformation and the time gap to start the transformation is the corresponding incubation period for nucleation to become stable for growth. It would be noted that incubation period or start of transformation time is high at higher temperature initially, coming down to minima at around 550°C, and increases again till it reaches the martensite transformation temperature. Start of martensite transformation has no incubation lag and it starts instantly when temperature during cooling crosses the martensite start temperature (M_S), which is 300°C for the 0.60%C steel as given in Fig. 2-6. Apparent reason for absence of incubation period for martensite is that nuclei for martensite is that nuclei for martensite are the strain-induced embryos and dislocation tangles in the matrix, which are plenty at that lower temperature, if cooled rapidly to suppress carbon separation or carbide precipitation. Such strained martensite nuclei are ready to grow quickly due to high energy associated with them in a strained matrix. The martensite transformation, as revealed by this type of study, grows very fast and they are not time dependent but temperature dependent i.e. it is an 'athermal' process in contrast to isothermal process of other phases. The athermal character of martensite is a consequence of very rapid nucleation and growth, so rapid that the time taken for growth can be neglected. This implies that once nucleated, growth of that martensite nucleus is not time dependent. But, for further supply of nuclei for martensite, austenite has to be cooled to lower temperature. With lowering of temperature, more austenite will transform to martensite, completing at a temperature known as martensite finish (M_F) temperature. Thus, martensite formation will start at the M_S temperature, and more martensite will form only on further cooling of austenite; reaching 50% transformation to martensite at around 190°C in Fig. 2-6 and 100% on reaching M_F which is about 80°C in the case of 0.6%C steel. Thus, the study of austenite transformation under isothermal condition provides the basis of our understanding about:

1. What are the phases/structures that will form from the transforming austenite, if the holding temperature is changed?
2. How much holding time is required for desired phase transformation?
3. How fast the steel should be cooled to avoid any separation of undesirable phase? Specifically, incubation time for different phases/structures at different temperatures and the ways of avoiding such undesirable phase separation. For example, ferrite separation when one is aiming for bainite transformation or martensite transformation.
4. How fast and at what temperature the steel must be cooled, to avoid any bainite separation when martensite supposed to be the desired phase?
5. At what temperature the steel must be cooled for completion of 100% martensite (e.g. the M_F)?
6. What could be the approximate strength (or hardness) of different phases that form at different temperatures and their overall nature—as indicated on the right side of the Fig. 2-6)?

The TTT diagrams are different for different compositions. By studying the TTT diagrams of various compositions of steels (e.g. plain carbon steels and alloy steels) one can obtain the above mentioned information about the respective steel, and thereby making it possible to design a fool-proof heat treatment or rolling operation. The other time temperature transformation diagram is the continuous cooling transformation (CCT) diagram as depicted in Fig. 2-7. The TTT and CCT diagrams of respective steels are mostly available, as published material for common grades of steels in books of atlas of transformation diagrams. Some references have been provided at the end of this chapter.

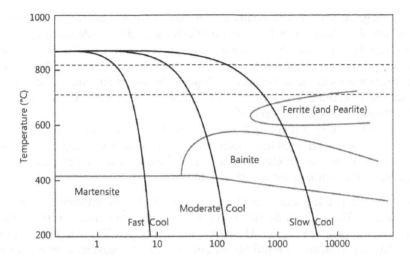

Figure 2-7. A typical CCT diagram of low-alloy steel, illustrating the possibility of different structure formations under different cooling rates. The cooling rate is assumed to be at the centre of the bar. In practice, cooling rate will vary in a bar of given diameter from the surface area to the centre.

Most industrial cooling is continuous in nature, giving rise to different transformation products when they cross different phase-fields as shown in corresponding CCT diagram of Fig. 2-7. Since this type of diagram is obtained by studying the phase transformation under continuous cooling condition, it is called 'continuous cooling transformation' (CCT) diagram. While thermodynamics of phase transformation in CCT diagram remains more or less similar as in TTT diagram, the kinetics of transformation changes due to heterogeneous structure that gets developed inside the steel body during such continuous cooling process. Continuous cooling not only produces continually differing morphology of a structure in the matrix, but also can produce different structures across the steel section. This is because different points across the section of the steel piece will cool at different rates due to thermal conductivity problem. For example, the centre will cool more slowly than the surface due to thermal conductivity lag.

Considering the practical situation of steel rolling and heat treatment, where mass effect on cooling cannot be avoided, CCT diagrams offer more realistic basis for designing steel heat treatment cycles or rolling processes. But, for interpretation of CCT diagram and to know what structures might be present and their nature, TTT diagram of the respective steel is very helpful. Characteristically TTT and CCT diagrams differ a bit as regards information about start and finish of a phase. A TTT diagram always indicates both 'start and finish' of transformation process (because the steel is held isothermally at a given temperature for the phase transformation to be completed, whereas CCT diagrams often show only the 'start' of transformation, because due to continuous cooling, different structures corresponding to the cooling rate, will form and the end point of transformation of a particular phase/structure may not be distinctly pin pointed (see the features of Figs. 2-6 and 2-7 in this regard). Separation of two C-curves in the CCT diagram of Fig. 2-7, splitting them into two distinct parts is characteristic of some alloy steels, where due to easy carbide precipitation (or their presence) ferrite formation curve moves upward, and thereby makes bainite or martensite formations easier.

A given CCT diagram can be conveniently used for understanding how the transformation progresses following an industrial cooling curve. It is more particularly used for understanding what should be the cooling rate for transformation of 100% martensite in a given section of the steel, though the diagram can also provide information about whatever phases/structure form under different cooling rates. In fact, a CCT diagram maps out the formation of all possible phases/structures under the coordinates of temperature and time of cooling. However, both TTT and CCT diagrams are specific to a specific steel composition i.e. they are 'composition specific'.

For example, the CCT diagram in Fig. 2-7 illustrates the formation of different structures following three cooling rates: slow cooled, medium cooled, and fast cooled. Following these cooling curves, the information regarding phase fractions/structures that can be obtained from this diagram (assuming that the cooling represents the centre of the mass) is as follows:

1. Slow cooling (the cooling curve farthest on the right) will produce essentially ferrite plus pearlite structure. The curve indicates that transformation reaction will be complete within the ferrite-pearlite temperature region. However, the ferrite-pearlite structure so produced will be finer than that is produced by equilibrium cooling, due to reasons explained earlier based on N&G process.
2. An intermediate cooling rate (marked moderate in Fig. 2-7) will produce 100% bainite structure, if the transformation can get completed within the bainite formation zone. However, due to faster cooling in this lower temperature region, the bainite transformation might not get 100% completed, producing some martensite along with bainite in the structure. The transformed bainite will be partly upper bainite and partly lower bainite, because the cooling curve enters the bainite formation region above the nose of C-curve for bainite. Had the cooling rate been little faster, all bainite would have been lower bainite.
3. Faster cooling rate (the first curve in Fig. 2-7 marked as 'fast cooled') will produce 100% martensite, as it enters the martensite start temperature without intersecting any other field. Martensite formation will continue with the cooling till temperature reaches M_F temperature, which is generally about $150^\circ C$ below the M_S temperature (this is only a ball-path figure).

If cooling is disrupted in between M_S and M_F temperatures, transformation of remaining austenite will stop. This austenite, however, will not transform to any other phase/structure than martensite, when taken out of the cooling bath or cooling is resumed. This is because at that lower temperature no other phase formation is possible from the remaining austenite, except some diffusion of carbon atoms out of austenite lattice and forming fine carbides, if held longer. However, by very nature of continuous cooling, there will be temperature lag in the section of the steel and, consequently, phases/structures formed at different points within the steel body will have different morphology (microscopic structural details).

A summary of cooling conditions for producing different microstructures in steel has been further illustrated in Fig. 2-8, including the condition of formation of upper bainite, lower bainite and martensite. The diagram shows the transformation 'start' and 'finish' curves in the form of C-curves, whereupon different cooling conditions have been superimposed. Nature of microstructures obtainable from such cooling conditions has been also indicated in the diagram. For formation of bainite or martensite, cooling should be such that the falling cooling line should not touch or cross the C-curve 'start' line at any point. Figure 2-8 diagram sums up the way steels can be cooled and the corresponding structures to be expected with reference to the C-curves. In application, this type of transformation diagram, superimposed with cooling curves, guides about how to control or design the cooling conditions during heat treatment for tailoring the microstructures for a specific application.

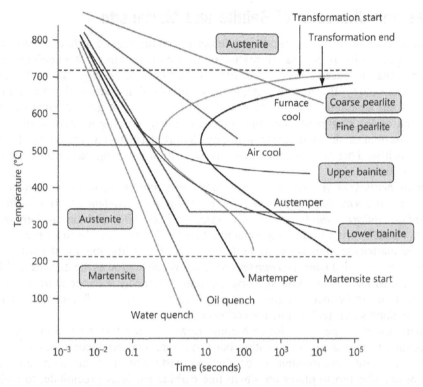

Figure 2-8. An illustrative CCT diagram with superimposed cooling curves to demonstrate the condition for formation of different structures from austenite decomposition. It also shows 'austempering' and 'martempering' regions of heat treatment process, which are now extensively used for heat treatment of alloy steels for enhanced properties.

Thus, with the help of TTT and CCT diagrams of steels, following information, relating to either applicable heat treatment cycle or the obtainable microstructures and their nature, including hardness, can be obtained:

1. What should be the cooling condition and cooling rate for obtaining the aimed microstructure such as bainite, martensite etc.?
2. What would be the nature of those structures?—would it be lower bainite or a mix of lower and upper bainite? 100% martensite or a mix of bainite and martensite or with traces of pearlitic carbides called toorstite?
3. What steps should be taken (i.e. how long or at what temperature the steel needs to be held in the heat treatment bath) so that martensite transformation becomes 100%?
4. How quenching stresses can be minimised during heat treatment by interrupting the quenching below a safe temperature?
5. What is the nature of transformed martensite under a given condition? The nature of martensite in alloy steel under water quenching and oil quenching (having different quenching severity) differs; water quenching will produce finer needles of martensite than oil quenching.
6. Hardness of corresponding phases that form during cooling at different temperature levels.

2.5. Nature and Character of Bainite and Martensite

The CCT diagrams of steels set the conditions under which martensite or bainite can transform from the decomposing austenite. In general, formation of martensite requires faster cooling than the bainite. The cooling rate that produces 100% martensite is called *critical cooling rate* in the terminology of heat treatment technology. Because of fast quenching involved in martensitic transformation, carbon atoms in the parent austenite from which martensite forms do not get any time or opportunity to diffuse out, making the martensite structure super-saturated with higher carbon than it can hold in equilibrium. This makes the martensite structure distorted (distorted tetragonal ferrite structure) but strong and brittle. This 'as-quenched martensite', which is strong but brittle, is not suitable for industrial applications.

To eliminate brittleness in such as-quenched structure, martensite has to be tempered at a higher temperature than its formation temperature for allowing some carbon atoms to diffuse out of the martensite and precipitate in the form of Fe_3C or alloy carbide. Alloy carbide can form only if the steel contains some carbide forming alloying elements, like Cr, V, Mo, etc. In this process of tempering, the as-quenched martensite loses some strength but gains in ductility, making the structure strong and tough. This structure is called *tempered martensite*, and it is considered as the most desirable structure of steel for many industrial applications. By using appropriate tempering temperature and time, tempered martensitic structure can be made to give wide ranging properties of different strength and toughness combinations, making such steel unique for various applications.

The second best structure is the *lower bainite*. Lower bainite forms when cooling rate is slower than that is required for martensite formation but faster than that produces *upper bainite*. Figure 2-9 depicts few representative micrographs of bainite and martensite. It should be noted that bainite is a combination of acicular ferrite plates on which fine carbide particles precipitate, to make it stronger and tougher. If the carbide precipitates are on the plate boundaries, it is called *upper bainite* and if the precipitates are all over the ferrite plate itself, it is called *lower bainite*, refer the notes and display in Fig. 2-9. *Lower bainite* is stronger and tougher than *upper bainite*, because of difference in carbide precipitation pattern as well as higher dislocation density present in the lower bainitic ferrite plates that form at relatively lower temperature than upper bainite. Upper bainite shows carbides mostly on ferrite plate boundaries and some might be on plate itself, but lower bainite shows finer carbides mostly on the plates and some on the boundaries.

Compared to lower bainite, tempered martensitic structures are a bit finer and superior in the scale of strength and toughness. However, lower bainitic structure will be generally free from internal strains (because of higher formation temperature than martensite) and sometimes preferred for some special applications like ship building, where internal strain can influence the 'ductile-to-brittle transition' temperature, making the ship structure vulnerable to cold weather cracking. An example is the use of 9% nickel steel plate for ship structure. 9% nickel steel shows predominantly lower-bainitic structure with very good low-temperature toughness.

However, if a structure is a mixed one with martensite and bainite, the strength and toughness will suffer with increasing bainite content—especially with upper bainite which is known to be poorer in elongation and notch toughness along with its lower strength. Table 2-1 gives some qualitative indication of relative properties of bainite and martensite. Considering the overall aspects of relative properties and likelihood of presence of bainite in the cross section of steel specimen due to continuous cooling character, martensite with small amount of lower bainite in the structure can very well be acceptable for applications requiring high strength and toughness.

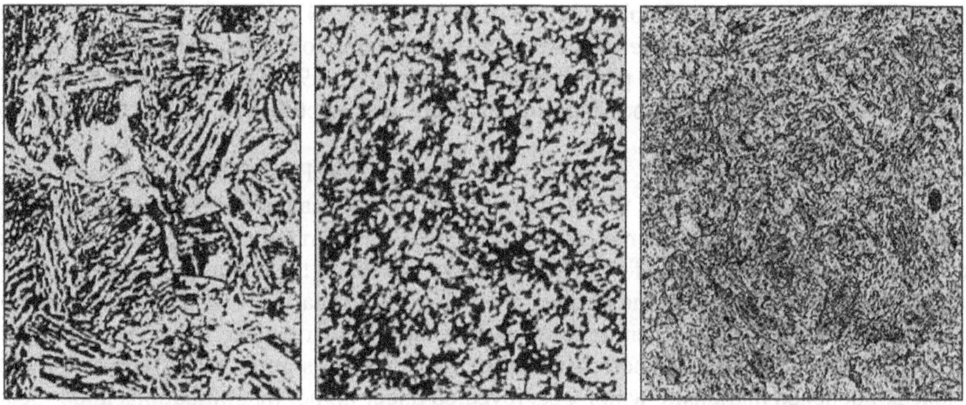

| Upper Bainite | Lower Bainite | Tempered Martensite |

Figure 2-9. An optical micrograph of upper and lower bainite and tempered martensite at ×200 magnification. While upper bainite structure is quite distinct from lower bainite and tempered martensite, the latter two structures can be often confused, unless examined very carefully under higher magnification.

Differences in properties between martensite, lower-bainite and upper-bainite arise due to their difference in structural morphology (i.e. their microstructural details), which, in turn, arise largely from their temperature of formation and partly from their alloy contents. Although the formation of these phases have been discussed and illustrated earlier, their structural features based on temperature of formation are relevant for further discussion. These are reiterated in the following:

Upper-bainite forms at relatively higher temperature, but below the pearlite formation temperature (see Fig. 2-7). Upper bainite forms at the upper range of bainite formation temperature range (see lower C-curve in the Fig. 2-7). The lower C-curve in the figure has a nose. When bainite forms at temperature range above the C-curve nose, it is 'upper-bainite'. At this temperature range, mobility of carbon atoms is good and the extra carbon that comes out from acicular ferrite plates can travel

Table 2-1. Outlines the difference in properties between martensite, lower bainie and upper bainite.

Properties	Martensite (Tempered)	Lower Bainite	Upper Bainite
Strength/Hardness (see Fig. 2-6)	Very high to high (700–750 VPN)	High (550–700 VPN)	Medium to high (450–550 VPN)
Yield strength	High	High	Medium
Toughness	High	Medium to high	Medium
Elongation	High to medium	High	Medium to poor
Reduction in Area (%RA)	High to medium	High	Medium to poor
Notch toughness	High	High	Medium to poor
Fatigue strength	Very high	High	Medium to poor

and get precipitated on the ferrite plate boundaries (see Fig. 2.4), plate boundary being the site of extra energy. Thus, the *upper-bainite* structure is an agglomerate of coarse feathery ferrite plates with carbides precipitated mostly on the plate boundaries (see Fig. 2.9). The presence of brittle carbides at the plate boundaries renders upper bainitic structure less ductile and tough, compared to lower bainite or tempered martensite.

Lower bainite forms below the nose of lower C-curve (see Figs. 2-7 and 2-8) and continues to form until the cooling curve touches the M_S temperature line. Because of relatively lower temperature of formation, carbon mobility in the matrix at the bainite formation temperature is limited. This causes the carbon atoms coming out of the ferrite lattice to precipitate on the site at the nearest vicinity i.e. on the ferrite plate itself as well as on the boundaries whichever is nearer. As such, carbide precipitates in lower bainite are all over; on the bainite plate and on the boundaries (see Fig. 2-9). This structure of lower bainite closely resembles tempered martensite. But, there is significant difference between lower bainite and as-quenched martensite which gets revealed under higher magnifications under Scanning or Electron microscope, see Fig. 2-10.

Martensite forms on faster cooling of steel (quenching) when the cooling condition is such that, the austenite does not get time to transform to any other phase like pearlite or bainite. This can happen when cooling curve misses the nose of C-curves for bainite formation. If the cooling rate is somewhat slow during quenching (at any part of the steel), some amount of lower bainite formation is possible inside the steel section under continuous cooling condition, giving rise to mixed structure. Since the cooling rate is very fast for martensite formation, carbon in the untransformed austenite does not get time to diffuse out to keep pace with lower solubility of carbon with lowering of temperature by changing to other phases or by precipitating out carbides. Hence, entire carbon in the austenite gets trapped inside and transforms on further cooling to highly-strained 'acicular ferrite plates', (see Fig. 2-10).

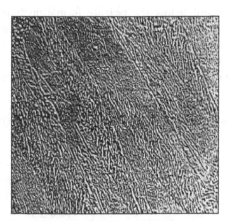 

Lower Bainite (×5000) (Electron microscopy—replica technique)

Martensite Plates (untempered) (×10000) (Electron microscopy—thin foil technique)

Figure 2-10. High-magnification micrographs of lower bainite and martensite plates (untempered). Dark areas in lower bainite are fine precipitates of carbide over the ferrite plates; darker areas in martensite are dense dislocation tangles in the strained ferrite plates supersaturated with carbon.

As-quenched martensitic structure, if held at room temperature for long, might induce some fine cracks in the matrix, arising from the tendency of relieving internal stresses of the strained ferrite lattice with time. *Hence, as-quenched martensite requires immediate tempering to relieve some strain in the matrix.* This takes place by heating the steel (called tempering), which induces some mobility for carbon atoms to diffuse out of strained ferrite plates and allow precipitation of carbides in the matrix. Thus, tempered martensite structure is also characteristically an aggregate of acicular ferrite and carbides that precipitateout from supersaturated ferrite on tempering.

Tempering of martensite will relieve some strain, precipitate out carbide particles, introduce some ductility and reduce some strength as well, depending upon the temperature and time of tempering. More about quenching and tempering of steels will be discussed in the chapter on 'Heat treatment', but some elaboration is necessary for emphasising the structural characteristics of martensite.

In general, tempered martensite structure will have fine carbides precipitating all over the matrix of acicular ferrite. In this respect, it is close to lower bainite structure after tempering and their properties could overlap. Yet, martensite in alloy steels are often very fine plates of ferrite or needles of ferrite with many internal twins over which very fine alloy carbides get precipitated during tempering, making this structure stronger and tougher than lower bainite. Fig. 2-11 illustrates the following features of martensite tempering in two major types of martensite i.e. in lath (a special type of thin plate-let type martensite) and needle types.

Upper set of photographs, pertaining to lath type martensite, illustrate that:

- Lath martensite has dense dislocation tangles associated with as-quenched state (a).
- On tempering at around 300°C or below, some fine carbide precipitates out, thickening the plate boundaries (b).
- On tempering at about 500°C, the dislocation structure recovers and forms thick cell-like structure sharing the plate boundaries (c).
- On tempering even at higher temperature, say around 600°C, the plate like structure of martensite tends to re-nucleate ferrite, and carbide precipitation occurs on the ferrite boundaries which thickens and balls up with time of holding as globular spots (d).

In Fig. 2-11 the lower set of photographs, pertaining to needle shaped martensite, illustrate that:

1. Needle type martensite (belonging to high carbon or some nickel bearing alloy steels) is highly twined (e).
2. When this structure is tempered at lower temperature between 100°C to 200°C, very fine transitional carbides ($Fe_{2.4}C$) precipitate across twin boundaries (f). This type of carbide is called 'epsilon carbide' which on further heating coalesces and changes to regular carbide.
3. Tempering at around 200°C–300°C causes the fine carbides to dissolve and re-precipitate as normal carbides (either cementite or alloy carbide) along the twin plate boundaries (g).
4. When tempered at around 400°C–600°C, twin structure breaks down and the carbides spherodise to prominent globular form, known as spherodised structure (g).

These structural changes in martensite with tempering temperatures allow the steel properties to be exactly tailored for an end application. However, temperatures mentioned for different tempering stages are only a ball-path figure for illustration; exact temperature depends on the composition of the steel and required strength and toughness.

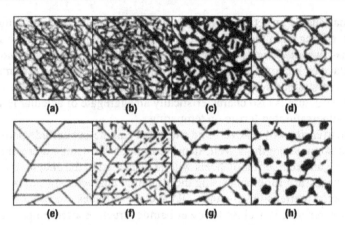

Figure 2-11. Schematic presentation of stages of microstructural changes with tempering temperature in lath (plate type) and needle type martensite, present in some low carbon plain steel and high carbon plain steel or alloy steels respectively.
Figure 2-11 (a), (b), (c) and (d) represent the stages of change in lath type martensite and (e), (f), (g) and (h) shows the changes in needle type twined martensite. Twining in needle type martensite is observed due to strain-induced parallel facets created on quenching the steel. *Source: Rollason, E.C. Metallurgy for Engineers. London, 1973.*

2.6. Uniqueness of Martensite: Utility and Utilisation

Martensite is one of the strongest phases in steel (except carbides), but it is hard and brittle in as-quenched condition. However, characteristically, martensite quickly improves and gains toughness with tempering—which can be carried out over a range of temperatures below the A_1 temperature. Martensite after tempering is called *tempered martensitic* structure. Higher the tempering temperature of martensite, lower is the strength but higher is the toughness. At relatively lower temperature of tempering, carbon that comes out of martensite plates is very fine (called epsilon carbide $Fe_{2.4}C$), but with increasing tempering temperature more carbon diffuses out of martensite plates and coalesce together to from coarser precipitates of normal carbide (Fe_3C). Nature of carbide precipitate and their size determines the strength and toughness of martensite; alloy carbide provides more strength and toughness than plain iron carbide and finer carbide precipitates provide higher strength. Strength of martensite can also vary with composition of the steel; it increases with increasing %C and alloying. Fig 2.12 depicts the variation of martensitic hardness (in VPN scale) with carbon content in plain carbon and alloy steels. This is *as-quenched hardness* of martensite.

Figure 2-12 indicates the following:

a. Martensite hardness in plain carbon steel steadily increases up to carbon percentage of 0.83%C (eutectoid carbon), but the rate of increase drops thereafter. This is because of the possibility of 'toorstite' formation in steels with higher carbon content. Toorstite is a very dense spot of pearlite which forms in high carbon steels due to the presence of very fine carbides in the matrix that act as nuclei for pearlite formation. This type of pearlite (toorstite) is sometime observed after hardening case-carburised parts containing high surface carbon.

b. Martensite hardness in alloy steels peaks at about 0.70%C and thereafter decreases because of the tendency of alloy steels to retain some untransformed austenite (*retained austenite*) upon fast

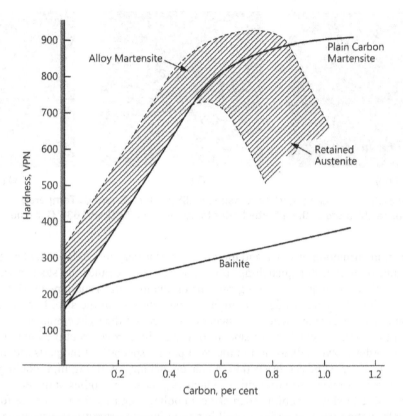

Figure 2-12. Variation of hardness of martensite and bainite with carbon content.

quenching. However, this retained austenite in the martensitic structure mostly gets transformed to martensite upon tempering, by the process of 'diffusion of carbon atoms' out of the austenite lattice and thereby, making the austenite leaner in carbon and unstable. However, small amount of finely dispersed retained austenite is known to improve the *toughness of steel*, making the steel more fracture resistance. This is because pools of retained austenite can act as 'arrester of crack growth' in the matrix.

c. Figure 2-12 also reveals that steels with %C less than 0.30% cannot be effectively hardened as the martensite of such steels is not hard enough. For martensite (as-quenched) to be considered hard enough for further processing or tempering, a hardness value of 580–600 VPN is considered minimum; and to reach to this value, a minimum carbon of about 0.32%–0.35% is necessary in plain carbon steel (see superimposed indicator lines). Hence, all hardening grade steels require sufficient level of carbon, or carbon and alloy mix, in order to get right type of martensite with right hardness level. Because of this fact, heat treatability of steels is related to the factor called 'hardenability', which implies the ability of the steel to effectively harden up to a required depth.

It is, therefore, obvious from Fig. 2-12 that *by choice of carbon and alloy in the steel composition, martensitic hardness can be manipulated* so that upon tempering, the steel provides the desired combination of strength and toughness. In Fig. 2-13, A illustrates the nature of as-quenched martensite and B and C show the nature of same martensite after tempering at different temperature.

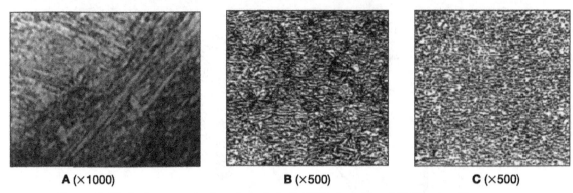

A (×1000)   **B** (×500)   **C** (×500)

Figure 2-13. A: As-quenched martensite in 0.5%C steel. **B** and **C:** Tempered martensite structure of 0.5%C steel after tempering at 500°C and 630°C, respectively.

Martensitic transformation in steels and their properties is very much dependent on the cooling rate during quenching, such as water quenching, oil quenching, etc. Leaner the steel in carbon and alloy content, higher will be the required cooling rate for producing 100% martensite. But, higher cooling rate during quenching also tends to develop high *residual stresses* in the steel body and, thereby, will cause distortion of the steel piece. Hence, it is necessary to choose the right composition of steel (carbon or carbon and alloy content) which will respond to martensite formation at medium cooling rate (e.g. oil quenching) in order to avoid distortion in the steel parts, especially if the parts are thin and slender. However, if the steel part is heavy in section, water quenching involving high cooling rate might be unavoidable; otherwise some non-martensitic products (e.g. pearlite noodles or upper bainite) will result inside the section due to slower cooling than critical cooling required for martensite formation. Thus, martensite transformation process and the flexible tempering opportunity of martensite structure offer many advantages for its use and utility in industrial applications.

Most heat treatment processes of steels are based on these features of martensite formation and tempering. They are based on the availability and flexibility of quenching and tempering operations, leading to the development of required combination of strength and toughness. Various combinations of chemical composition and cooling conditions can be adopted for getting the exact type of tempered martensitic structure that suits an end-application. Some such conditions of cooling have been shown in Fig. 2-8.

In addition to these advantages, martensitic heat treatment is also central to 'surface hardening' by case carburising or induction/flame hardening. In surface hardening, the aim is to produce fully martensitic structure on the surface, leaving the core structure to remain tougher. Such a combination of high surface hardness and tough core allows application of the steel for both wear and fatigue resistance. Tougher core structure can be produced by controlling the hardening process after case hardening of carburised jobs or giving prior hardening and tempering operation for flame or induction hardened jobs.

Martensitic structure is very effective for engineering applications involving wear and fatigue. For wear resistance the structure must have high hardness and for fatigue and fracture resistance the structure must have good strength and toughness, which can be produced by appropriate quenching and tempering. *Tempering at 450°C or above for 45 min/sq in area has been found necessary for alloy steel martensite to attain greater toughness. However, for plain carbon steels, this temperature would be somewhat lower.* At times, lower tempering temperature can be used to retain some more hardness, if the service condition calls for, but that would be possible only by sacrificing some toughness.

Because of these merits of martensite, it is the most sought after structure in all hardening grade steels requiring high strength and toughness for fatigue type applications or high hardness for wear resistance. But, it is not so for forming or bending applications, because of its limited ductility. Martensite is not easily workable because of its lack of ductility (see Table 1-1). Hence, applications of fully martensitic steel plate for structural applications are rare. Where high strength and high toughness steel plate is essential for applications, the steel plate must be well tempered or more ductile 9% nickel steel plate— which upon quenching and tempering produces lower-bainitic structure with lots of 'retained-austenite' spots—can be considered, if cost permits.

The process of formation of martensite (i.e. heat treatment process) is relatively simple once the right composition of steel has been chosen—one of the many standardised cooling method can be adopted—e.g. oil quenching, water quenching, martempering, etc. In such a cooling process, care must be taken to make the cooling rate fast enough to avoid touching the C-curve nose of the chosen steel. Once martensite has formed, it needs tempering as early as possible in order to avoid any chance of stress crack. For high alloy steels, where internal strain in the martensite is high and can cause cracking of the steel parts, often recourse is taken to thermally hold the transforming martensite into an intermediate thermal bath, called 'martempering bath'—maintained above M_F temperature but below M_S—in order to lower the strain in the martensite and prevent either cracking or distortion due to high residual stresses arising from fast quenching.

Therefore, the depth and breadth of utility of martensite comes from the fact:

- It's *as-quenched hardness* and strength can be chosen and manipulated by choosing the right combination of carbon and alloy in the steel and appropriate cooling method;
- Required level of application-specific hardness, strength and toughness can be easily achieved by controlling the tempering temperature and time
- Additional special properties (like low temperature property) can also be induced in martensitic structure by adopting appropriate alloy steel and special thermal treatment like 'martempering', and
- Mechanism of tempering of martensite is also a thermally activated process; and hence can be easily controlled by adjusting the temperature and time of tempering.

The tempered martensitic structure, despite its higher strength and toughness, is amenable to good machining and grinding for its hardness level. This feature adds to added attractiveness of martensite in fabrication process involving machining, which is a major manufacturing process in industries.

In sum, martensite is the most sought after structure in steel, except where normalising and annealing are required for both economy and applications. Therefore, most heat-treatment operations are essentially focused, to getting as much martensite in the steel structure as possible and improving the nature and character of martensitic microstructure by alloying and tempering—so that properties like wear resistance, fatigue resistance or fracture resistance can be easily obtained. For example, *alloy steels containing chromium, molybdenum* and *nickel* can be chosen—in preference over plain carbon or carbon-manganese or many other common alloy steels—to produce tougher martensitic structure for high duty applications like high-speed automobile gears or aircraft landing gears. Martensite structure of this type of steel allows formation of complex carbides by tempering at relatively higher temperature (e.g. 550°C and above), producing a stronger and tougher tempered martensitic structure, suitable for applications against high dynamic fatigue load and vibrations. Martensite in the microstructures of such alloy steels will have fine needles (twined martensite) that are extra tough. Additionally, the matrix of such structure can be made to contain some spots of *retained austenite*—in the form of small islands and finer precipitates of carbides—which make the steel further resistant to growth of cracks leading to fracture. Small islands of retained austenite act as crack arrester due to their highly ductile character. More about the martensite and its tempering process will be discussed in Chapter 8: Heat Treatment and Welding of Steels.

2.7. Comparison of Structures of Ferrite, Pearlite, Bainite and Martensite in Steel

Foregoing discussions on 'austenite decomposition' establish that there are four basic microstructures that form from austenite on cooling. They are *ferrite, pearlite, bainite* and *martensite*, in order of descending temperature of formation from austenitic region. In the light of discussions so far, it may appear that the structures like pearlite, bainite and tempered martensite are a kind of aggregates of ferrite (α) and carbide (Fe_3C or alloy carbide), which are the thermodynamic phases of steel. While this could be a macro-view of the situation, in reality there are sharp differences in the morphology of these structures due to temperature of formation and mechanisms of formation. It is these morphological differences between them that produce different properties in steels.

Carbides are chemical compounds; they form from austenite of a given composition during the phase separation under cooling. The form of carbide depends on the temperature of formation and the kinetics i.e. N&G factor. For instance, carbide that forms during pearlite transformation at the lower critical temperature is lamellar in form, with long carbide-ferrite interface, whereas carbides that precipitate out from austenite during bainite formation (below the pearlite formation zone) are particulate in form, with short interface, appearing on the grains or grain boundaries. Carbides that form on tempering martensite, on the other hand, are very fine globular precipitates, characteristic of formation at lower temperature, and form all over the matrix. Particle interface area of globular carbide is very small compared to pearlite or even bainite. These differences in nature and character of carbides considerably influence the properties of steels.

Ferrite (α)—the other structure in the aggregate—forms either on equilibrium cooling in the form of equiaxed grains or under non-equilibrium cooling, when the shape and size of ferrite can be acicular or plates/needles, as in martensite. The presence of ferrite either in different grain sizes or in other non-equilibrium shapes and forms makes considerable difference in the mechanical properties of steels.

The foregoing discussion outlines how these two structural aspects of steels can individually influence the properties of steels. When they exist in physically combined forms in a structure, their behaviour can change further according to their structural synergy. It is in this respect that some more structural characteristics of different phases of steels will be outlined and their uses and utility will be highlighted.

Depending on the cooling condition and formation temperature, structure and properties of ferrite will differ. For example, slow equilibrium cooling (i.e. annealing) and higher temperature of formation will produce nearly equiaxed ferrite of coarser grains (see Fig. 2-1), whereas faster air cooling (i.e. normalising) will produce finer grained ferrite due to higher nucleation sites available under faster cooling. The latter structure is stronger and tougher (as could be measured by %elongation) than the former structure. Hence, normalising by air cooling is resorted to when higher strength and ductility is required for a given application. But, if the steel is of high carbon where softness to be introduced for bending, forming or machining, annealing by slow furnace cooling can be resorted to produce ferrite with coarser grains, which is softer than fine grained annealed steel. However, unless the steel is having extra low carbon, structure will not be 100% ferrite—either by normalising or annealing—it will be mixed with *pearlite* (a lamellar combination of ferrite and carbide (Fe_3C)), see Figs. 2-3A and B. If the cooling rate is higher, as in normalising, there will be a tendency for formation of more pearlite and finer pearlite than when the cooling is slow like annealing. These faster cooled finer structures of ferrite-pearlite aggregate have higher strength and ductility than coarse grained ferrite that is formed by high temperature annealing.

However, what happens when the cooling rate is even higher than the air cooling used for normalising? As illustrated in the CCT diagram in Fig. 2-7, formation of pearlite and ferrite is still possible, but their characteristics will be different from the one formed under the equilibrium cooling conditions discussed earlier. Under these cooling conditions, pearlite, if produced, will be finer and dense, and ferrite will be feathery and coarser. Structurally, finer pearlite means narrower lamellar spacing between the ferrite stringers and carbide stringers within a pearlite colony (see Fig. 2-3A and B). Such fine pearlitic structure forms due to the effect of faster than air-cooling rate which induces large number of nuclei for pearlite formation due to higher under-cooling during faster cooling.

If the cooling rate is made even faster than above, bainite and martensite will form (see Figs. 2-7 and 2-8). Here again, morphology of bainite and martensite will change with the actual cooling rate being attained (see right hand axis of Fig. 2-6); indicating the nature of transformation products and their hardness levels. Even, martensite formed by water quenching and oil quenching a similar steel, will give rise to different martensitic morphology, and their tempering characteristics and the final properties will be different.

A closer examination of the detailed features of microstructures of pearlite, bainite and tempered martensite can be summed up as:

1. Pearlite is composed of alternate layers of ferrite and carbide lamella (see Fig. 2-3B).
2. Bainite is also composed of ferrite and carbide, but ferrite is present in acicular/platelet form and carbide in fine precipitates form—carbides precipitating on the ferrite plate boundaries or all over the ferrite body and boundary.
3. Depending on this carbide precipitation nature, bainite is divided into 'upper bainite' and 'lower-bainite'(see Fig. 2-9).
4. Tempered martensite is also composed of fine ferrite plates/needles with fine carbide precipitates all over the matrix (see Fig. 2-9).
5. Ferrite platelets in martensite are not only finer, but also have more dislocation density than ferrite in bainite.
6. Generally, martensite is tempered at temperature lower than bainite formation (B_S) temperature, in order to keep the tempered martensitic structure fine with evenly dispersed carbide precipitation.

Thus, *pearlite, bainite* or *tempered martensite* are *aggregates of ferrite* and *carbide*—the two primary thermodynamically stable phases of steel at room temperature. Despite this, there are wide differences in properties amongst these phases, mainly based on the nature of ferrite, size of ferrite and formation, distribution and composition of carbides, giving rise to different morphological characters in the microstructure of steel.

Table 2-2 gives the possible combination of morphological changes that might result from differing cooling conditions. An understanding of the influences of structure and its morphology on the properties of steels is at the centre of heat treatment of steel. Efforts to develop various special heat

Table 2-2. Influence of cooling conditions on microstructures of steels and their morphology.

Phase	Cooling Condition	Nature/Morphology of the Microstructure
Ferrite	Equilibrium cooling	Nearly equiaxed Ferrite grains—mostly present along the prior austenite grain boundaries. Within this, slower cooling (e.g. furnace cooling) will produce coarser grains and faster cooling (e.g. air cooling) will produce finer grains
Ferrite	Faster than equilibrium cooling	Feathery ferrites, at time resembling to Widmansttaten structure type
Pearlite	Equilibrium cooling	Lamellae of ferrite and carbide (see Fig. 2-3) interspaced closely, formed due to 'eutectoid' reaction of carbon-rich austenite at A_1 temperature
Pearlite	Faster than equilibrium cooling	Denser lamellae of ferrite and carbide; closely interspaced, spacing decreases with further increase of cooling rate or carbon and alloy content.
Bainite (upper)	Initial fast cooling of the austenite till the nose of C-curve or the B_S temperature of the steel (whichever is higher), then can be held isothermally or continuously cooled (see Fig. 28).	Feathery ferrite plates with fine carbide precipitation, dominantly over the plate boundaries (Figs. 2-4 and 2-8)
Bainite (lower)	Fast cooling till below the nose of C-curve of the steel or below the B_S temperature, then slower cooling or isothermal cooling (see Fig. 2-8).	Fine feathery plates of ferrite with fine carbide precipitation all over the matrix; see Figs. 2-9 and 2-10 Lower bainite structure can closely resemble tempered martensite structure if formed at the lower range of temperature, but above the M_S temperature.
Martensite (tempered)	Fast cooling, which should be fast enough to avoid intruding into other higher temperature phase areas of CCT diagram of the steel e.g. pearlite or bainite formation areas (see Figs. 2-7 and 2-8).	Fine ferrite plates or needles with dense dislocation tangles and very fine carbide precipitation all over the matrix (see Figs. 2-10 and 2-13A). For revealing dislocation density very light etching and high magnification examination by transmission electron microscopy should be used. Needle shaped martensite can have signs of twinning inside, especially in alloy steel, making the steel stronger and tougher. The carbide can be iron carbide or alloy carbide, if alloy steel of appropriate composition is used.

treatment processes are mostly guided by, the consideration of how to modify structural morphology for improved properties. Development of various thermo-mechanical heat treatment processes—like sub-critical or inter-critical thermo-mechanical treatment (TMT) or thermo-mechanical controlled processing (TMCP) are few examples. They are all directed for developing special properties by inducing morphological improvements in the structures of steel. In steel, structure is the essence of its properties.

2.8. Structure–Property Relationship in Steel

Throughout this chapter, the focus of discussion has been to demonstrate how different structures can be developed in steel, by playing through the combination of composition and cooling rate. Steel draws a special place in the world of materials due to its structure-sensitivity of properties. Slightest variation in the structure or structural morphology leads to variation in properties, especially the mechanical properties. Though this has been discussed repeatedly in the chapter, importance of this subject, demands a brief review and reiteration of how structure influences the mechanical properties of steel.

Structure of steel is dependent on the composition of the steel and the way the steel is processed and/or heat treated. For example, a 45C8 steel (equivalent to SAE 1045) when hot rolled or forged, the structure will be a combination of (ferrite + pearlite) with relatively coarser structures. But, the same steel when normalised after forging/rolling, the structure will be more uniform and fine with relatively higher strength and elongation, than as rolled/forged steel. If this steel is further subjected to hardening by quenching and tempering, the structure will be essentially tempered martensite having much higher strength and toughness. Even in hardening, if the quenching is carried out in water as quenching medium, depth of hardening and volume of martensite will be higher than when quenching is carried out in oil; the former method producing better mechanical properties than the latter due to higher volume of martensite in the structure.

Therefore, for a given steel composition, microstructure that gets produced through a processing route and/or heat treatment determines its properties, especially the mechanical properties. There could be mild influence of structure on other types of steel properties (e.g. chemical and physical properties), but mechanical properties are solely dependent on microstructure. Table 1-4 in Chapter 1 has outlined how various microstructural constituents in steel influence different aspects of mechanical and forming properties. Corrosion, an important chemical property of steel, is also said to be influenced by the structure, such as by grain size and inclusions.

The subject of structure-property relationship in steel can cover a very wide range of studies involving all sorts of physical, chemical and mechanical properties, but that is not the purpose of discussions in this book, where the focus is on mechanical properties. Influence of different steel microstructures—such as ferrite, pearlite, bainite, martensite, carbide, etc. which are the products of austenite decomposition—on the mechanical properties will be highlighted here. Influence of grain size and inclusions on mechanical properties will be discussed in Chapter 3.

Characteristics and nature of microstructures produced by austenite decomposition under different cooling conditions have been outlined in Section 2.6 in this chapter. All transformation products—like ferrite, pearlite, bainite and martensite—are very sensitive to the change of composition and cooling conditions. Slightest change in the composition or cooling rate can cause changes in the morphology of these structures; consequently influencing its properties due to structure sensitivity. Discussions here will focus to sum up, what structural changes influence what properties of steels, including forming properties.

Figure 2-14 illustrates the basic wrought structure of steels with increasing carbon, illustrating increasing volume of pearlite in the structure with increasing carbon. The diagram also indicates the change in mechanical properties—like hardness, strength and elongation—with increasing pearlite in the structure. Perhaps, structure property relationship in steel can be best discussed by starting with this diagram.

Figure 2-14 shows how increasing carbon leads to increase in pearlite content in the steel, resulting in higher hardness and tensile strength but lowering of the elongation. Lower elongation implies lower ductility in the steel. Therefore, high carbon wrought steel, where pearlite content is high and free

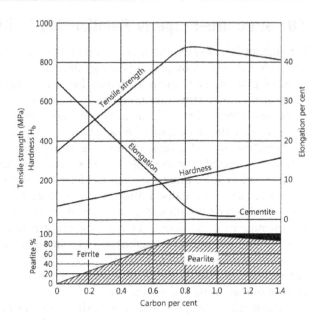

Figure 2-14. Change of microstructure (volume fraction of ferrite and pearlite) with increasing carbon content and consequent changes of properties in plain carbon wrought steel. Wrought steel refers to cast and rolled steel to shape, but without heat treatment. (1 Kgf/mm^2 = 9.85 Mpa.)

cementite (Fe$_3$C) starts appearing (see Fig. 2-14, starting of cementite formation at carbon level above 0.80% in the steel) loses normal room temperature workability due to lack of sufficient ductility. This type of change of properties in steel with increasing carbon is due to its structure, having high volume of pearlite plus some free cementite.

However, if the same structure is given a heat treatment like prolonged annealing or spheriodising, there will be changes in the structural morphology. Free carbides in the structure will break down and ball up along with spheriodisation of pearlitic structure, producing large numbers of carbide spheroids of different sizes in a matrix that is essentially ferrite. Such a structure will vastly improve the cold workability of the steel, especially machining. Thus, by appropriate heat treatment a steel which was difficult to machine can be rendered machinable through changes in structure. Annealing or spheriodisation of high carbon steel will reduce its strength but improve the ductility, thereby rendering the steel more workable.

If, on the other hand, a medium carbon steel (e.g. SAE 1045 steel with carbon content of 0.45%) requires improvement in strength and ductility for—cold forming, machining or any other end-use specific application—normalising heat treatment can be given to the steel. Normalising will produce fine grained structure with finer pearlitic spacing, adding to ductility with some improvement of strength as well. Limitation of normalising is that it is more effective for change in strength and ductility in lower and medium carbon steels than in high carbon steel. In low and medium carbon steel, normalising will not only refine the structure, but will also change the relative volume fraction of ferrite and pearlite in the structure compared to annealing due to relatively faster cooling rate than annealing; thereby further influencing the properties. However, normalising can still be used for homogenising and modifying the structure of higher carbon steel, in order to bring uniformity in properties or for uniform response to heat treatment after rolling/forging. Thus, properties of steels can be altered through structural changes brought about by different heat treatment, like annealing, spheriodising, normalising or any other heat treatment process that is considered appropriate. Important part is to know what structure or structural changes are required in the steel for a given set of properties.

It may be recalled from earlier discussions that cementite (Fe_3C) in most commercial steels exists in intimate mixture with pearlite, bainite or even tempered martensite, but rarely as free cementite. Free cementite is more detrimental to workability of steel than the carbides in pearlite, which exist in the form of lamellar carbide. Lamellar carbides in pearlitic structure are interspaced between ferrite stringers, and this type of structural morphology may help in cold drawing of the steel where part of deformation load is taken by the ferrite stringers in between the carbide lamellae. Therefore, cold drawing of high carbon wire is not uncommon in industries.

Improvement of structure by annealing or spheriodisation might render a higher carbon steel more ductile and suitable for further working (e.g. bending or machining), but the steel will lose its strength. Strength is an important property for structural and load bearing applications. If the load bearing application involves static load, as in structural applications, steel of suitable composition can be subjected to normalising process for improvement of strength and ductility. But, if normalising of the steel does not produce enough strength and ductility for meeting the property requirements of an application, there could be other methods of inducting higher strength and ductility in the steel. One such method is controlled rolling of micro-alloyed steel for getting very fine structure and precipitates of carbides and nitrides. Thus, opportunities for microstructural adjustment in steel for meeting different properties are plenty. However, amongst various techniques available for microstructural adjustment, hardening of steels occupies bulk of the technological space for improving properties of steels.

Hardening of steel drastically changes the structure and properties of steel by formation of martensitic structure, which is very hard and brittle in as-quenched condition. But, this structure can be easily tempered for improving the toughness with some sacrifice of strength. Tempering of hardened steel can be done over a range of temperature and time, for adjusting the final mechanical properties as required for specific application. Therefore, if the load bearing application involves dynamic load (e.g. fatigue load), the steel could be given a hardening and tempering operation for developing structure, that is stronger and tougher than other methods discussed here.

Heat treatment for hardening of steel is a very versatile option, which can be applied for 'bulk hardening' (also called through hardening) or 'surface hardening'—where hardening is confined to a specified surface area of the component. These processes are used for producing martensite based structures (i.e. tempered martensite) that are very strong and tough and capable of meeting very demanding service conditions involving wear, fatigue and fracture. The structure and properties obtained after hardening and tempering are primarily the functions of chemical composition of the steel, cleanliness of the steel, and the hardening and tempering process. Example of composition influencing the steel structure and properties are plenty from the austenite decomposition study (refer Sections 2.2 to 2.4), which demonstrates that steel structure and its properties can be easily tailored by choosing right composition of steel and cooling condition.

Choosing cooling condition from the standard set of cooling media, each having its characteristic cooling rate, may not be difficult, but choice of composition for an end application requiring some specific properties could be a challenge. Because, there are hundreds of steel grades available for hardening, and identifying which one is most appropriate is often difficult. For example, strong and tough tempered martensitic structure can be obtained from a number of combinations of alloying elements containing Mn, Cr, V or Mo, but if the properties so obtained call for highest toughness at the strength level, then choosing an alloy steel with appropriate percentage of Ni in combination with Cr and Mo might be necessary. Combination of Ni with Cr and Mo is believed to induce extra toughness in the tempered martensitic structure—due to their synergistic effect on the structure that gets developed—e.g. heavily twined martensite with lots of very fine and evenly dispersed alloy carbide on the twin boundaries. In addition to such judicious composition control, if these steels are made with very low inclusion content by using clean steel making technology, the toughness property in the steel can be made even better. Thus, various combinations of steel parameters—like the composition, cleanliness and grain size—along with different options for heat treatment of steel allow a wide choice for tailoring the steel structures and properties befitting any specific application.

There might be a number of mechanisms for strengthening of steels (see Table 1-2), but hardening of steel by producing martensitic structure is by far the most effective and widely used method. Martensite so produced allows stress relieving/tempering for exactly adjusting the properties for a given application. Tempering induces ductility in the martensite structure, by promoting migration of extra carbon from the strained martensitic structure and forming fine carbide precipitates. The precipitated carbides could be plain iron-carbide or alloy carbide as per composition. Stages of tempering of steel and associated changes in the structure during tempering have been described in the Section 2.4 in this chapter. Figure 2-15 further illustrates a schematic tempering behaviour of martensite with temperature and time for helping to understand—what happens to martensite structure during tempering and how to control the ultimate properties.

Figure 2-15 is only a qualitative illustration of how hardness (strength) can change with different tempering temperature and time. The loss of hardness (or the strength) of martensite with tempering is due to change in the structure. With tempering at a given temperature and time, martensite precipitates out fine carbides, formed by the carbon atoms coming out from the super-saturated and distorted ferrite lattice. Initially, these fine carbides are transitional carbides ($Fe_{2.4}C$), which coalesce with each other with increasing temperature forming normal Fe_3C. If the steel contains some carbide forming alloying

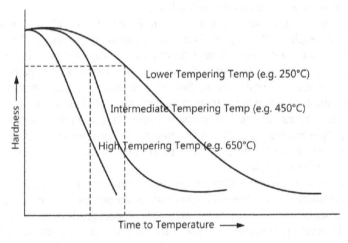

Figure 2-15. Qualitative illustration of drop in hardness of martensite with tempering temperature and time.

elements, in addition to iron carbide, some alloy carbides (chromium carbide, molybdenum carbide, vanadium carbide etc.) will also form. But, the rate of formation of alloy carbide will be lower than the plain iron carbide due to lower diffusion rate in the alloy matrix. For a given tempering temperature and time, alloy carbides will be finer in size compared to plain iron carbide.

Difference in strength and toughness between tempered plain carbon martensite and alloy steel martensite are mainly due to the carbide type and size distribution, and partly due to acicular nature of ferrite. Shape, size and distribution of precipitated carbides are significant factors in determining the tensile properties of steels. Highest strength in tempered martensite is obtained when the carbide precipitates are very small and coherent with the matrix structure, but that might not be the toughest. For toughness, martensite with very fine and even dispersion of carbide is preferred. As such, surface hardened martensitic structure, which is required with high hardness to withstand wear as well as fatigue, is generally stress relieved or tempered at very low temperature (below 200°C) in order to control the structure for highest strength (hardness). But, bulk hardened martensitic structure, which is used for high bending and fatigue load, and less for wear resistance, is generally given a tempering operation at relatively higher temperature (above 400°C) for inducing more toughness through even dispersion of fine carbide.

Structural changes brought about by tempering or re-heating of initial martensite or bainite or pearlite are essentially concerned with carbide precipitation and controlling the carbide size in the microstructure. They all aim to bring down the strength and increase the ductility or toughness of the steel, with the exception of 'secondary hardening' steel. Steels containing molybdenum may exhibit 'secondary hardening' (i.e. increase in hardness with tempering above 550°C), and thereby increasing the strength of the steel, instead of lowering, with tempering at around 600°C. This happens due to very fine precipitation of complex molybdenum bearing alloy carbide. Such steels are, therefore, preferred for application involving higher temperature of application (e.g. creep), because complex alloy carbides produced by secondary hardening of Mo-bearing steel will facilitate structural stability and high temperature strength of the steel for creep applications. A rule for deciding appropriateness of structure for higher temperature applications is that the structure must be stable at the operating temperature. And, in this respect, general rule is that a structure is stable upto the temperature where it had originally formed.

Thus, mechanical properties of steel are strongly influenced by the type of structures and their exact morphological character. When ferrite is acicular and carbides are fine discrete precipitates—as in lower bainite and martensite—strength of the structures is remarkably high compared to lamellar structure of ferrite and carbide in pearlite. Strength data (as depicted by VPH in Fig. 2-6) indicate that pearlite can have maximum hardness of 450 VPH when very fine as against hardness of about 700 VPH for lower bainite and ≥750 VPH for martensite in 0.60%C steel. This drastic increase of hardness in bainite and martensite is partly for the acicular or needle shaped ferrite and mostly for the fine dispersion of carbides in the matrix. Nonetheless, it must be appreciated that martensite must be tempered or at least *stress relieved* before put to service. Stress relieving of martensite is generally carried out below 300°C, in order to retain the near-original structure with minimum drop in strength. Typically, medium carbon low alloy steel will lose about 2–3 HR_C hardness by tempering at around 250°C–300°C.

Thus, close relationship of structure and mechanical properties in steel opens up the vista of many ways steels can be designed, processed, treated and finished for developing the necessary structure and properties. And, in this direction, rules and principles of austenite decomposition, discussed in this chapter, act as the guiding post.

SUMMARY

1. The chapter focuses on the process of development of different microstructures in steel and the character and nature of those microstructural products through the austenite decomposition under different cooling rates. The chapter covers the process of austenite decomposition—by illustrating the effect of N&G, equilibrium cooling, isothermal cooling, and continuous cooling conditions.

2. Formation of ferrite, pearlite, bainite and martensite, and their morphological differences and character based on cooling rate have been pointed out and the reasons discussed. Bainite and martensite formation has been discussed with reference to both TTT and CCT diagrams of steels. Difference between TTT and CCT diagrams and their uses have been pointed out. With reference to an illustrative CCT-diagram, arising from different microstructural phases in steel and the corresponding cooling rate has been shown.

3. Because of the importance of bainite and martensite structure in heat treated steels, nature, character and differences in structure and properties between these two critical structures in steels have been discussed; differences in their properties have been compared (see Table 2-1). Different forms of bainite (e.g. upper and lower bainite) and martensite (lath and needle types) and the conditions of their formation and difference in their properties have been discussed.

4. Martensite being the most important of all structures in steel, the reasons for its unique properties have been highlighted. Importance of tempering of martensite for obtaining the right strength and toughness in steels has been pointed out.

5. For further clarity, finer aspects of microstructural features of pearlite, bainite and martensite with reference to the distinguished nature and form of carbides, have been discussed and their contributions to mechanical properties have been highlighted.

6. Finally, structure–property relationship in steel has been discussed and illustrated.

FURTHER READINGS

1. Reed-Hill, Robert E. *Physical Metallurgy Principles,* 2nd edition. New York: Van Norstrand, 1973.
2. Dieter, George E. *Mechanical Metallurgy.* New York: McGraw-Hill, 1986.
3. Rollason, E.C. *Metallurgy for Engineers.* London: The English Language Book Society & Edward Arnold (publishers) Ltd., 1973.
4. Hume-Rothery, W. *The Structure of Metals and Alloys.* London: Institute of Metals Monograph, 1962.
5. Raghavan, V. *Physical Metallurgy: Principles and Practice,* 2nd edition. New Delhi: Prentice Hall of India, 2006.
6. Atlas of Isothermal Transformation Diagrams, 2nd revised edition. Pittsburgh, United States Steel Corporation, 1974.
7. American Society for Metals. *Atlas of Isothermal Transformation and Cooling Transformation Diagrams.* Ohio: Metals Park, 1977.

Influence of Grain Size and Inclusions on the Properties of Steels

The purpose of this chapter is to highlight the importance of grain size and inclusions in steel and their influence on mechanical properties and behaviour of steels. Contents of the chapter cover the formation, characteristics and effects of grain and inclusions on the properties of steels, highlighting the importance of interfacial character of these entities in the steel matrix. The chapter also covers the methods of determination and measurement of these micro-constituents for the purpose of reference and specification.

3.1. Introduction

Grains and inclusions are a part of microstructural features, which significantly influence the behaviour of steel. Table 1-4 in Chapter 1 focussed on the qualitative influence of grain size and inclusions on various mechanical properties and forming behaviour of steels. While the influence of inclusions is limited to the mechanical and forming properties of steels, influence of grain size extends beyond the mechanical properties; it also influences transformation behaviour of steels as discussed in Chapter 2. In general, grains and grain sizes can be used with advantage for tailoring mechanical and forming properties in steels, but inclusions are held as deleterious for similar purposes, except in the case of machining of steels.

Grains and inclusions are integral to any steel structure, excepting in the case of single crystal. No matter what the conditioning of steel solidification process is or control over cooling rate during austenite decomposition, grain boundaries consisting of atomic-level structural disorientation will form, delineating the individual grains of different crystal orientation. Similarly, whatever could be the precautions in steelmaking process; there will be some internal gas-metal reaction or presence of external impurities originating from slag-metal reaction or metal-refractory reactions or carryover from unused chemical additions, giving rise to inclusions of different chemistry. These reaction products or the trapped particles fail to get floated-off to the surface of the steel bath during steelmaking—due to viscosity of liquid steel and the poor buoyancy of such small particles. Hence, these particles get entrapped in the solid steel and remain as inclusion. Therefore, grain boundaries and inclusions are an integral part of solid steel, and by being present in the steel structure, they exert considerable influence on the mechanical and forming behaviour of steels. The exact nature of their influence will, however, depend on the shape, size and distribution of these physical entities.

Grains and grain boundaries formed during steel solidification is not unchangeable; their shape and size can be changed by subsequent hot working and/or heat treatment, to make them favourable for particular uses or application. For example, hot-working and normalising of steel will make the grain size very fine, rendering the steel favourable for applications requiring resistance to deformation, under load at normal or below the room temperature. However, the same steel with fine grain size will not be good for high-temperature creep deformation, because deformation under creep load (creep is a slow high temperature deformation process) gets accelerated by grain boundary sliding, where finer the grains more is the grain boundary area for sliding. Hence, coarser grains with lesser grain boundary area are preferred for application under creep. Similarly, finer grains with higher grain boundary areas are bad for machining due to higher strength and ductility in the steel, leading to difficulties in chip breaking in fine grained steel. Thus, by changing the grains and grain sizes in steel with the help of mechanical deformation and/or thermal treatments, application specific properties of steels can be improved or areas of weakness can be overcome—e.g. machining fine grained steel. Therefore, control of grain size in steel should be done with discretion of what is good and what is bad for the uses and applications.

Similarly, inclusions, which originate from steelmaking process, are always present in steel in varying degrees; no matter what the precautions in steelmaking practices are. They can be minimised but not totally eliminated. Inclusions of any kind are considered bad for steel properties, but their detrimental influence can be lowered by making them finer, globular and evenly dispersed. Again, there is exception to this rule and that exception relates to machining. For machining, hard inclusions of any size and distribution are detrimental, because they cause wear and erosion of cutting tool tip, but softer 'sulphide inclusions' are desirable due to their lubricating effect that helps to reduce friction between metal and cutting tool. Other than such special effects, inclusions, in general, have detrimental effect for all load bearing applications of steels where components can fail by fracture under the applied stresses e.g. fatigue fracture, shear fracture etc.

Thus, as regards mechanical properties of steel e.g. strength, toughness, ductility etc.—which are the major considerations for selection and application of steels—both grain size and inclusions play important roles. For majority of applications, fine grained steel with minimum inclusion content is desirable. Therefore, understanding of the role of grain boundaries and influence of grain size, on the one hand, and the origin and effect of inclusions in the behaviour of steel, on the other, are necessary for efficient use and effective application of steels.

3.2. Formation of Grains and Grain Boundaries

Formation of some kinds of grains (or fibres) is a natural phenomenon in all materials, except in the case of glass. Therefore, all metals and alloys that we use, excluding normal glass, are polycrystalline solids, consisting of many constituent tiny single crystals (i.e. grains) which are disorientated with respect to each other and meet at internal interfaces. These tiny single crystals are called 'grains' and the internal interface where these tiny crystals (i.e. grains) meet is called 'grain boundaries'. Thus, grain boundaries are the interfaces between two grains in polycrystalline material. A schematic arrangements of atoms and the crystals (grains) and their interfaces is shown in Fig. 3-1A, and the actual grain boundaries that result from such atomic arrangements in metals is shown in Fig. 3-1B (revealed by polishing and very light chemical etching of the meal surface).

Most solids such as metals and alloys, therefore, have a crystalline structure where atoms are arranged in an orderly manner in the crystal lattices in three-dimensional periodic manner, forming crystals, and these crystals meet each other at inter-crystalline interfaces. The orderly arrangement of atoms in the crystal lattice gets broken down at the inter-crystalline interfaces, forming the crystal

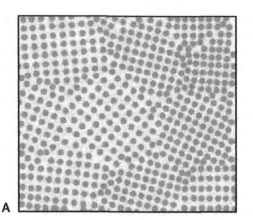 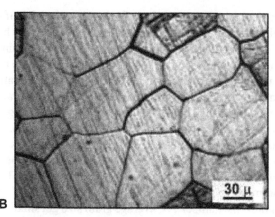

Figure 3-1. A: Schematic outline of crystal boundaries in polycrystalline material, and B: Micrograph of real grain boundaries in metals revealed by acid etching.

boundary (i.e. grain boundary) where atoms lose their orderliness of arrangement and the interface (boundary) structure becomes full of defects with atomic voids.

The crystal interfaces (i.e. the boundaries) are, however, generally planar, having two-dimensional periodic atomic structure. This situation implies that in a polycrystalline cube of size 1 cm, with grains of 0.0001 cm in planar diameter, there would be 10^{12} crystals, resulting in grain boundaries of several square meters. Hence, any influence of grain boundaries on any property—either mechanical or chemical or physical—would significantly change with grain diameters i.e. sizes.

At atomic level, boundaries between crystals (grains) are jagged interfaces with ledges and atomic level disorder. This makes the grain boundary structure defective, having higher interfacial energy. Thus, grain boundaries are considered as 'defects' in the crystal arrangement, and they tend to influence the *physical properties* like thermal and electrical conductivity; *chemical properties* like the corrosion behaviour, and the *mechanical properties* like strength and elongation by their presence. Grain boundaries obstruct the dislocation glide (necessary for deformation to progress) through them, influencing the strength of steel. Further, these boundaries, having higher interfacial energy, can act as preferred sites for—impurities to segregate, precipitates to occur and nucleation to start for new phase formation.

During the formation of grains, the shape is governed by its neighbours, and they are never sphere. It generally takes the shape of irregular polyhedron that when packed together completely fills the space. Depending on the way grains have formed, they could be equiaxed, elongated or pancake shaped when viewed on planar surface; equiaxed when characteristic dimensions are nearly same in different directions; elongated when dimensions are not the same, one direction is much longer than the others, and pancake structure is the one where two dimensions are much higher than the third dimension. The elongated grains are also called the columnar grains, which are common in steel solidification due to chemical inhomogeneity and impurities causing disturbances in uniform crystal growth. The size and shape of crystal growth is also influenced by the way heat is being extracted out from the liquid during solidification; faster directional cooling gives rise to directional columnar growth.

However, such initial grains of different sizes and shapes can be subsequently changed by hot-working and heat treatment in order to get the desired size and shape of grains for specific applications. For instance, in the case of steel, the steel can be heavily hot-worked and recrystallised near the lower critical temperature to form fine equiaxed grains of smaller diameter (giving more number of grains and higher grain boundary areas in a given space) for higher strength and ductility.

As regards mechanical properties of steels, grain size influences 'yield strength' (YS), tensile strength, elongation, impact toughness properties, and creep deformation. Except in cases of creep deformation and machining operations, in all other areas of applications finer grains (having higher grain boundary area in a given volume) are preferred for their favourable influence on the deformation and plastic flow of metals. As such, there is a need for describing the grain sizes in metals and steel in a standardised manner for referring to the effect of grain size variation. Recognising this necessity early, American Society for Testing of Metals (ASTM) came out with a standardised measure of grain sizes in steels, assigning numbers from 1 to 9. The ASTM grain size number "N" is defined by: $n = 2^{N-1}$, where n is the number of grains per square inch when viewed at ×100. The formula expresses the average grain diameter with permissible departure within the individual field of examination. This formula implies that with increasing 'grain size number', N, average grain size get smaller and the grain boundary areas exponentially increase. More about the application related properties and behaviour of grain sizes will be discussed later.

3.3. Formation of Grains in Steel and their Structure

Steel is an alloy of iron and carbon, where other application specific alloys like Manganese (Mn), Chromium (Cr), Nickel (Ni) etc. could be present in the steel composition along with some incidental or residual elements like Sulphur (S), Phosphorous (P) etc. Along with chemical composition, liquid steel also contains many suspended particles. Therefore, the liquid steel from which grains form on solidification is often a non-homogeneous liquid with many suspended particles. In such a liquid, uniform nucleation and uniform growth in all directions is difficult; instead solidification proceeds largely with columnar or columnar-dendritic growth, especially due to non-equilibrium solidification under directional cooling. Figure 3-2 schematically depicts the stages of grain and grain boundary formation during such non-equilibrium solidification, showing the formation of dendritic structure and merger with each other.

As the molten steel starts cooling, some nuclei form at random due to under-cooling effect. Some of these nuclei which can quickly acquire a 'critical mass' become stable and act as points of growth for solidification. This solidification process generally proceeds from these points by formation of dendritic pattern, growing in direction opposite to the direction of heat extraction in the solidification mould. The dendrites so formed grow directionally and tend to merge with each other. As a consequence, distinct boundaries form where two dendritic arms meet (see Fig. 3-2). The grains and grain boundaries so formed at high temperature during the solidification process are relatively large and coarse, and the structure is termed as 'as-cast dendritic structure'.

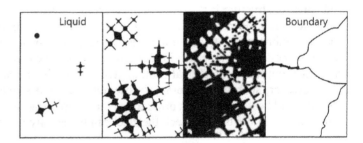

Figure 3-2. Different stages of nucleation, dendrite formation and grain boundary formation in metals on solidification. *Source: Rollason, E.C. Metallurgy for Engineers, 1973.*

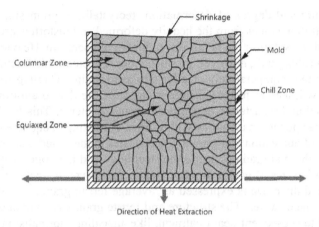

Figure 3-3. An illustrative mix of solidification pattern in a mould. Arrows show the direction of heat extraction due to mould walls.

In practice, metals are made to solidify inside a mould and the shape, size, wall thickness and material of mould, exert considerable influence on solidification pattern by influencing the way heat is extracted out of the liquid metal. Hence, in practice, the solidified structure of steel is a mixed one—there would be a zone of chilled small equiaxed grains due to chilling effect of mould wall, followed by columnar dendritic zone growing opposite to the direction of heat extraction, and finally a central zone of coarser equiaxed grains arising from slow cooling at the centre of the mould. An illustrative mix of solidification pattern is given in Fig. 3-3.

An important feature of steel solidification is segregation of solute elements like C, S, P etc. Segregation takes place between the solidifying arms of dendrite due to freezing of solute-enriched liquid in the inter-dendritic spaces. Similarly, impurities will get entrapped in between the dendritic arms. Thus, the as-cast structure of steel contains grains of different morphology and zones of segregation and impurity concentration in the inter-dendritic arms. Such cast structures are weak and are not favoured for engineering applications. Therefore, cast structure is to be further reheated and hot worked for breaking down the structure, for producing more uniform grain structure with uniform properties. This is done by giving heavy hot-working to cast steels by rolling or forging. During re-heating and hot working of steel, dendrites present in cast structure break down into smaller grains due to: (a) *allotropic transformation* i.e. phase changes and (b) by *recrystallisation*.

When as-cast steel is heated to higher temperature (above A_3) and soaked, the entire as-cast structure should allotropically transform to austenite (γ) with grain size characteristic of the re-heating temperature and soaking time at that temperature. Impurities and segregation present in the steel migrate and take positions in the newly formed austenite grain boundaries. But, this structure is also coarse, non-uniform and not equiaxed, because of limitation of holding time. Therefore, the structure needs further working for refinement, which is accomplished by hot-rolling or forging. Hence, hot-working by forging or hot-rolling follows the soaking of the steel at higher temperature and the steps of deformation process are such that, the structure breaks-down and causes formation of fresh grains by recrystallisation process. Recrystallisation takes place during hot-working (rolling) of steel itself. Hence, by controlling the rolling temperature and degree of deformation, recrystallised grain size can be controlled. The energy for recrystallisation comes from the heavily deformed and distorted structure of austenite under rolling and the rate of recrystallisation is fast due to higher temperature. Heavier deformation and lower temperature of finish rolling will, however, produce more refinement of structure by lower temperature recrystallisation, forming comparatively smaller equiaxed grains. Cooling of steel from the rolling/forging temperature will lead to further change in the structure due to allotropic transformation from austenite, as discussed under austenite decomposition in Chapter 2. This final structure, arising from rolled/forged austenite upon normal cooling will be finer, consisting of equiaxed ferrite and finer pearlite due to large increase of nucleation and growth (N&G) sites in the rolled matrix.

However, due to thermal gradient inside the steel body and heterogeneity of steel structure, there will still be variation in sizes of grains in steel. Hence, grain size in steel is not one single size—it is mixed to an extent and their size is expressed as 'average ferrite grain size', in terms of ASTM grain size number, N mentioned earlier. The structure and ferrite grain size may undergo further changes, if the steel is subjected to subsequent heat treatment, like annealing, normalising etc. It is this final grain size and structure which ultimately influences the workable properties of the steel. Following the logic of nucleation and growth for different structure formations in steel (refer Chapter 2), these grain sizes will be influenced by the parent 'austenite grain size' from which they form, and are popularly termed as ASTM grain size.

It is this parent austenite grain size that participates in the process of austenite decomposition/transformation as per the rules of N&G process. Hot-worked austenite will have higher N&G sites for its allotropic transformation; consequently, pearlite produced from such austenitic structure will be finer with fine lamellae of ferrite and carbide and having higher mechanical properties. Because of this unique feature of steels, where its grain sizes and structure can be controlled—by controlling the rolling, forging or transformation conditions—considerable advantages can be derived towards making steel more fit for a given end use.

Grains formed by equilibrium isotropic transformation or by recrystallisation are generally strain-free. But, if the steel has been quenched or cold deformed and heated, there could be dislocation cells and sub-grains of low-angle grain boundary within the distorted grains or crystals. Such structure is highly strained and requires further heating for recovery from the strain. This step refers to 'tempering' for quenched steels and 'recovery' for cold-worked steels—e.g. wire drawing, bright drawing, cold rolling, etc. Recovery after cold-deformation is generally carried out at relatively lower temperature than full recrystallisation. Recovery involves stress relieving (i.e. lowering of energy) by rearrangement of dislocations which gets produced in the matrix due to deformation, forming dislocation cells; sub-grains of lower disorientation (low-angle sub-grain boundary); and high-angle grain boundaries. In well recovered matrix, generally sub-grains within the grains are present. These sub-grains of low-angle boundary are equally effective in stopping the dislocation movements necessary for continuation of plastic flow under deformation by dislocation glide. Hence, materials with presence of sub-grains give rise to higher strength than that of only equiaxed recrystallised grains. This situation can, therefore, be gainfully used for increasing the formability and strength of cold rolled or cold drawn steels, where sub-grain boundaries also act as barriers for dislocation glide.

Grain boundaries along with its sub-boundaries act as obstacles to plastic flow of metal while under deformation. This happens due to the fact that any angular disorientation of boundaries and sub-boundaries will obstruct the movement of dislocations through them, causing more stresses to be applied to overcome those obstacles for the progress of the plastic deformation by dislocation movement. Hence, higher the area of grain/sub-grain boundaries, higher is the resistance for plastic flow i.e. finer the grain size higher is the strength—resulting in higher grain boundary area. Hence, by controlling the grain size and its sub-structure, mechanical properties of steels can be gainfully altered for different applications.

However, for consistent and effective control of grain size in steels, just hot-working and control over allotropic transformation may not be good enough for industrial practice. This is due to rapid growth of unpinned grain boundaries above certain re-heating temperature. Hence, steels are made with Al-killing where aluminium reacts with and removes oxygen from molten steel on the one hand and combines with nitrogen present to form aluminium nitride (AlN) particles, which are stable at a high temperature. The stable AlN particles then precipitate and pin the grain boundaries. Pinning of grain boundaries by AlN ensures that the grains are not free for growth in subsequent re-heating. Hence, for fine and stable grains, steels are recommended to be produced by Al-killing; refer Chapter 4 for more information.

3.4. Influence of Grain Size on Mechanical Properties of Steels

Grains and grain boundaries have wide ranging influence on all three types of properties of metals, namely physical, chemical and mechanical. They offer resistance to free flow of electric current and heat at the grain boundary junctions, provide high energy sites for corrosion to start at the grain boundaries, and influence the deformation and plastic flow by offering resistance to dislocation movements and glide. With reference to steels, influence of grain size and grain boundary is most prominent in the areas of mechanical properties of various descriptions. The following table summarises the qualitative influence of grain sizes on different mechanical properties and corrosion of steel.

Finer the Grain Size	Fatigue Strength	Fracture Strength	Impact Strength	Machining	Cold-Forming	Creep Strength	Corrosion and Pitting Resistance	Yield Strength	Tensile Strength
	↑↑	↑	↑↑	↓	↑	↓↓	↓	↑↑↑	↑↑

Most fundamental influence of grain size is in improving the YS of steels with decreasing grain size. Figure 3-4 illustrates the change of YS with grain size, where d is the grain diameter.

Variation of YS with grain size follows the Hall–Petch relationship:

$$\sigma_y = \sigma_0 + k_y\, d^{-1/2}$$

where σ_0 is a material constant for the starting stress for dislocation movement, k_y is the strengthening coefficient (a constant unique to each material) constants and d is the average grain diameter. This experimental figure shows the possibility of increase in YS from about 150 MPa (about 15.4 kgf/mm^2) to about 500 MPa (about 52 kgf/mm^2) with increasing grain refinement in ferritic structure. However, producing steel with extremely fine grain size (ferrite grain size above 12) is a challenge, because during re-heating austenite grains are known to grow in exponential relationship with soaking temperature. For producing fine grains in standard steels, growth of austenite grains at the soaking temperature is controlled by using Al-killed steel, so that fine AlN particles can pin the grain boundaries from further growth. During subsequent heating, growth of pinned grain boundaries becomes restricted, avoiding grain growth and ensuring formation of fine grain structure by following the rules of N&G process

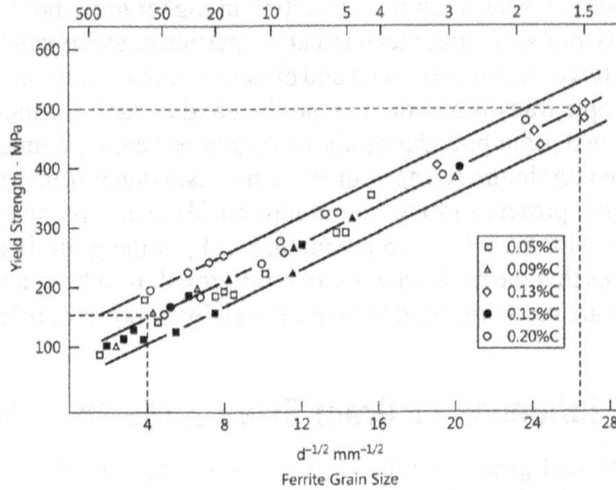

Figure 3-4. The relationship between yield strength and mean ferrite grain size in micro-alloyed HSLA steel. *Source: HSLA Steel. Brussels: International Iron and Steel Institute, 1987.*

during cooling (as discussed in Chapter 2). There are other methods available for refining steel grains—common methods are normalising or cold-work and recrystallisation. However, grain refinement by these methods is also limited to about ASTM number 5–9. Hence, gain of YS improvement from grain refinement by these traditional methods is also limited.

Appreciable improvement in YS of steel is made when grain sizes become finer than ASTM 10 or 12 and go up steeply thereafter, see Figs. 3-4 and 3-5. Considering the enormous possibility of gain from YS increase, techniques have been developed to produce steel structure with very fine micron level grain sizes by special micro-alloying—using niobium (Nb), vanadium (V), titanium (Ti) etc.—and thermo-mechanical rolling (refer Chapter 4 on steelmaking and rolling for more details). Extra fine grain size is achieved in these grades of steels by thermo-mechanical rolling where special type of micro-alloying (e.g. Nb and V) is used for delaying recrystallisation of austenite during early stages of rolling and thereafter controlling the finish rolling temperature and rolling schedules to below 900°C. Not allowing recrystallisation of austenite at upper temperature and finish rolling below 900°C with heavy deformation produces very fine recrystallised grains of ferrite, where mean ferrite grain size often exceeds 12–16.

These are called high strength low-alloy (HSLA) steels where YS improves with extra fine grain sizes, taking the YS to the level of 400–500 Mpa (42–52 kgf/mm²). HSLA steel is used extensively for manufacturing automobile frame parts and in structural construction requiring high YS. In HSLA steels, special micro-alloying with Nb, V and Ti is done for strengthening the steel by very fine precipitates of carbides and nitrides of these elements. Fine precipitation is achieved by thermo-mechanical rolling at an intermediate temperature between A1 and A3 (e.g. a typical rolling temperature of 875°C is used for extra low-carbon Nb-treated HSLA steel) which promotes fine carbide precipitation of micro-alloyed elements on partially recrystallised austenite grains. By controlling the finish rolling temperature and cooling, ferrite grains coming out of the partially recrystallised austenite can be made even finer.

Scope of improvement of YS by grain refinement clearly demonstrates that either the load bearing capacity of a steel part can be increased by refining the grains or the cross-section of the steel (i.e. specific weight of the steel) can be reduced for a specific application. Hence, steels with finer grains, especially HSLA steels are preferred for applications in structural load bearing parts (e.g. long-members and cross-members of a chassis frame, column, beams, angles etc.) where it offers significant commercial

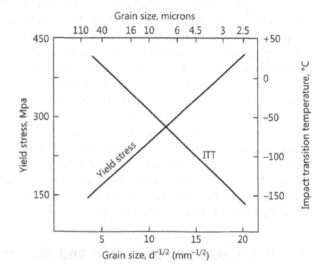

Figure 3-5. Effect of ferrite grain size on yield strength and impact transition temperature of steels.

advantage due to higher strength to weight ratio. A popular example of such choice is the selection of HSLA steel plates for main load bearing members of vehicles—long members and cross members of the vehicle frame. High YS is also necessary for improved springiness of steel members.

Effect of grain size on tensile strength, which also increases with decreasing grain size, is not so drastic. The reason for contribution of grain size for increased strength is attributed to grain boundaries acting as barriers for deformation to progress as they offer resistance to dislocation movements. During initial yielding, grain boundaries can act more effectively to stop the dislocation movement through them. But, as the plastic deformation starts all over the matrix upon crossing the YS, stress build-up ahead of moving dislocations becomes high and role of grain boundaries becomes less effective. Hence, the influence of grain boundaries on tensile strength—which is not related to yielding but relates to strength of resisting plastic deformation and ultimate fracture—is rather limited.

Other significant contribution of grain size is its ability to improve *toughness* especially, *notch toughness*. Figure 3-5 shows the influence of grain size on yield stress strength and impact transition temperature (ITT) of steel—a measure of toughness requirement in applications under fluctuating temperatures and impact stresses i.e. sudden loading at low temperatures.

Impact transition temperature refers to the point where failure mode of steel changes from *ductile* in nature to *brittle* and the part can fail suddenly without showing any signs of deformation and elongation. In plotting Figure 3-5, impact transition temperature has been considered as the temperature where energy absorption on impact test is 27 Joules i.e. *if the steel absorbs less than 27J on fracture, that fracture is essentially brittle; and if energy absorption is 27J and above, the fracture is considered ductile.* This temperature at which a change of fracture mode takes place is called *ductile-brittle transition temperature.* In order to ensure safety of operations under seasonal variation of temperature and fluctuation of load, all engineering applications—like industrial structures, high-sea platforms, rigs and tables, welded joints, pressure vessels, externally exposed auto body parts etc.—call for this fracture transition temperature well below the room-temperature. Fine grained steel structures can assure this requirement.

Amongst various strengthening systems available for steels—namely grain boundary hardening, solid solution hardening, precipitation hardening and transformation hardening—grain boundary hardening is the least expensive and has holistic effect on all other hardening/strengthening methods. Thus, grain size and grain boundaries, which are constitutional characteristic of steels, can be very gainfully used in the applications of steels. To take advantage of this situation, metallurgical specifications of conventional forging and engineering grade steels call for 'Al-killed fine grained steel' within ASTM grain size of 5–8. For grain sizes finer than ASTM 8–9, additional measures for suitable micro-alloying and controlled rolling are specified.

However, fine grains can pose problems for machining (see Table 1-3). Hence, to overcome machining problem of Al-killed fine grained steel, full 'annealing' or 'sub-critical annealing' treatment can be given to steel before machining, in order to either coarsen the grains by annealing or spheriodising the pearlite by sub-critical annealing. Sub-critical annealing treatment brings down the hardness and brakes down the fine lamellar pearlite of high carbon steels to globular carbides and ferrite, making the structure amenable to good machining.

3.5. Grain Size Determination and Measurement

Grain boundaries are generally planar with two-dimensional periodic atomic structure as against array of three dimensional atomic arrangements inside the grains. Therefore, grain sizes—bounded by grain boundaries—are best measured by planar distance across the boundaries as revealed under microscope after flat polishing and chemical etching. ASTM standard E112, which deals with the measurement and standard expression (for comparison) of grain size, refers to average planar grain size. The test method measures grain size with a microscope by counting the number of grains within a given area, by determining the number of grain boundaries that intersect a given length of random line. The average grain diameter d can be determined from measurements along random lines by the equation:

$$d = \frac{l}{n},$$

where l is the length of the line and n is the number of intercepts which the grain boundary makes with the line. ASTM grain size, N is then expressed by using the relationship:

$$n = 2^{N-1},$$

where n is the number of grains per square inch and N is the ASTM grain size number. The method measures 'average grain size', which is reported in Table 3-1 which shows the average grain diameter for each of common ASTM grain size number, N.

The intercept method of measuring average grain size requires counting of grains and intercepts, which is somewhat cumbersome due to statistical rules. Hence, an easier estimation has been developed by ASTM based on examination of similar field as used for intercept method, but reporting the number of grains per square inch. The comparison procedure does not require either counting grains or intercepts—it involves selecting a field under magnification ×100 and compares the field with a standard chart approved by ASTM.

Figure 3-6 shows a standard ASTM comparison chart of different grain sizes. Due to size variations within a field, the comparison method may not be exact, but still best serves the purpose to know and compare—what the average grain sizes are and the properties expected from the steel. Repeatability and reproducibility of comparison chart is generally within ±1 grain size number.

The comparison chart shows how many grains are there per square inch under a magnified field ×100. With decrease of grain size per square inch, ASTM number goes up and number of grains per square inch increases exponentially (as it should be from theoretical considerations discussed earlier). Thus, while *intercept method* gives average grain diameter in size (μm), *comparison chart* gives number of grains per square inch at ×100. In comparison chart, ASTM size 1 implies coarsest grain size (1–1½ grain/sq in) and ASTM size 8 implies finer grain size with over 96 grain/sq in, within the accuracy of ±1 grain size number. Such variation of grain sizes in the same field of examination occurs due to variation of steelmaking process and heterogeneity in steel. Impurities, inclusions, solute concentration etc. influence the grain formation in steel not only during solidification but also during allotropic transformation in the solid state.

ASTM grain sizes 1–4 are considered coarser and sizes 5–8 are considered finer.

Table 3-1. Average grain diameter for each of the common ASTM grain sizes, N.* *(Source: ASM Metals Handbook, ASM, Ohio, 1948)*

ASTM Grain Size No. (N)	Grain Diameter (μm)
0	359
1	254
2	180
3	127
4	90
5	64
6	45
7	32
8	22.4
9	15.9
10	11.2
11	7.94
12	5.61
13	3.97
14	2.81

* 1 μm = 10^{-3} mm.

Benefits of fine grained steel with enhanced mechanical properties have been discussed earlier. However, for consistently obtaining fine grains of sizes 5 and finer, the steel needs to be degassed and Al-killed. If the steel is not degassed and killed, the steel is termed as 'rimmed steel' where rims of sub-surface gas bubbles entrapped in the steel during solidification can be observed. Such steels generally have coarser grains, because of unrestrained grain growth during heating or cooling in the absence of any grain boundary pinning effect. There is another category of steel, namely semi-killed or balanced steel, where the gases are partly removed by Al-Si treatment and rest remains in the liquid for compensating volume shrinkage during solidification (refer Chapter 4). Semi-killed steel grain sizes are also coarse and non-uniform in size.

Grain sizes in steel influence many mechanical properties of steel, which includes—hardness, YS, fatigue strength, ductile-brittle transition temperature and susceptibility to environmental embrittlement like hydrogen enbrittlement. The influence of grain size on strength properties is best understood through the famous 'Hall–Petch Equation', which has been explained under Fig. 3-4.

However, there are applications where fine grained steel is not the choice. Such applications are creep, machinability and hardenability. Influence of grain sizes on creep and machinability has been indicated earlier. As regards hardenability of steel, coarser grains tend to delay the phase transformation of steel by pushing the transformation C-curve to the right (see Fig. 2-7) and, thereby, effectively help in getting more martensite transformation on hardening. More about the influence of grain sizes on hardenability has been discussed in the Chapter 8 on 'Heat Treatment'.

ASTM-1: (1-1/2 grains/sq in), ASTM-2: (1-1/2 to 3 grains/sq. in), ASTM-5: (12 to 24 grains/sq. in)

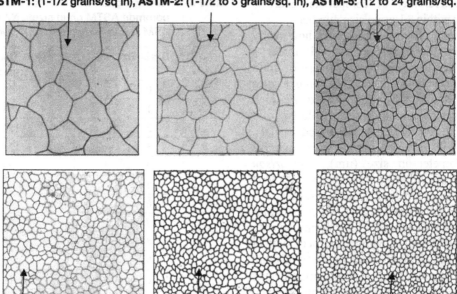

ASTM-6: (24 to 48 grains/sq in), ASTM-7: (48 to 96 grains/sq. in), ASTM-8 (>96 grains/sq. in)

Figure 3-6. A photograph of ASTM grain size chart. Note the exponential increase of grain numbers per square inch with ASTM number. For full chart, see ASTM standard E112.

3.6. Formation of Inclusions in Steel

Inclusions are non-metallic chemical compounds that are present in steel by virtue of their formation—either by chemical reactions or contamination or physical effects (erosion and entrapment) during high-temperature steelmaking process. These inclusions are categorised by their origin—i.e. *endogenous* or *exogenous*. Endogenous inclusions form within the steel melt due to various chemical reactions e.g. oxide inclusion, sulphide inclusion, nitride inclusion, etc. These inclusions are generally finer. Exogenous inclusions come from external sources, such as: entrapment of non-metallic particles from slag, fluxes, refractories, etc. These inclusions are generally larger in size than endogenous inclusions. Inclusions can be also grouped into: (1) micro-inclusions and (2) macro-inclusions. Inclusions formed by gaseous reactions (i.e. endogenous inclusions) are generally fine and called micro-inclusions. Inclusions originating from erosion and entrapment (i.e. exogenous inclusions) are coarser in nature and are called *macro-inclusions*. These two types of inclusions can co-exist by sharing their mutual interfaces.

Four major ways by which most non-metallic inclusions form in steel:

- By oxidation and re-oxidation reactions;
- By interaction between liquid steel and liquid slags in the steelmaking vessel;
- By erosion and corrosion of refractory ladle during pouring of liquid steel; and
- By inclusion agglomeration due to clogging during steel pouring (clogging is caused by deoxidation of the liquid steel during pouring due to exposure to air; refer Chapter 4 on steelmaking).

For example, as deoxidisers (Al, Si, Ca, Mg, etc.) are added to the steel, oxygen dissolved in the melt reacts and that is how the stable oxides of these metals form. Re-oxidation can occur due to: (a) contact of liquid steel bath/stream with air, (b) contact with slags containing high levels of FeO, MnO and Silica, and (c) reaction with refractories. These reactions give rise to oxide inclusions of various chemistry e.g. alumina (Al_2O_3), silicate (SiO_2), calcium oxide (CaO), and magnesium oxide (MgO) etc. Similarly, there would be excess nitrogen in liquid steel, forming nitrides (e.g. AlN, TiN etc.) by reacting with aluminium, titanium etc. if present in the liquid steel. If impurities like sulphur are present in steel—which is always the case due to sulphur in-put through pig-iron or steel scrap feed to steelmaking vessel—it will form sulphide of metals like FeS, manganese sulphide (MnS), CaS, MgS etc. Even the presence of 1 ppm (parts per million) oxygen or sulphur gas in steel can give rise to many minuscule particles of oxide or sulphide inclusions in steel.

Thus, from various sources of origin of inclusions, various types of non-metallic inclusions can form. To assess their influence and effect on steel properties and applications, non-metallic inclusions are divided based on their chemistry (non-metallic inclusions always exist in the form of chemical compound of fixed composition). The divisions are:

- Oxides—simple: FeO, MnO, Cr_2O_3, TiO_2, SiO_2, Al_2O_3 etc.—or some complex oxide compound;
- Sulphides—FeS, MnS, CaS, MgS, Al_2S_3 etc.—or some complex sulphide compound;
- Nitrides—TiN, AlN, ZrN, CeN etc.—or complex carbo-nitride like Nb(C,N), V(C,N) etc., which can be found in alloy steels with strong nitride forming elements like Al, V, Ti etc.; and
- Phosphides—Fe_3P, Fe_2P etc.

Oxides and sulphide inclusions can again exist in different shapes and sizes (e.g. elongated, broken and globular) in the hot rolled steels. Figure 3-7 illustrates some of these forms.

Figure 3-7 illustrates those shapes and forms of inclusions that are typical of their chemical character, and these can be used for identification of inclusion types. If the inclusion type is plastic at the rolling temperature of the steel, it will elongate due to higher rolling in one direction. Longer the inclusion length, higher is its harmful effect, especially for bending and forming.

Study of inclusion contents and inclusion types in commercial steels generally focuses on oxide and sulphide type inclusions, due to their prevalence and effect on different types of forming and application behaviour of steels. Presence and effect of other two inclusion types, namely phosphides and nitrides, are comparatively limited. Nitrides in commercial steels may come from Al addition to liquid steel for killing and degassing, where AlN particles have their own role. Otherwise, nitride can be present in special steels, like stainless steels, tool steels, ball-bearing steels etc., which can contain higher percentage of alloying elements that have strong affinity for nitrogen, e.g. Cr and V to form nitrides. Such nitrides are very fine and the role of such nitrides is more for strengthening the steel rather than acting as harmful inclusions—except for ball-bearing applications involving contact fatigue—where nitride inclusions are especially harmful.

Thus, the presence of non-metallic inclusions is a part of steel—they will be present to a certain degree irrespective of steelmaking precautions. Their shape, size and distribution can, however, be changed by special steelmaking process, hot-working and deformation, and also by special chemical treatment in some cases, e.g. control of sulphide morphology by calcium treatment.

Inclusions of all types, especially the sulphides and oxides, are harmful for most steel uses and applications; but the degree of harmfulness depends on the shape, size and distribution. Removal of inclusions during steelmaking and tapping is an expensive operation too. Hence, without attempting complete removal of inclusions for all applications, it is economical and technically judicious to control

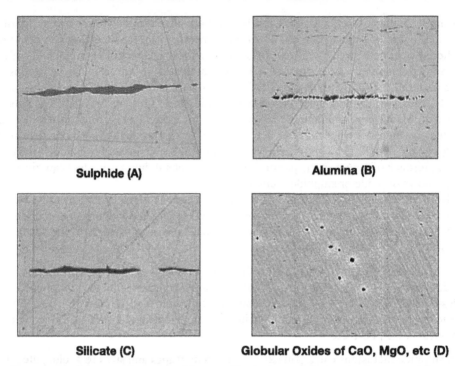

Sulphide (A)	**Alumina (B)**
Silicate (C)	**Globular Oxides of CaO, MgO, etc (D)**

Figure 3-7. Different shapes and forms of inclusions and chemical nature of those inclusions.

the level of inclusions and their size and distribution based on needs for specific application. The ASTM standard E45 deals with the testing and specification of steels with permissible inclusion levels for different applications.

Control of inclusions in steel can be very effectively dealt with in most modern steelmaking processes and the steel with tolerable level of inclusions (as per ASTM chart) can be produced with consistency. For this reason, steels are specified in at least two qualities, one *air melted* quality that may contain inclusions as per a standard and standard chart, and the other as *vacuum melted* quality steel containing very low level of inclusions and are meant for critical applications in forming and load bearing, and specified based on exact needs for the uses and applications. Choice of steelmaking process and route is, therefore, adopted based on inclusion limits and specifications. Developments of *ladle metallurgy treatment* or *vacuum degassing* of steels are steps towards this direction. If high grade low inclusion content steels are required—as for the high duty ball-bearing application or high forming automobile body parts—ladle furnace (LF) treatment or vacuum degassing (VD) of steels should be specified. More about inclusion control in steelmaking will be discussed in Chapter 4: Elements of Steel-making and Rolling for Quality Steel. For other standard applications of steels, air melted quality can be specified, but limiting the inclusion contents as per applications. As such, all steel specifications mention the inclusion limits, because of its harmful effects.

3.7. Determination of Inclusion Types and Ratings

Definition wise, non-metallic inclusions are chemical compounds of metals and non-metals carried over from molten steel to solid steel. For example, Al_2O_3, the aluminium oxide, forms by oxidation reaction at high temperature, where Al is the metal part and O_2 is the non-metal part, forming the chemical compound Al_2O_3. A *major source of inclusion is the dissolved gases like O_2, N, S etc. in the molten steel, which reacts with some metallic present in molten steel and form the inclusion type.* As reported earlier, presence of 1 ppm oxygen or sulphur in steel can give rise to many minuscule particles of oxide or sulphide inclusions. Hence, no steel is absolutely clean; there will be some inclusions, and the task of steel users is to know how much can be tolerated for the specific end use or application and how to estimate and express that in the steel specification.

There could be two sizes of inclusions for estimation—macro-inclusions and micro-inclusions requiring macroscopic or microscopic examination of a field. Macroscopic examination involves examining large areas and is generally done by opening a fracture through the inclusion (for example, 'Blue Fracture' test) or by 'Step Down' test where new surfaces are opened up by step machining. Inclusion lines or patches, revealed by such tests, are examined visually or under a magnifying glass, and with or without the aid of some physical method of testing such as magnetic particle testing for cracks (inclusions behave like a surface crack under magnetic particle testing). A blue-fractured sample with streaks of macro-inclusions (light grey lines in dark background) is shown in Fig. 3-8. More about this test will be discussed in the chapter for testing and evaluation of steel quality.

The other important method of macro-examination is the *sulphur print*. Sulphur print is carried out by dipping or swabbing the flat steel surface with dilute hydrochloric acid. Hydrogen from the acid then reacts with sulphur of the steel and causes evolution of hydrogen sulphide gas. This hydrogen sulphide can then be made to react with photographic bromide paper which gets stained and gives rise to a pattern of the macrostructure of the steel exhibiting the sulphur segregation spots. More about this test has been described in Chapter 5 under metallographic tests (see sulphur print macrograph of Fig. 5-13).

Macroscopic examination is not, however, suitable for smaller inclusions because of limitation of detection below a size (e.g. inclusion of size below 0.5 mm is difficult to observe macroscopically). Hence, stress of inclusion rating method is on microscopic examination, which characterises inclusions by shape, size, concentration and distribution rather than by chemical composition or sources of their formation.

Microscopic test method for inclusion rating as per ASTM E45 attempts to characterise inclusions that form endogenously, such as by deoxidation during steelmaking (microscopic method is not meant for exogenous inclusions formed by entrapped slag or refractory). Though this characterisation does not directly attempt to identify the chemical composition, the ASTM methods place these endogenous

Figure 3-8. A blue-fractured test sample showing presence of macro-inclusion streaks (light blue streaks).

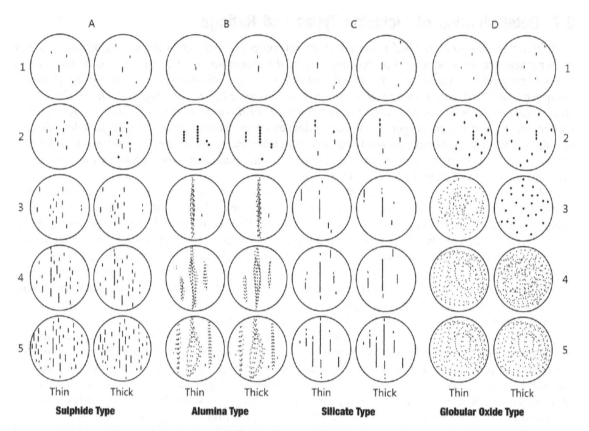

Figure 3-9. Standard comparative chart for determination of inclusion content in air-melted steels. Classification and specification of inclusions are made in terms of A, B, C and D types in grade thin or thick by referring to chart *vis-a-vis* the field being examined from the steel sample.

inclusions into one of several composition related categories, such as sulphide, alumina, silicate and globular oxides (see Fig. 3-9).

Inclusion levels estimated by ASTM E45 are expressed in the form of a chart, which is also referred to as *JK-chart* for inclusions in steel. The ASTM chart deals with four types of inclusions—namely A (Sulphide), B (Alumina), C (Silicate) and D (globular oxides) sub-dividing them into 'Thick' and 'Thin' categories. The globular oxides (type D) often originate from special treatment of steel by rare-earth metal compound or calcium treatment for some special applications. Figure 3-9 exhibits the standard ASTM inclusion chart.

Examination and estimation of micro-inclusion requires careful preparation of microscopic samples as per clauses of ASTM E45 for revealing the inclusions in a sample. The standard also calls for examination of number of fields and reporting result on the 'worst field'. Worst field reporting is

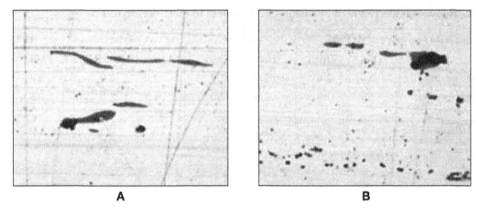

Figure 3-10. Inclusions in steel: (A) sulphides in light grey with elongated shape; (B) oxides in dark grey with broken chains, including some mixed area of sulphide and oxide at right hand top corner.

recommended for safeguarding the chance of under-rating the inclusion level and jeopardising the steel applications.

The ASTM method is based on similarities in morphology and not necessarily their chemical identity. For chemical identity of these oxide inclusions or inclusions arising from carbide, nitride, boride and their carbo-compounds, more rigorous tests—like atomic image analysis, energy dispersive X-ray spectroscopy, etc., can be used. Nonetheless, the simple comparative method of micro-inclusion examination and rating system of ASTM (as shown in Fig. 3-9) provides a very effective means of inclusion rating in steels. Therefore, this method is widely used in industries. Many a time, inclusion of one type may co-exist with the other type e.g. oxide and sulphide mixed together. Figure 3-10 shows one such example, where oxides (darker) are present with elongated sulphide (grey). Hence, such situation calls for careful sample preparation and closer observation.

Determination of inclusion type and level by ASTM E45 applies to wrought steel after rolling or forging, which is generally the starting material for further processing of steel. During rolling or forging, soft sulphide inclusions tend to get elongated along the direction of final rolling whereas harder and brittle oxide inclusions tend to get fragmented. Thus, oxide inclusions often show up as thin broken stringers, rendering them more harmful than sulphides, see Fig. 3-10. Harmfulness of broken oxide inclusions mainly come from the brittle interfaces between the hard oxide particles and the steel, as well as due to increase in interfacial area due to fragmentation and stringer formation.

Steels cannot be made totally inclusion free, especially of sulphide and oxides. Hence, users have to choose their steel quality based on end uses and applications. In this regard, Table 3-2 provides a general acceptance plan for inclusions in steels. This table includes generally acceptable level of micro-inclusions for steels made through air melted, ladle metallurgy and vacuum degassing routes. The latter two types of steels are used for more challenging applications. For closer bending radius in forming or for applications under high fatigue load—steels with much stricter inclusion control are preferred.

However, the table refers to micro-inclusions that are detectable under normal magnification. Very fine inclusions often escape the chance of detection by normal microscopy, because of the limitation of detection by traditional microscopic methods. Figure 3-11 illustrates the different techniques that can effectively detect and determine the size and size frequency distribution of inclusions. Thus, more

Table 3-2. Recommended level of inclusions with respect to steelmaking route.

Acceptance level / Quality	A-Type		B-Type		C-Type		D-Type	
	Thin	Thick	Thin	Thick	Thin	Thick	Thin	Thick
Air Melted quality	2.5	2.0	2.0	2.0	2.0	2.0	2.0	2.0
Ladle Treated/Argon Purged	2.0	1.0	2.0	1.0	1.5	1.0	1.5	1.0
Vacuum Degassed	1.0	0.5	1.0	0.5	0.5	0.5	1.0	0.5

Source: ASTM chart in Fig. 3-9 for types and thickness.

Figure 3-11. An illustration of inclusion size frequency and size ranges that can be observed by different methods. Different methods of estimation are required for accurate detection of different sizes, which are not amenable to measurement by all processes. (Note the inclusion size range that can be determined by optical microscopy: under Light Optical counting)

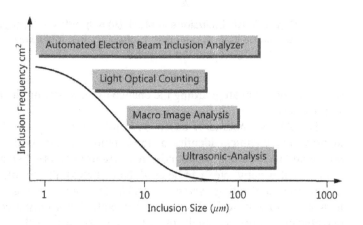

accurate inclusion determination will depend on the technique being used for revealing and measuring the inclusions, especially very fine inclusions. Micro-inclusions which are less than one micron in size are considered harmless in use, and most harmful ones are in sizes above 20 μ. Larger inclusions of size above 20 μ are quite prevalent in air-melted steels.

3.8. Effects of Inclusions on the Properties of Steels

Inclusions are regarded as harmful for all mechanical and forming properties of steels, with the exception of machining where soft sulphide inclusions are said to be helpful for providing lubricating effect at the cutting tool tips. But, that apart, inclusions are harmful for all applications—only difference between them is that some types are more harmful than others with respect to some specific applications. Adverse effects of inclusions on mechanical properties of steels spread from—tensile strength and reduction of area to impact strength, fatigue strength and overall fracture toughness. Inclusion affects the heat-treatability of steel by making steels susceptible to cracks during quenching and hardening. Inclusion also affects the forming properties of steels, like the bendability, stretchability and drawability. Thus, adverse influence of inclusions ranges from affecting processing of steel to the applications of steel.

While all inclusions are harmful, sulphides and oxides come to general notice first—because of their common presence and occurrence in steels. Sulphides are soft and mostly deformable, taking elongated shape along the rolling direction. Elongated shape of sulphide inclusions gives rise to anisotropy in properties between transverse and longitudinal directions. Sulphides, which are mostly present as elongated inclusions, cause transverse cracks in cold forming of steels when bent along the rolling direction (i.e. axis of bending is parallel to the rolling direction). However, same sulphide may not create that much problem when the part is bent across the rolling direction (i.e. at 90°). Common example of elongated sulphide inclusion is MnS. Sulphides also lower the through thickness transverse notch toughness of steel for engineering applications. Figure 3-10(A) shows the general appearance of sulphide inclusion in steel.

Oxide inclusions are mostly present as hard and irregular in shape, often in broken chains, see Fig. 3-10(B). Oxides are hard and not easily deformable at the rolling temperature. Hence, under the rolling pressure, these hard oxide inclusions get fragmented and lie as broken chains. Because of the nature of hard oxide inclusions, these inclusion spots generally cause high stress concentration at the inclusion and metal interface, facilitating crack formation and growth under the applied load; be it static or dynamic load like fatigue. The effect is more pronounced when the oxide inclusions are coarser in size and angular in shape than when oxide inclusions exist in fine globular form. Hard inclusions are also difficult to machine and forge. During forging or rolling, hard oxide inclusions of larger sizes often cause rupture and split of metal being rolled.

Not all oxides are hard; oxides like CaO, MgO etc. are softer and generally occur in globular form. Globular oxides are not as harmful as hard alumina (Al_2O_3) inclusions for applications, but they are bad for pitting resistance—required for bearing and contact load applications—due to their deformability. These inclusions are present in calcium or rare-earth treated steels. *Silicates* are the other oxide type inclusions (e.g. SiO_2 or mixture of SiO_2 with Fe, Mn and other oxides in the steel) and they form a large part of the inclusion population. They are glassy type and brittle at room temperature but plastic at the rolling temperature; and hence occur in elongated form. They are grouped as a special type in the ASTM chart to indicate quality of deoxidation of steels. They are equally harmful for bending, forming, machining etc. like other hard oxides.

Though all inclusions are harmful, their degree of adverse influence however changes with the type of inclusion, their size and distribution on the one hand and the strength of steel matrix, and on the other; rising with increased strength of steel due to increasing notch effect of the matrix-inclusion interfaces. Thus, adverse effects of inclusions on steel properties are influenced by any one or a combination of following parameters—shape, size, quantity, distribution, interspacing between two particles and interfacial strength of inclusions, on the one hand and physical properties of the steel matrix, on the other. Individual assessment of effects of each of these factors for different uses and applications is difficult to quantify, but overall qualitative effect of inclusions on different application specific properties can be summed up as shown in Table 3-3.

Effects of inclusions on steel properties are primarily due to weakness of the interface between inclusion and the steel. Inclusion interfaces in steel matrix are incoherent in nature without any natural bonding. Hence, the interfacial strength is low and interface is the source of micro-void and crack formation under the applied load. The situation gets aggravated when the inclusion size is coarser and shape is irregular, leading to the possibility of high stress concentration at the interfacial points. Due to high stress concentration, micro-cracks can grow from the inclusion interfaces under the applied load, even when the steel has enough ductility. Under dynamic fatigue load—where fatigue crack can grow even below the YS of steels with the help of stress concentration—this effect of inclusions assumes greater significance.

Though fracture and failure of steels can be associated with any type of inclusions, exact influence of inclusions depends on the hardness and shape of the inclusion concerned. In general, harder oxides

Table 3-3. Qualitative effect of inclusions on different application specific properties.

	Fatigue Strength	Fracture Strength	Impact Strength	Machining	Cold-Forming	Hot-Forming/Forging	Corrosion and Pitting	Yield/Tensile Strength.
Effect of Inclusion (Harder has greater effect)	↓↓	↓↓	↓↓	↓ *	↓	↓	↓	↓

* Improves with sulphide inclusions.

Note: Harder and irregular shaped inclusions have more adverse effect than softer and spherical inclusions, irrespective of type.

inclusions like—Al_2O_3, silicates and spinel, which are brittle and less deformable—can cause higher risk for micro-cracking leading to failure. This is because these hard inclusions are likely to get fragmented under rolling pressure, producing irregular shape and size, which are more harmful. Sulphides, like MnS, are deformable and get elongated along the rolling direction, causing anisotropy in properties, but are not so detrimental as regards micro-cracking. Thus, harmfulness of inclusions depends on types, size, shape and the application requirements. However, brittle alumina or oxides are considered more harmful than ductile sulphides in the uses and applications of steels, in general.

Influence of inclusion on fatigue strength of steel is, perhaps, the most widely studied area. While the inclusion population increases, the fatigue strength of steel decreases. Inclusions have incoherent interfaces with steel matrix—as such these interfaces can act as sources of voids and micro-crack formation for initiation of fatigue crack. However, in general, finer inclusions of smaller diameter have less adverse effect than the coarser inclusion of larger sizes. Figure 3-12 shows the variation of fatigue strength in alloy steel with different inclusion sizes. Finer inclusion sizes were produced in this steel by vacuum degassing and coarser inclusions belonged to 'air-melted' steel quality. Air-melted steel is known to give larger and coarser inclusion sizes compared to vacuum degassed steel. This is due to chance of higher contact with outside air and less efficient inclusion removal system in air melting process. The figure shows that presence of larger size inclusions drastically brings down the fatigue strength of steels. Therefore, incidence of fatigue can be considerably reduced by selecting steel with lower inclusion content and by ensuring that, inclusions are finer in size and even in distribution.

Sparsely distributed inclusions with wider inter-spacing are found to be much less harmful than unevenly distributed or clustering of inclusions of similar sizes. This is the reason why VD or LF treated steels along with high degree of hot-deformation (i.e. high hot-rolling or forging reduction ratio) are preferred for all critical applications. This is because these treatments lead to—(a) much reduced inclusion level in steel and (b) make the inclusions finer and sub-microscopic. In general, fine inclusions with spherical interfaces have higher interfacial strength and less harmful effect on mechanical properties of steels, including fatigue strength.

Inclusions must have a critical size for acting as effective site for nucleating fatigue crack. But, the size effect also depends on the inclusion shape i.e. if they are spherical or angular. Even small angular and brittle inclusions are harmful for fatigue, especially if the steel matrix is of high hardness. For example, in ball bearing applications, brittle oxide inclusions, which often take the shape of fine chains

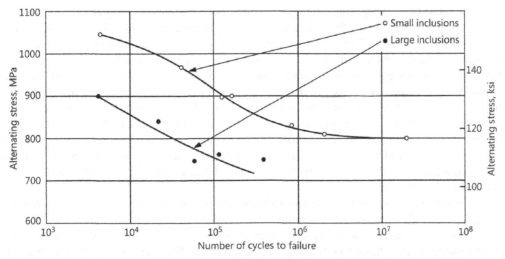

Figure 3-12. Influence of inclusions and inclusion size on fatigue strength of SAE 4340H grade steel. *Source: ASM Handbook, Vol. 1. Properties and Selection: Irons, Steels, and High-Performance Alloys. ASM Handbook Committee, 1991, pp. 673–688.*

of broken down particles, are found more harmful for fatigue crack initiation than soft sulphide or oxy-sulphides. However, even softer and globular oxides (e.g. CaO) are also objectionable for contact fatigue in ball bearing. Because, these softer oxide particles yield under the compressive load of the bearing elements and give rise to local pitting that acts as centre for contact fatigue failure.

Inclusions also significantly influence the 'ductile-brittle' transition temperature in steels. The ductile-brittle transition (i.e. the temperature where fracture mode suddenly changes from ductile mode to brittle mode) temperature is influenced by inclusions and inclusion spacing—more inclusion spots with closer inclusion spacing lowers the impact energy (energy absorbed on fracture in impact test) and widens the range of transition temperature. The net effect of this is the lowering of ductile-brittle transition temperature—i.e. the temperature corresponding to 27J impact value which, in practice, is considered as the transition impact value point for ductile to brittle transition.

In sum, major effects of inclusions are through its shape, size, and distribution, which are, in turn, dependent on the hardness and deformability of the inclusions. In this respect, oxide inclusions of Al_2O_3 and silicate types appear to be more harmful than sulphide of MnS type. Sulphide inclusions improve machinability of steels, but they cause anisotropy in mechanical and fatigue properties due to their deformable nature and elongated profile. Steel containing high sulphur (i.e. higher sulphide inclusions) shows sharp drop in transverse ductility and impact value compared to longitudinal value. Transverse impact energy value of high sulphur steels can be 25%–40% lower than the longitudinal value of the same steel. Thus, anisotropy of properties in high sulphur steel is a problem. Therefore, for applications relating to high stresses in transverse loading—high sulphur steels are often treated with 'rare-earth meat' (e.g. cerium) to convert the sulphide inclusion morphology from elongated to globular. Rare earth treatment in liquid steel converts the deformable sulphide inclusions to non-deformable globular sulphide inclusions—which are less harmful either for transverse bending or for transverse toughness. This effect is also valid for fatigue crack initiation and growth.

Thus, inclusions have very wide ranging influence on the mechanical properties of steels—affecting the tensile strength, fatigue strength, impact strength, toughness, and formability amongst others. Major influence of inclusions comes from the interfacial character of the inclusions in the steel matrix—which acts as points of stress concentration and void formation—initiating micro-crack and crack growth. Coarser in size and irregular in shape makes the inclusions more harmful than small and spherical inclusions. Adverse effects of inclusions also depend on the strength of steel matrix where the inclusions are present; higher the strength of steel more adverse is the effect of presence of any particular inclusion type. Therefore, tolerance level of inclusions and inclusion type has to be decided on the strength of steel being used and the intended applications. Since inclusions cannot be totally eliminated from the steel, the best way to deal with it is—to take measures for minimising the volume fraction, size and evenly dispersing them in the matrix. This is essentially the task of steelmaking and rolling, which will be discussed in the next chapter.

The other deleterious effect related to grain boundary and segregation of impurities is *hot shortness* and *cold shortness*. Hot shortness is generally caused by the sulphur; when sulphur is present, it can segregate to grain boundaries and can form iron sulphide, which is low melting point constituent. During re-heating of steel these low melting iron sulphides at the grain boundary can melt and cause loss of cohesion between grain boundaries, leading to brittleness in the steel while being worked at higher temperature. Remedy of this problem is to fix the sulphur present with Mn to form MnS, which has higher melting point than FeS, and prevents hot-shortness. It is also believed that, segregation of low-melting tin and copper to the grain boundary areas can also lead to grain boundary melting and hot-shortness in the steel. Hence, steels are often specified with strict control of residual level of copper and tin. Cold shortness is caused by phosphorus segregation (when P is present in steel in excess of 0.10%), reducing the ductility of the steel and leading to cracking while being cold-worked. Hence, steels with high level of P are susceptible to become very brittle—especially under low temperature condition—and leads to cold cracking in the steel. Cold shortness is associated with substantial increase in YS—caused by blocking of the dislocation movements for glide and sharp drop of ductility—which also shows up in impact testing of the steel. Cold shortness is of particular importance for structures used in cold region, under deep sea or for any cryogenic application. Degree of cold shortness can be reduced by the elimination of harmful impurities, heat treatment, grain size control, and by controlling alloying elements in the steel, such as by increasing the nickel content and reducing any alloying element that has body centred cubic structure (e.g. chromium).

SUMMARY

1. The chapter highlights the importance of grain size and inclusions in steel and their influences on mechanical properties and behavior for application of steels. This chapter covers the formation, characteristics and effects of grain size and inclusions on steel properties, highlighting the importance of interfacial character of these (non-crystallographic) phases in the steel matrix. The chapter also covers the methods of determination and measurement of these phases for the purpose of reference and specification.

2. Grain boundaries in metals arise due to the fact that orderly arrangement of atoms in the crystal lattice gets broken down at the inter-crystalline interfaces. Such inter-crystalline meeting points are few atoms thick and have planar arrangement of atoms with many atomic voids.

3. Though grains are tiny crystals inside the metal, they occupy appreciable area inside the material. In a polycrystalline cube of size 1 cm, with grains of 0.0001 cm in planar diameter, there would be 10^{12} crystals (i.e. grains), resulting in grain boundaries of several sq. meters. Hence, any influence of grain boundaries on any property—either mechanical or chemical or physical—would significantly change with grain diameters i.e. grain sizes. Therefore, considering the importance of grain size effect on properties of materials, American Society for Testing of Metals (ASTM) came out with a standardised measure of grain sizes in steels, assigning numbers from 1 to 9. The ASTM grain size number "N" is defined by: $n = 2^{N-1}$, where n is the number of grains per square inch when viewed at ×100. The formula expresses the average grain diameter with permissible departure within the individual field of examination.

4. Discussing the grain boundary formation in steel, it has been emphasised that grains in as-cast steel are highly inhomogeneous and of different shapes and sizes, which are of little practical use. It is important to break down this structure into more homogeneous and equiaxed grains by heating and hot-working the steel for allotropic transformation and recrystallisation that leads to formation of fresh and more uniform grains sizes.

5. It has been pointed out that grain boundaries—along with their sub-boundaries, if any—act as obstacle to plastic flow of metal while under deformation. This happens due to the fact that any angular disorientation of boundaries and sub-boundaries will obstruct the movement of dislocations through them, causing more stress to be applied to overcome those obstacles for the progress of the plastic deformation. Hence, higher the area of grain boundaries, higher is the resistance for plastic flow i.e. finer the grain size (resulting in higher grain boundary area) higher is the strength. Thus, by controlling the grain size and its sub-structure, mechanical properties of steels can be gainfully altered for different applications.

6. Based on this nature and character, influence of grain size on different mechanical properties of steels—such as YS, tensile strength, elongation, impact toughness, low-temperature properties, and creep deformation—have been discussed, tabulated and illustrated.

7. Sources of inclusion formation in steel have been discussed and the nature (e.g. endogenous and exogenous inclusions) and types of inclusions (e.g. oxides, sulphides, silicates, nitrides, etc.) have been identified and their effects discussed. Difference between quality of steel made through 'air-melting' and secondary metallurgy treated or 'vacuum degassing' route have been highlighted.

8. Effect of inclusions on mechanical properties of steels based on nature and character of different types of inclusions and their shapes and sizes have been discussed and illustrated. It has been shown that larger inclusion size with closer interspace has more adverse effect on properties than finer and evenly distributed or sparsely distributed inclusions of same type.

9. The chapter also outlines the ASTM methods of inclusion determination and rating, including macro and micro examinations. Based on ASTM chart, the recommended level of acceptance of inclusion levels from different steelmaking routes has been mentioned.

FURTHER READINGS

1. Reed-Hill, Robert E. *Physical Metallurgy Principles*. New York; Van Nostrand Reinhold Co., 1973.
2. Dieter, George E. *Mechanical Metallurgy*. McGraw-Hill, 1961.
3. Bhadeshia, H. and Honeycombe, R. Steels: *Microstructures and Properties*. Butterworth-Heinemann, 2006.

4. Smallman, R.E. *Physical Metallurgy and Advance Materials.* Butterworth-Heinemann, 2007.
5. Grain Boundary. Wikimedia Foundation, Inc. 12th May, 2013. *http://en.wikipedia.org/wiki/Grain_boundary; http://creativecommons.org/licenses/by-sa/3.0.*
6. Leslie, W.C. *The Physical Metallurgy of Steels.* McGraw-Hill Hemisphere Publishing Corporation, 1981.
7. Kiessling, Ronald. *Non-Metallic Inclusions in Steels.* UK: Iron & Steel Institute, 1965.
8. Nyar, Alok. *Testing of Metals.* New Delhi: Tata McGraw-Hill Education, 2005.
9. Metals Handbook, Vol. 1. *Properties and Selection: Irons, Steels, and High Performance Alloys.* ASM International, Metals Prak, Ohio; 1991.
10. High Strength Low Alloy Steels. Brussels: International Steel Institute, 1987.

Steelmaking and Rolling for Quality Steel Production

Τ he purpose of this chapter is to highlight the means and methods of making and shaping quality steels for different applications. The chapter describes and discusses the stages of steelmaking, necessary precautions for making quality steels, various methods of producing cleaner steels with stricter control of inclusions, and precautions necessary in the casting of steels for making defect-free steel. Role of good rolling practice for the production of quality steel has been highlighted, along with the description of some features of hot-rolling and cold rolling of steels.

4.1. Introduction

At the centre of steel technology is the steelmaking and rolling for producing quality steel. Therefore, understanding steelmaking and rolling is essential for taking any decision about selection and application of steels. Though steelmaking is an expensive operation, modern developments in this area offer much flexibility for making steels as per exacting needs. The make and break of good quality steel lies in good steelmaking along with good rolling. Rolling of steel is equally important for good quality steel, as it is said that—*bad steel cannot be made good by good rolling, but good steel can be made bad by bad rolling*. This chapter, therefore, aims to highlight various steelmaking and rolling options and processes that would help the users of steel to select the optimum process or specify the right steelmaking and rolling route for economical uses and applications.

STEELMAKING

4.2. Introduction to Steelmaking Processes for Quality Steel

Steels are made by oxidising 'charge materials' in a suitable high temperature furnace, followed by refining the molten steel and then casting the molten steel into a pre-designed shape e.g. ingot, billet, slab, thin strip etc. *Charge materials* could be liquid iron (popularly called *hot metal* from the blast furnace route—a liquid pig iron with higher carbon, approx. 4.0%, silicon (Si), phosphorus (P), and other elements which do not at that stage represent the desired composition of the steel to be made), steel scarps, sponge iron or any other iron bearing material, and some chemical fluxes. Figure 4-1 illustrates the schematic and diagrammatic presentation of iron making and steelmaking upto the rolling stages.

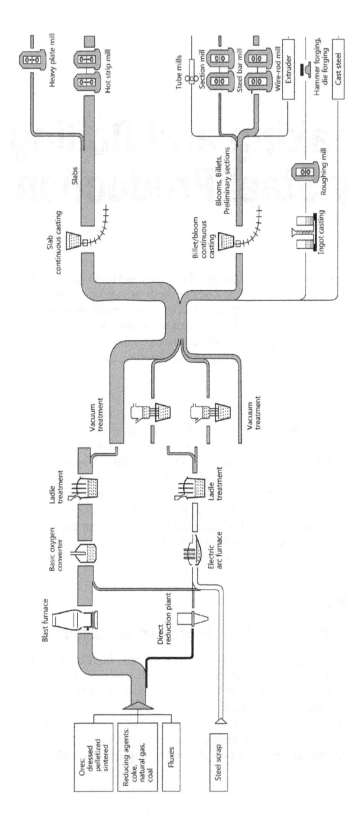

Figure 4-1. A diagrammatic presentation of steelmaking and rolling at one place—starting from blast furnace hot metal, their treatment, steelmaking processes and rolling to different products/shapes.

These step-wise processes have been further elaborated in Fig. 4-2 where all actions required for clean steelmaking have been listed. Important steps for clean steelmaking in this diagram are:

- *Injection treatment* of hot metal for desulphurising and deposhporising the hot metal from the blast furnace, which is generally high in sulphur (S) and phosphorus (P);
- *Bottom stirring* in the Basic Oxygen Furnace (BOF) (e.g. LD furnace);
- *Argon flushing* for removing inclusions and uniformity in temperature;
- *Injection treatment* for lowering of sulphur and inclusion modification; and
- *Ladle furnace/vacuum treatment* for low hydrogen and sulphur control, in specific, and inclusion removal in general.

The diagram also shows the subsequent steps and choices of casting and the preferred route for final shape by rolling. While blast furnace iron making is essentially a reduction process—where iron

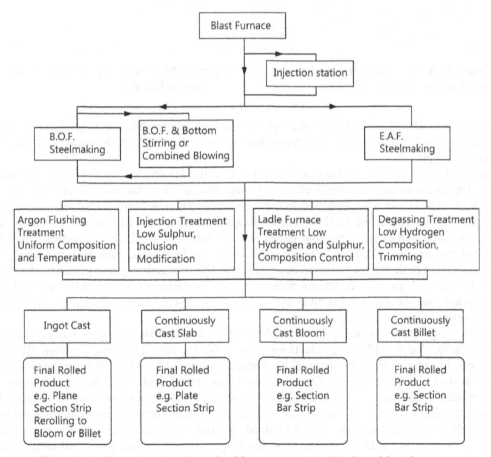

Figure 4-2. Flow diagram of steelmaking process starting from blast furnace.

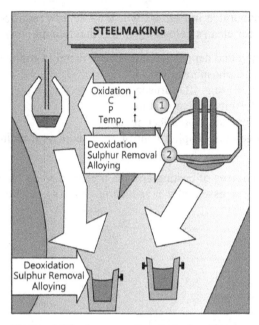

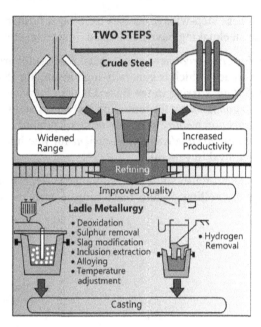

Figure 4-3A. A conventional steelmaking set-up using LD or Arc furnace.

Figure 4-3B. A two stage modern set-up for improved quality.

bearing ores are reduced in blast furnace at appropriate high temperature—converting this iron to steel is dominated by oxidation, followed by deoxidation process, in order to remove/control all undesirable elements and impurities.

Functions of steelmaking process can be further visualised from the Figs. 4-3A and B. Figure 4-3A depicts a typical single step conventional steel making process—popularly called 'air melting' process. Figure 4-3B depicts more modern two-step processes—involving 'ladle metallurgy'—that deploys additional step of refining the steel for improved quality; especially with regard to homogeneity in composition, inclusion content, residual gases, and accurate control of casting temperature for better integrity of cast product. For more effective removal of residual gases, especially hydrogen, vacuum degassing (VD) of molten steel is carried out (see Fig. 4-3B). Vacuum degassing is particularly used for removal of hydrogen gases using stream degassing or recirculation-degassing (RH), where the molten steel is sucked into the vacuum chamber and continuously re-circulated. More about these processes will be discussed later in this chapter.

For carrying out the oxidation of extra carbon in the charged materials, high purity oxygen is used for steelmaking. The deoxidation products are then allowed to float off or made to float off by special mechanisms in order to improve the quality of steels. Along with carbon, other trapped elements in the molten steel also get oxidised. Therefore, for clean steelmaking, the primary task is to remove these deoxidation products as completely as possible. One of the popular modes of removing the deoxidised products is by ladle furnace (LF) treatment. In ladle furnace treatment, high purity argon gas (or high purity nitrogen gas) can be bubbled through the molten steel, where argon gas being inert gas does not react with the molten steel but purges out the impure gases and inclusions out of the molten steel. The purging takes place as the rising argon bubbles rise through the melt and any particle or impurity that comes on the way gets attached to the bubble due to surface tension effect and gets carried on to the surface (see Fig. 4-4). Slag chemistry on the surface is so maintained that these oxidised products are quickly absorbed into the top slag.

Ladle furnace is not effective for removing hydrogen from the steel. Hydrogen removal from the liquid steel requires VD. Methods of vacuum ladle degassing utilise the reaction of deoxidation by dissolved carbon in the steel according to the equation:

$$[C] + [O] = \{CO\}$$

where, [C] and [O] denote carbon and oxygen dissolved in liquid steel and {CO} denotes gaseous carbon monoxide.

Vacuum treatment of molten steel decreases the partial pressure of CO, which leads to shifting equilibrium of the reaction of carbon oxidation, giving rise to large number of bubbles of carbon monoxide. Hydrogen dissolved in liquid steel then diffuse into the carbon bubbles and is evacuated by vacuum system. Vacuum degassing also helps in further removal of inclusions, which agglomerate due to bath agitation under vacuum and gets floated up and absorbed in the slag. Thus, steels refined by vacuum are characterised by low gases, low inclusions and homogeneous structures.

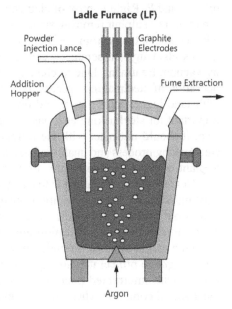

Figure 4-4. A schematic view of ladle furnace used in steelmaking and its functions. *Courtesy: www.substech.com.*

Sources of hydrogen in molten steel are primarily moist air, wet refractories, wet fluxes and ferro-alloys used in the steel making process. Thus, vacuum treatment of steels is frequently specified, especially for heavy section or for critical applications, e.g. ball and roller bearing steels, rotor forging steel, etc. Figure 4-5 shows different traditional VD processes used in treating steels.

Most basic functions in steelmaking are the removal or reduction of carbon, sulphur, phosphorus and various gases and impurities in the liquid steel. These are accomplished by stages, starting with removal

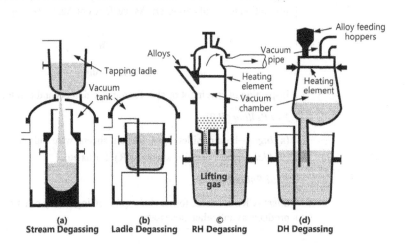

Figure 4-5. Different methods of VD process in use. *Source: Rollason, E.C. Metallurgy for Engineers. London: Edward Arnold, 1973.*

of C, S and P. Right control of slag chemistry and temperature is necessary for carbon and phosphorus removal. For sulphur removal, special desulphurising agents are added in the ladle. Impurities in steel—which come not only from the charged material but also from the erosion and wear of refractory lining of the vessels at high temperature—can be effectively removed by ladle metallurgy, and gases can be removed by vacuum treatment. Due to chances of erosion of refractory lining material during high temperature operation of steelmaking process, care is necessary in the use of high quality refractories, which can withstand higher temperature and do not vigorously react with the slag layer. Slag chemistry is very important for steelmaking as it has to absorb and retain the oxidised products, including the impurities; also, it should not react with the refractory wall. Thus, the quality of steel produced is dependent on steelmaking process and quality of slag and refractory practices. The latter practice may differ from plant to plant, and thereby making the difference in intrinsic quality of steel produced by different plants.

Next step in steelmaking is to cast the molten steel into a primary shape—like ingot, billet, slab etc. Casting is also a critical operation for quality steel making, because when the molten stream of steel comes out of the casting ladle, it can react with the atmosphere and again pick up oxygen and gases, which forms inclusions and pin-hole porosity in the subsurface areas of the solid metal. Hence, the molten metal is directed through a nozzle tube to cover the liquid stream or the *stream is shrouded by an inert gas* to protect it from exposure to environment. The casting method can be conventional ingot casting with interrupted pouring in between the ingots lined up in a pit or *continuous casting* under mechanised condition where the cast length moves over a set of rollers under water spray (for cooling)

Table 4-1. Flow diagram of steps and actions in steelmaking and rolling.

A. Steelmaking

Steel melting, decarburisation, deoxidation, sulphur removal, slag modification, inclusion removal, tramped-element removal, alloying and degassing

B. Casting

Ingot casting in pits or continuous casting—to blooms, billets or slab—with precaution to avoid corner cracks, surface cracks, tears, slag entrapment etc.

C. Rolling

Inspection of cast ingot or slab for surface defects and conditioning of defects.

Hot Rolling (HR):

(a) *Ingot*—bloom, billet, bars, section, sheets, wire rods etc.

(b) *Slab*—plates or coils, sheets (HR), forms etc.

Cold Rolling (CR):

Sheet (CR), sections, forms etc.—galvanizing or any other coating of CR products as and when necessary and the casting is automatically cut after the mass has completely solidified at an 'ear-marked' distance from the casting table. A schematic flow diagram of principal steps in steel making and rolling is shown in Table 4-1.

Though there had been many steelmaking routes in vogue, now-a-days steelmaking route is limited to either 'Oxygen process' or 'Electric Arc process' (see Figs. 4-3A and B). Electric arc furnace is a very high energy consuming route and entails high cost of energy. Hence, electric arc furnace (EAF) route of primary steelmaking should be selectively used—such as for steelmaking using high percentage of steel scraps or for special grade steels which cannot be easily made through oxygen process. Therefore, mostly 'oxygen steelmaking' route is preferred because its economical, unless the steelmaking becomes difficult by the oxygen process due to problems in alloy recovery, e.g. stainless steelmaking—where recovery of chromium will be erratic due to oxygen blowing.

Popular oxygen process is the 'LD converter' process, which is carried out in a special vessel where high purity oxygen is blown through the top or bottom tuyers (see Fig. 4-3B). The process of oxidation is exothermic, which generates enough heat of reaction to sustain the steelmaking reaction velocity in the LD converter. Reaction rate in LD converter is very high and hence the process requires sophisticated control mechanisms. Nonetheless, whatever process is used main steps in the process of steelmaking are as have been indicated in the Figs. 4-2, and 4-3A and B.

4.3. Salient Process Features for Quality Steelmaking

From the preceding discussion on the steelmaking processes, it can be seen that salient process features of steelmaking steps through LD or EAF route are:

Decarburisation: This is carried out by controlled oxygen lancing of the liquid steel. During this process carbon in the liquid steel reacts with the oxygen to form carbon monoxide (CO), which escapes from the bath as combustible waste gas. If any other solid oxides form in reaction with fluxes etc. present in the bath, those reaction products are taken into the slag by appropriate slag composition control to absorb them.

Deoxidation: Deoxidation of steel melt is done by adding aluminium (Al), silicon (Si), ferro-manganese (FMn) etc. which are oxidising elements to the impurities or gases present in the liquid metal. Of these, aluminium deoxidation (popularly called Al-killing) is preferred for making the grain size inherently fine grained and resistant to coarsening during subsequent hot-forming or heat treating. Silicon killing is carried out for some special grades where electrical property or coarser grains are preferred. The mechanism of grain size refinement by Al-killing involves reaction of nitrogen gases of the molten steel with aluminium and precipitation of fine aluminium nitride (AlN) particles on solidification, which pins the moving grain boundaries from growth. Aluminium nitride being stable at high temperature, it is effective in stopping grain growth during subsequent heating and rolling.

Alloying: Alloying of the steel is carried by adding measured quantity of respective ferroalloys or metallic alloys at the appropriate temperature during steelmaking, with precautions to ensure correct recovery of alloying elements. Care should be taken during alloying to avoid loss of expensive alloying elements due to oxidation or entrapment in the top slag layer. Final adjustment of alloying is done by trimming addition after degassing.

Degassing: Degassing can be done by vacuum treatment or by inert gas (argon) bubbling at controlled conditions and temperature during steelmaking—when rising argon bubbles carry the adsorbed gases and inclusions along with it to the surface layer. For *full degassing*, vacuum tank is used, and argon-bubbling is used for *partial degassing* and inclusion removal.

Temperature: Temperature adjustment before casting is necessary in order to regulate the shrinkage inside the cast product, macrostructure of cast product—which is influenced by the cooling rate from the casting temperature—and tendency to form thermal crack at the points of contacts with the mould or constraints. Use of *ladle furnace* (LF) is a popular method for adjusting and controlling the final casting temperature.

Casting: The process involves casting to shape with care to ensure surface quality and internal soundness. Liquid steel has to be cast appropriately. It is cast into either ingot or slab, but with enough precautions during casting for ensuring good surface and soundness of cast product in order to obtain defect-free rolling and good quality rolled products. Any deficiency in the cast quality needs to be countered by taking additional steps for ingot or slab inspection and conditioning the defective areas before rolling. In continuous casting of slabs for flat product rolling, slab casting and slab rolling processes are integrated with each other for productivity and economy. Hence, for such integrated 'Hot Strip Mills', input quality of slabs has to be ensured for quality output from the hot strip mill—especially if those hot strip mill coils are to be further rolled by *cold-rolling process*.

In sum, making good quality steel is a very complex job which depends on—the capability of the adopted processes, quality of input materials and refractories, precision of temperature controls, and efficiency of various process operations like degassing, alloying and casting. A case of bad casting can spoil all efforts of good steelmaking and hence, extreme care is necessary to ensure that casting conditions are correct in order to produce high quality castings—free from pin-hole and porosity, blow holes, segregations and internal or external cracks. However, every step in steelmaking costs money, which has to be carefully examined and balanced in order to off-set any down the line cost of rejection during processing. While an expensive steelmaking process need not be adopted for common grade of steels, due care must be taken to improve the overall yield of good metal from the processes.

4.4. Features of Steelmaking by using Ladle Furnace and Vacuum Degassing

Due to special requirements for different applications, there are many steelmaking-specific tasks to be performed by the ladle furnace technology or by VD. These processes are not interchangeable as they are distinct and have their own advantages and associated cost, which should be taken into consideration while making a choice between them. Depending on the specific tasks, there are again different methods within each of the processes that work best for a given objective. For example, Vacuum Oxygen Decarburization (VOD) is reported to work best for stainless steel making, but Recirculation Degassing (RH) in vacuum chamber works well for low-carbon slab casting for deep drawing grade steels. Common types of LF and VD treatments are:

Ladle Refining: Ladle Furnace method, Ladle Furnace Injection method—Powder or Cored Wire Injection system.

Vacuum Ladle Degassing: RH recirculating degassing method, Ladle Tank degassing (VD), VOD or RH-OB—recirculating degassing with oxygen top lance.

Technically, their major benefits are:

Benefits of Ladle Furnace treatment: Ladle furnace is used for refining a wide variety of steels where degassing for hydrogen removal is not required. The main advantages are:

- Deep sulphur removal (desulphurisation)—by injection technology;
- Good temperature control and provision for controllable reheating by electric power;
- Convenience in alloying with good alloy recovery under controlled condition;
- Thermal and chemical homogenisation of the melt before casting;
- Very effective non-metallic inclusions removal; and
- Economy—compared to VD.

Benefits of Vacuum Degassing: Vacuum degassing can be carried out by following any one of the methods mentioned under vacuum ladle degassing. Benefits of VD are very similar to LF treatment, but with additional capability of better hydrogen removal. Benefits of individual VD processes are summarised as follows:

- *Benefits of Recirculation Degassing* are—its capability for hydrogen removal by degassing, oxygen removal by deoxidation, carbon removal by decarburisation, sulphur removal by desulphurisation, precise alloying, plus non-metallic inclusion removal and temperature homogenisation for casting. Ladle tank degassing uses vacuum treatment and simultaneous argon purging through porous bottom plug at the ladle back, and thus stirring is more intense.
- *Benefits of VD* are—carbon removal, hydrogen removal, oxygen removal, deep sulphur removal, precise alloying, inclusion removal and temperature equalisation and obtaining chemical homogeneity.
- Vacuum oxygen decarburization is a special process for stainless steel making where under normal operating conditions, chromium—the main and expensive alloying element added to stainless steels—tends to get oxidized along with carbon. *Benefits of VOD process are*—low losses of chromium along with deep decarburisation, sulphur removal, scope of preheating for temperature control and homogeneity, and also inclusion removal.

Hence, major advantages of VD are deep decarburisation and removal of hydrogen gases from the steel; presence of dissolved hydrogen in steel often creates problem for heavy section forging by inducing hydrogen flaking and cracking inside. However, any VD process is more expensive than LF treatment. As such, for most other grade of steels, which do not require deep degassing and hydrogen removal, LF route is more economical and preferred route.

A schematic view of LF in operation has been shown in Fig. 4-4. Similarly, various processes of VD have been shown in Fig. 4-5. The vacuum process involves churning of liquid steel under vacuum in a closed chamber so that all absorbed gases get sucked out of liquid pool and removed. VD is a must for steels with low hydrogen requirement. In a low pressure environment produced by vacuum, all chemical reactions relating to gaseous products are enhanced, and the products are taken out from the liquid steel pool. Thus, the steel made by VD process should be cleaner than the LF-route steels. However, VD process would cost more—hence, for economic reasons this route should be chosen where it is necessary. Many a time, LF treated steels are good enough for most forging steel applications, whereas for low-carbon sheet steels, especially the one meant for cold rolling, VD-route of steelmaking is preferred.

4.5. Making Clean Steels for Critical Applications

Earlier, clean steel meant steel with low level of oxide and sulphide inclusions. But, with increasing demands on steel for superior properties as well as for critical formability—as in the case of automobile body parts—the term 'clean steel' now relates to steel with low level of sulphide and oxide inclusion contents as well as low levels of hydrogen, nitrogen, phosphorus, and sometime even carbon. Extra low-carbon Interstitial-free steel with %carbon (%C) less than 25 ppm—which are in demand for critical automobile parts formation—is an example of the latter. Table 4-2 shows limits of some of these elements that are demanded now-a-days for certain end-use applications. The Table 4.2 gives an idea as to how the definition of clean steel will depend on the criticality of end use. Cleanliness level of steels is to be chosen as per the end-use requirement due to its impact on cost and availability.

It is further expected that purity level of such ultra clean steel will further grow with the sophistication of applications and criticality of end use. In addition to the limitations of above elements, clean steels also frequently demand control of other residual metallic elements in steel e.g. copper, chromium, tin etc.—that can cause some detrimental surface quality or affect forming properties for some applications, e.g. hot shortness in steel affecting hot workability.

Therefore, clean steelmaking involves:

- Removal of oxide and sulphide impurities in steel;
- Removal of gases like the hydrogen and nitrogen; and
- Controlling other residual elements from the steel as per application requirement or agreed supply specification.

Impurities in steel mean anything that remains trapped inside the solid steel and cause deterioration of properties, like physical, mechanical, chemical (e.g. corrosion), forming etc. Impurities are generally the products of de-oxidation process, dissolved gases and suspended non-metallic that might come from refractories and fluxes used in steelmaking. On cooling into ingot or slab, these impurities remain trapped in the steel and cause dirtiness in the steel—that act as discontinuities in the steel matrix and impair mechanical properties, machining property, forming property etc. Dissolved gases—like oxygen, nitrogen, hydrogen, etc.—on cooling segregate and might cause 'ageing', 'flaking', 'hydrogen cracking' etc. that create internal flaws and loss of certain properties.

Table 4-2. Purity level of clean steel required for some specific applications.

Element (Metalloids)	Product	Content Limitation (ppm)*
Oxygen	Steel chord Wire Rod	≤10
Sulphur	Sour gas pipe line steel	≤10
Hydrogen	Rail steel	≤1.5
Nitrogen	Electric grade steel	≤20
Phosphorus	Off-shore steel (drilling rig)	≤80

* 1 ppm = 0.0001%.

To fulfil such stringent cleanliness requirements, steelmaking technology is getting continuously improved by optimising each stage of steel production. These stages or steps are:

- Iron making (pig iron),
- Iron treatment (hot metal treatment),
- Oxygen steelmaking process (e.g. LD or EAF),
- Tapping,
- Ladle treatment,
- Tundish treatment, and
- Continuous casting (CC).

Clean steelmaking calls for precautions to control the aforesaid elements in all stages of production by appropriate means.

Some essential steps for clean steelmaking are:

1. Sulphur control: Major source of sulphur in steel is the pig iron (hot metal from the blast furnace). Hence, hot metal desulphurization is carried out by using sulphide-forming agent like calcium compound (e.g. lime injection) in a reducing atmosphere inside the treatment vessel. Calcium carbide with some additive is preferred for its comparatively lower cost. However, during oxygen steelmaking (e.g. the LD process) molten steel encounters high oxidizing atmosphere inside the vessel and again some sulphur gets picked up. Therefore, some extra precautions are taken during steelmaking and tapping the steel to control sulphur as well as oxygen. Some common steps include:

- Slag-free tapping i.e. slag is a source of oxygen—general course is to bubble some inert gas through the molten steel to float off all slag to the surface and then add some protective reducing agent;
- The use of vessel lined with basic refractory lining e.g. dolomite;
- Adjusting pouring temperature of the liquid steel for better sulphur removal;
- Use of desulphurising agent e.g. injection of calcium silicide (CaSi) wire;
- Pouring the liquid steel into the mould or tundish (in case of continuous casting) under protective shroud cover (or nozzle), to avoid oxygen pick-up from the surrounding air.

In addition to control of sulphur, shape of sulphur—which is generally lenticular—can also be controlled by using CaSi wire, when the shape of sulphide becomes more spherical and less harmful.

2. Phosphorus control: Phosphorus control in steelmaking requires virtually opposite of steelmaking conditions used for sulphur control. Lower steel temperature and higher oxidation potential in the melt is the right condition for phosphorus removal, which is applicable to both oxygen process and EAF process. Lower phosphorus in LD steel is achieved through a special blowing process during steelmaking called *combined blowing* where top surface is highly oxidise by controlled oxygen blowing on the top slag for making phosphorus partition to the slag easier, and simultaneously blowing inert gas through bottom tuyers of LD vessel to ensure intimate contact of melt with the oxidising top slag.

Further, phosphorus control requires use of low phosphorus fluxes and ferro-alloys that are to be added, if required, subsequently. And, strict slag removal measure is also needed to take care that no oxidising high phosphorus slag is left over in the melt after tapping. In general, slag free tapping, use of active slag cover of the melt, and inert gas bubbling from the bottom (see Fig. 4-4)—considerably improves the cleanliness of steel with regard to oxide inclusions as well.

3. Homogenisation of steel melt: This is another step in clean steelmaking. This is necessary not only for chemical cleanliness and uniformity, but also for uniformity and consistency of alloy recovery at the later stages of alloy adjustment and better control of casting temperature. Homogenisation of the melt is largely attended by stirring effect of the inert gas blown from the bottom tuyers (see Fig. 4-4). Control of casting temperature is largely achieved at this stage by cooling the melt by adding select-scraps or by additional heating if required, for which provisions are generally provided in the ladle metallurgy process (see Fig. 4-3 particularly the electrodes at the top).

4. Vacuum degassing treatment: This is a popular process for improving the cleanliness of steel, especially the alloy steels. The process involves churning of liquid steel under vacuum in a closed chamber so that all absorbed gases get sucked out of liquid pool and are thus removed. Figure 4-5 shows the setup of VD by different techniques that can be adopted as per process requirement. Vacuum degassing is a must for steels with low hydrogen requirement. In a low pressure environment in the vacuum chamber, chemical reactions that produce different gaseous products in the liquid steel get accelerated and the vacuum system takes those gaseous products out of the liquid steel pool. Such a vacuum treatment can for example be combined with the other options of secondary steelmaking as well in a tank degasser, as indicated already in earlier figures. The low pressure causes a remarkable reduction in hydrogen and nitrogen content of the steel.

Thus, the VD process has the versatility to remove hydrogen, nitrogen, and other atmospheric gases, on one hand, while removal of volatile impurities like tin, lead, arsenic etc. that may come from scrap used for steelmaking on the other. Vacuum degassing also helps in reduction of metal oxides, removal of deoxidation products, and control of alloy recovery and adjustment to close limits. Hence, for steels with narrow specification range, high degree of cleanliness requirement, and freedom from any trapped oxides—such as bearing steel—VD is a necessary process.

Low carbon stainless steel, ball bearing steels, high quality forging steels for critical applications like turbine blades, steel for ship hulls etc. are made through VD process. The clean steel produced through VD route is of consistently high quality, with highly improved properties in transverse to rolling direction. Amongst the VD processes, RH-degasser technique is very popular for mass production—like production of extra low carbon stainless steel—due to its flexibility for handling large tonnages and fast processing time.

5. Casting/continuous casting: This is the stage of solidification of liquid steel into a mould (ingot mould) or to the shape of billet or bloom, etc. Steels, made clean by following the processes discussed earlier, may become dirty again if adequate care is not taken for protecting gas absorption—like oxygen, nitrogen, and hydrogen—by the liquid stream while being cast. Besides, liquid steel can also get contaminated by the erosion of refractory being used in the tundish or the clogged oxide particles at the casting nozzle tips.

Since the liquid steel is made low in oxygen and of other gas contents, there would be a tendency to absorb these gases again from the environment, if not:

- The vessel carrying liquid steel is continued with mild stirring to float off any such oxidation product;
- The liquid pool at the casting point is covered; and
- The casting stream is shrouded from the atmosphere.

Hence, submerged nozzle and special types of casting powder are used in the tundish for continuous casting or for direct ingot mould casting. Ingot mould casting is getting out of favour now—due to lower productivity, cost and also for quality—unless the steel is required for very large forgings where continuous casting of such heavy bloom is a problem.

Standard technologies used for CC (continuous casting), therefore, include: application of submerged nozzles between the ladle and the tundish and between the tundish and the CC-mould, special technique for preventing the ladle slag entering the tundish, use of good quality casting powder, and preheating of tundish for exact control of casting temperature. Casting temperature has considerable influence on segregation and cast structure; hence the super-heat—necessary for ensuring smooth casting flow—of the liquid steel being cast needs control. In addition to tundish heating, there is a trend now-a-days to use dams, filters and porous plugs at tundish, for further improving the clean steel quality. Furthermore, for improved surface quality, the continuous casting process is run with oscillation and proper mould lubrication for better surface.

Thus, depending on the criticality of usages, clean steelmaking process can be chosen from number of alternatives—either individually or in combination—as described above. However, they are all characterised by the application of a sequence of metallurgical operations in separate vessels or in a combined vessel—to allow optimisation of certain basic reactions that remove the impurities and dissolved gases to make the steel cleaner. Care has to be taken to ensure that the processes chosen are capable of giving good yield for economic reasons, because the challenge to steel technology is not only of making clean steel but also making the steel the most cost-effective material for industrial and domestic applications. Role of secondary metallurgy with respect to quality and cost optimisation of steel grades is, therefore, becoming increasingly important and being adopted in steel plants across the world.

4.5.1. Secondary Metallurgy and Its Functions

Secondary metallurgy is a post-treatment operation in steelmaking, which is carried out after the primary steelmaking operations in the main vessel, in order to improve the quality of steels by carrying out certain special treatment. For example, LF treatment or argon purging of liquid steel, discussed earlier, are parts of secondary metallurgy. The entire steelmaking process is divided into primary and secondary steps—due to the necessity of more control for cleaner steels and for the sake of improved productivity from a steel melting shop.

In a nut shell, basic functions and aims of secondary metallurgy are:

- Slag carry over control to the next step of processing;
- Mixing and homogenising the melt;
- Decarburisation, desulphurisation, and dephosphorisation;
- Elimination of tramp elements (for improved cleanliness);
- Precise alloying and improved alloy recovery;
- Degassing e.g. by vacuum treatment;
- Deoxidation, and prevention of further oxygen pick up, and;
- Sulphide inclusion shape control for improved machining.

Secondary metallurgy is now being extended to continuous caster as well where liquid steel streams are shrouded in the tundish to prevent oxygen and hydrogen pick up. Many steel plant operators also employ 'electro-magnetic stirring' (EMS) in the CC-mould as secondary metallurgy practice for promoting sound cast structure. Secondary steel metallurgy has become an integral part of quality steel making for its capability to economically produce much better cleanliness and closer tolerances of alloying in the steel. With the advent of secondary metallurgy and many new developments of techniques in this area including—ladle and tundish metallurgy, quality demands of large verities of special steels can be fully met. Because of the versatility and flexibility of modern secondary metallurgical processes, the process, in some form or other, is adopted in all types of steel plants, namely integrated steel mill with BF hot metal making facility as well as mini steel plants using EAF technology for production of steels. Amongst the secondary metallurgy processes, argon purging to remove deoxidation products from liquid steel is most widely used for economically making various commercial grade clean steels.

4.6. Killing of Steel: Killed, Rimmed and Balanced Steels

Liquid steel has excess oxygen content, coming from the oxidation process during steelmaking which extensively uses oxygen to oxidise the impurities. Dissolved oxygen in the liquid steel is undesirable, because it will react with *carbon* to form gaseous carbon monoxide. Oxygen will also react with *alloying* elements to form oxides. Continuation of such reactions at the time of tapping the steel poses problem for bath composition adjustment for final tapping. Moreover, gases produced by oxygen reaction also cause foaming of the liquid steel bath and formation of gas-holes in the castings. Hence, it is necessary to reduce the oxygen level in liquid steel to negligible level before the final composition adjustment, if the steel is not planned for 'rimmed quality'. *The process of fixing the oxygen in the liquid steel is called killing.* Degassing or killing of steel is done during tapping by using special deoxidation agents which readily and preferentially combine with oxygen in the bath. These deoxidising agents are aluminium, ferro-silicon and ferro-manganese.

The name killing comes from the fact that presence of high oxygen in the melt causes foaming from the gasses formed by oxidation reactions, and the bath, or the mould, in which the melt is cast, becomes foamy. When deoxidising agents are added and oxygen gets readily drawn out of the bath, bath becomes quiet. This in olden days used to be called 'killing the bath', and the name is still continuing—though the process is technically known as *deoxidation process*. All deoxidation products have to be suitably separated from the liquid steel and absorbed into the slag—so that they do not lead to entrapment of oxides in the steel and make the steel dirty i.e. high in inclusion content.

Amongst the deoxidation agent, aluminium is most reactive and widely used. Though it is more expensive than silicon killing, aluminium killing is known to produce inherently fine grains in steel due to pinning and locking of grain boundaries by aluminium-nitride particles that form by reaction between nitrogen gas in the bath and added aluminium. All forging grade steels and fine grained steels for engineering and automotive applications specifically call for 'aluminium killing' because of finer and more stable grain sizes obtained by aluminium killing *vis-a-vis* silicon killing. Silicon killing is also effective in removing excess oxygen, but it is ineffective in controlling grain sizes or grain growth. Grains produced in silicon killed steel can undergo rapid grain growth process when re-heated for hot working or hot rolling. Such large grains are not desirable for good strength and ductility (as discussed in Chapter 1). Also large grains give rise to pancake structure after cold rolling and stretch forming, which is an objectionable surface defect in steel body parts. Therefore, killing steel by silicon only is not very common now-a-days, unless the steel is for electrical applications. Instead, steels could be partly killed by silicon and finished by adding aluminium i.e. Al-Si killing. However, for distinct advantage of using aluminium killed fine grained steels for automotive and engineering applications, the overwhelming trend is to go for aluminium killing, and there have been many technological advances to perfect the technique of adding or feeding aluminium wire at tap or during teeming of hot steel for better control of the killing process.

Sometimes, steel is purposely left without killing the oxygen in the liquid, because of the necessity for avoiding shrinkages and piping in ingot casting, where gas holes produced by the oxygen reaction with carbon compensate the shrinkage of volume with cooling. This type of steel is known as *rimming steel*. This type of steel casting is deliberately designed to produce a chill layer of relatively purer metal on the surface by partial deoxidation (mostly by small addition of aluminium) and controlling the casting condition to produce a thick chill layer. Rimming steel may not be exactly without any deoxidation, but very controlled deoxidation whereby foaming is stopped but enough oxygen is left to cause gas holes to compensate the shrinkage. Due to chilling of the surface, gas holes produced move away from the surface and towards the centre of the ingot. These gas holes are not oxidised; hence they mostly get welded up under the pressure of rolling or forging. Because of soft surface, rimmed steels are easy to form and cold draw. As such, its applications are found in areas of very low carbon steel where surface finish is not a big criterion, such as for nails and screw forming in high speed machines.

There is another type of steel called *balanced steel/semi-killed steel* where the melt is deoxidized by adding more aluminium than rimmed steel but just enough to leave some oxygen for gas formation and balance the shrinkage. Benefit of semi-killed steel is its higher solid metal yield due to less shrinkage, but the semi-killed steel could have more and larger inclusions than fully killed steels. Hence, these semi-killed steels are not preferred for any parts that may face fluctuating load in service i.e. for fatigue and fracture sensitive applications.

4.7. Steel Casting Routes

Liquid steel produced in steel melting shop has to be cast to some definite shape, size and weight for subsequent rolling or forging to finished product. Purpose of casting is to give the desired shape and size to cast steel—generally described as ingot, slab, bloom or billet—so that the cast steel can be further rolled (sometimes forged) to required finished product—like the bar, billet, strip, sheet, plate, angles, sections, beams etc. Towards this purpose, methods of casting generally take two routes: *Ingot (mould) casting* and *Continuous casting*. There is another method of casting steel, and that is 'sand casting' as followed for casting of steel component in foundries, but that is not in the scope of this study.

Ingot casting is the age old process and had been the only route of steel production in the pre-World War-II era. In 1950s, continuous casting process was first developed but its initial application was in casting of copper bars. Continuous casting of steel did not draw much attention until late 1960s. But, since then, this technology has rapidly gained ground and today over 70%–80% of all steels are cast through CC route. Main difference between ingot casting and continuous casting is in the condition of solidification. In ingot casting, solidification process is comparatively slow, producing coarser columnar structure (see Fig. 4-6) and other solidification and shrinkage-related defects, like—pipes, porosity, and segregation. The propensity of these defects is even more when the casting temperature is not adequately controlled. Figure 4-6 schematically depicts some such defects that are common in ingot casting.

Ingot casting can be single ingot or multiple ingots casting either by manually pouring the liquid steel into the individual ingot mould or by feeding through the bottom of ingot moulds—by pouring metal through a central pouring gate to number of moulds connected to the central gate by refractory tubes. Since the liquid comes in contact with the mould wall with rising level of liquid steel, and in the process heats up the mould wall, the cooling rate is not uniform along the mould length and

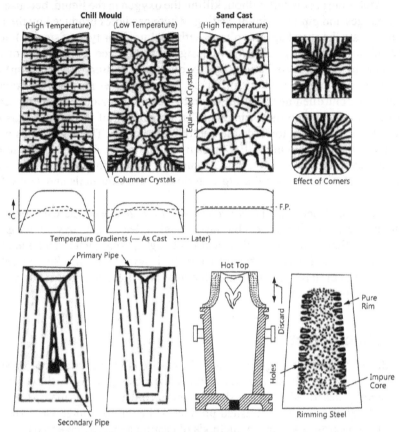

Figure 4-6. Schematic representation of ingot cooling, resultant structure and defects.
Source: Rollason, E.C. Metallurgy for Engineers, 1963.

solidification structure differs at spots due to temperature variation. Ingot casting also suffers from shrinkage—arising due to volume contraction from liquid to solid—segregation, and non-uniform structure along and across the ingot, containing many areas of defects that could not be used for downstream processing or uses. This necessitates use of hot-top practice to reduce top shrinkage, and cropping and dressing of defective areas of ingots—the latter causes yield loss. Further, to breakdown the coarser macro-structure of the ingot steel, heavy reduction during rolling is needed so that variation of microstructures along and across the rolled product can be minimised. These are few limitations of traditional ingot route casting.

Continuous casting, on the other hand, involves faster cooling rate, producing a better surface and more uniform macro-structure across the section being cast. Continuous casting is, in effect, an endless process of casting where molten steel is transformed into solid on a continuous basis through an open-ended mould. The mould has the facility for external cooling for controlling the solidification process, pattern and the rate. Shape of the solid steel produced by continuous cooling is determined by the shape of the mould, which could be spherical, rectangle, square etc.—only limitation being that the shape should allow easy withdrawal of the solidified steel by the withdrawal rolls of the caster placed underneath the mould. Figure 4-7 depicts a typical continuous casting arrangement.

Due to faster cooling rate, structure is also relatively finer than ingot, and segregation problem is less—except some tendency of central segregation which can be controlled to a large extent by proper design of cooling and providing a soft-reduction below the casting mould when reasonably thick solid shell has been formed and some liquid metal is trapped in the centre zone of the cast section.

In CC process, molten steel flows from the ladle through the tundish to the mould. Flow to and from tundish can be suitably covered or protected from the outside atmosphere to avoid oxygen pick up (this improves cleanliness of the steel due to lower oxide inclusion). Tundish can hold enough molten steel to ensure smooth flow and also acts as reservoir for facilitating exchanges of ladle and sequence casting of different heats (this improves metal yield and productivity). Tundish metal is protected from the atmospheric exposure by a slag cover, and offers many facilities and flexibility for controlling the quality of steel by, what is known as—specialized tundish metallurgy, including

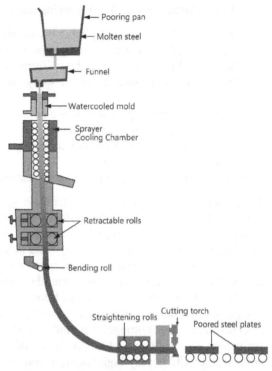

Figure 4-7. An illustration of a typical continuous casting arrangement for steel. *Note:* In this figure, 'Tundish' has been named as 'Funnel', and Retractable rolls represent 'below the mould soft reduction facility' necessary for reducing centre segregation.

refining, inclusion removal and homogenization. Tundish metallurgy, which is a part of secondary metallurgy, is fast gaining ground in the production of clean steel through CC-route.

From the tundish, liquid metal flow to the mould where a constant level of liquid pool is maintained for controlling the heat flow and prevention of rupture of frozen steel skin that has already formed in contact with the water-cooled mould wall. Moulds are generally bottomless copper moulds, cooled by an external water jet, and oscillated vertically in order to avoid sticking of the shell to the mould walls. Drive rolls that follows below the mould table and water spray zone, take over the task of slowly withdrawing the shell from the mould at a 'casting speed' that matches the flow of liquid metal to the mould, in order to ensure constant level of liquid pool in the mould. A special sensor device called 'mould level controller' is used for this purpose. Most critical part of the casting process in CC is the control of initial solidification below the meniscus level of mould-metal (shell) junction. This is where the surface of the solid mass starts forming and this surface must remain uninterrupted by ensuring proper mould level control and mould lubrication. If there is any disruption or puncture of the shell, there could be 'breakout' of the casting and total disruption of the casting process. Also, any defect of the shell skin arising at this stage will cause defective surface in the final stage.

With reference to Fig. 4-7, following measures and control are essential for good CC-casting:

1. Molten steel from pouring ladle to tundish and tundish to CC-mould must be shrouded to protect oxygen, hydrogen or any other gas pick up in the melt while being cast.
2. Spray cooling by water must be uniform and adequate for ensuring enough solid shell formation by the time casting comes out of the spray zone to withstand some 'soft reduction' to eliminate centre segregation.
3. Soft reduction roll pressure should be so controlled as not to put too heavy a pressure on the liquid core to cause its rupture.
4. Progressing cast-billet should be gently bent and directed towards the cooling table by using sufficient 'bending radius'.
5. On reaching the cooling bed, cast bars should be straightened while hot and cut to length.

For good casting, CC-billet or bar after leaving the mould shall develop at least 6–20 mm thick shell by the time the casting comes to the point of 'bending roll'. Task of bending rolls is to gently bend the semi-solidified shell (still with liquid steel inside) towards the casting table by exerting minimal pressure. Bending roll should make the casting bend at a radius just enough for it to enter the straightening roll set. Spray cooling continues at a graduated rate to ensure total solidification of the cast by the time it reaches up to the straightening rolls and the cutting torch table.

Thus, CC process offers lot of additional points for control and adjustments required to make clean and consistent quality of steels. Because of the chance to better control the solidification process in CC, steel quality is more homogeneous and with much less segregation and inclusions than ingot cast steel. Inclusions can be floated off from the liquid melt by proper tundish practice. However, the biggest advantage of CC is very high 'liquid metal to solid metal yield', which is more than 95—as against about 80% in ingot route steel casting. In CC, the liquid steel is cast into an appropriate size and shape that is ready for finish rolling in a modern continuous mill—this helps to avoid or minimise number of reheating for rolling and minimises scale loss and other rolling related losses. Thus, CC process saves some intermediate stages of reheating and rolling and produces higher yield as well. Therefore, the process has proved to be very economical and ideal for adopting in mass production. Though the initial capital cost of the CC set-up would be much higher than ingot casting, the operational costs however are much lower and the quality of steel is better. The advantages of CC can be summed up as follows:

- Saves energy and improves yield;
- Higher labour and capital productivity;
- Improved steel quality e.g. cleanliness, homogeneity, uniform cast structure etc.;
- Reduced number of process steps to finish product; and
- Better surface quality compared to ingot cast surface.

However, there are some technical difficulties in making all grades of steels through continuous casting route e.g. high alloy steels, especially with regard to centre line segregation. But, because of economy and high productivity through CC route, extensive amount of research work is going on in this field to overcome any such problem.

As a result, world steel production through continuous casting route is steadily growing and has currently touched over 70%. It has been reported that in Germany 100% of crude steels is produced through CC now. Furthering the development of CC, there has been extensive development of 'thin slab casting' through CC route that facilitates direct rolling in 'hot-strip mills' for production of stringent quality steel sheets, matching the requirements for critical automobile applications.

4.8. Steel Defects and Their Causes

Foregoing discussions on steelmaking and casting indicate that howsoever careful one could be, there will be chances of defects arising in the cast steel due to segregation, non-homogeneity in composition and structure, gas holes and porosity, and inclusions of both types—micro and macro-inclusion. These defects need to be prevented or minimised for ensuring the suitability of steel for a given application. Hence, sources and causes of such defects in steels should be examined.

4.8.1. Segregation in Steel and Its Causes

Composition of steel after casting is not exactly uniform throughout the section, and there occur some concentration of impurities as well as alloying elements, in various parts of the steel section. This non-uniformity of composition is due to 'segregation'. *Segregation is the outcome of separation of impurities and other soluble elements e.g. alloying elements present in liquid steel during solidification while being cast—both ingot casting as well as continuous casting.* During solidification process, there is always a temperature gradient across the casting. This temperature difference causes some solute elements in the liquid to get rejected from the solidifying mass, which then diffuses out to the melt ahead of the dendritic solidification front. As a result, there are areas of solid dendrites—with relatively low solute elements and areas of liquid pool—with higher solute elements, trapped in between the dendritic arms. When this part of liquid pool solidifies, the composition of this solidified steel is somewhat different from the earlier. Insoluble impurities—like oxide and sulphide inclusions or any slag particles present in the steel melt—also get trapped in these inter-dendritic regions. This phenomenon gives rise to what is called 'segregation'. Therefore, segregation may refer to chemical segregation or segregation of insoluble particles—which are trapped between the solidifying dendritic fronts.

Chemical segregation—i.e. segregation of soluble impurities and other soluble elements present in the steel—could be either micro-segregation or macro-segregation. In fact, micro-segregation is thought to be the beginning and source of macro-segregation. Micro-segregation refers to microscopic segregation of alloying/solute elements in the body of steel due to a thermally activated process called 'diffusion', which gives rise to formation of numerous small volumes of liquid enriched with the segregating solute elements. This leads to composition differential with the surrounding, and adjacent liquid solidifies—as dendrites having low solute concentration—resulting in chemical heterogeneity. State or degree of micro-segregation can be corrected to some extent by heavy hot-working, reheating the steel at higher temperature or by special heat treatment.

However, when the temperature gradient during solidification is steeper, micro-segregation can move further to coalesce to form larger macro-segregation, which can cause serious problem in uses of steel due to formation of banding, crack, uneven mechanical properties, machining difficulties etc. Macro-segregation, which is coarser in nature and associates with other insoluble impurities, cannot be easily removed if once formed. They are more harmful than micro-segregation. Macro-segregation generally takes place at the top part of ingot or central part of continuously cooled billets—which are the last portion to solidify. *Hence, extra care needs to be taken to ensure that the steel product is reasonably free from macro-segregation by using special techniques like hot-topping for ingot casting and control of cooling condition by water spray in continuous casting along with, what is called, 'soft reduction' of the solidifying billet, blooms, etc. from the caster.*

Earlier, when ingot route of steelmaking was predominant, macro-segregation was a major problem. With the advent of continuous casting technology, problem of macro-segregation has been somewhat less but not altogether free. Ingot route macro-segregation is generally characterised by their location of formation e.g. hot-top segregation at the top of the ingot, negative segregation at the bottom, V-segregation at the central zone, and A-segregation associated with the coarse columnar structure of the ingot steel. Typical illustrative macro-segregation locations in ingot are shown in Fig. 4-8.

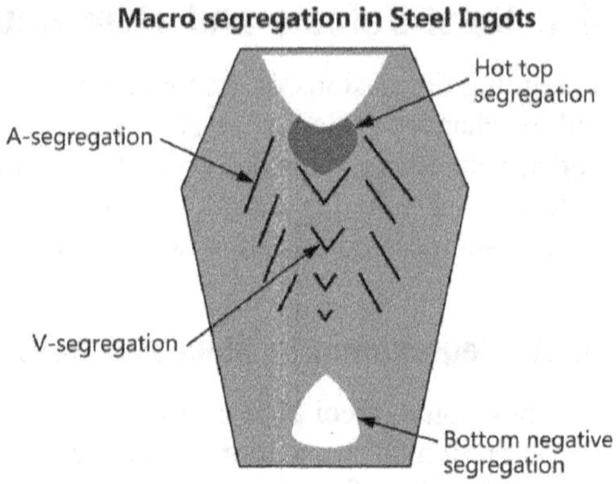

4.8.2. Other Steel Defects Arising from Solidification Process

Solidification is the physical-chemical process by which liquid metal transforms into solid mass involving— crystallization, segregation of impurities and some dissolved elements, liberation of dissolved gases from the solidifying liquid, shrinkage cavity due to thermal

Macro segregation in Steel Ingots

Hot top segregation

A-segregation

V-segregation

Bottom negative segregation

Figure 4-8. Illustration of the locations of macro-segregation in steel ingot. In continuously cast steels, V-segregation is often evident.

contraction and porosity formation by the trapped gas bubbles. Segregation phenomena which have been discussed earlier are a part of this solidification process and result. A typical schematic solidified structure is shown in Fig. 4-9, which has chilled crystals of purer metal on the very surface (small equiaxed grains), followed by columnar dendritic structure and the large equiaxed grains at the centre.

Such a pattern of solidification can give rise to various types of defects—by segregation, shrinkage, gas evolution, and inclusion entrapment. Thus, steel defects arising from solidification could be macro-segregation, shrinkages, gas porosity, and entrapped impurities and inclusions at the inter-dendritic region that solidifies last. These defects make the cast product unsound. Therefore, extra care is necessary to control the solidification process to minimise these defects.

Shrinkage is the contraction of volume caused by dropping temperature during casting as well as due to thermal contraction in the solid state itself. Shrinkage, in the form of cavity, forms when a large isolated region of liquid remains surrounded within solids produced during solidification. Shrinkage defect— in the form of macro-porosity—is generally associated with the central portion of the casting where wide temperature range occurs during solidification and liquid melt cannot penetrate down through the inter-dendritic regions and supply fresh liquid to fill the shrunk spaces.

Gas Porosities could be of two types: (a) *gas pores* (micro-porosity) caused by dissolved gases in the melt, mainly hydrogen, that gets forced out due to decreased solubility with cooling, and (b) *blow holes* caused by gases resulting from gaseous reactions during steelmaking (e.g. deoxidation

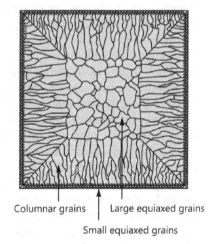

Columnar grains | Large equiaxed grains

Small equiaxed grains

Figure 4-9. Schematic solidified structure across a section and presence of columnar grains and coarser central structure.

reaction that produces carbon monoxide gas due to the reaction of carbon and oxygen in the steel) and remains entrapped.

Sometime, *surface blow-holes* may form due to mould metal and the liquid steel reactions. These blow-holes are confined in the sub-surface areas, and if not removed carefully can give rise to long surface seams during rolling. When the cast ingot/billet is reheated for rolling, thick scale forms on the surface, which when gets dislodged from the surface during rolling exposes the sub-surface blow-holes. These blow-holes can then get rolled along with the billet and form long streaks of seams—a narrow and shallow crack/discontinuity. Seams, if left without removal by scarfing and grinding, can lead to surface cracking during subsequent forging or forming. The other types of solidification defect is the *impurity entrapments* in the inter-dendritic arms, which give rise to coarse patch of heavy inclusions in the steel, rendering it bad for machining and any form of dynamic load bearing applications.

4.9. Constraints and Limitations of Inclusion Control in Steel

Inclusions in steel arise from various sources during steelmaking, such as chemical reaction products, physical suspension of insoluble particles, erosion of refractory lining in the steelmaking vessels etc. They all are some types of oxides, sulphides, nitrides, and phosphides, which are non-metallic in character and harmful for the processing and application of steels. The term 'inclusion' is synonymously used with 'impurities' in so far defining cleanliness of steel is concerned. Major sources of inclusions are the de-oxidation process associated with steelmaking and the erosion or reaction of refractory linings during steelmaking. Therefore, inclusions originate from: (a) dissolved impurities, and (b) suspended non-metallic—oxides, sulphides, refractories, etc.—in the liquid steel.

Complete removal of inclusions is nearly impossible; except by 'zone-refining' process which is not commercially or economically adoptable for mass consumed steel grades. Hence, complete inclusion removal in commercial grade steels is not attempted. Commercial steelmaking processes using special techniques for clean steel making (e.g. LF treatment, VD etc.) attempt to remove all coarser insoluble inclusions and gases that may react later to form inclusions or cause other defects e.g. hydrogen cracking. These special processes (LF, VD etc.) float off all large inclusions and also create a condition by reducing gas content in the liquid steel so that chances of gaseous reaction—like the oxidation of iron and alloying elements—are vastly reduced. However, even in clean steel making processes, steels are left with some amount of inclusions that are difficult to economically remove—either for process limitation or for longer cycle time to remove. But, whatever is left behind for economic reasons, those inclusions must be very fine, so that they are not harmful for subsequent forming and using. Therefore, the aim of tonnage clean steel production is not complete removal of all dissolved impurities and reaction products, but minimising them to a level that is not harmful for the end applications or for processing.

Therefore, what is required during steelmaking is to control the chemistry of the inclusions and their sizes. Chemistry of an inclusion determines its properties if the inclusion is brittle e.g. oxide inclusions or soft e.g. sulphide inclusions (refer Chapter 3). Depending on the end use or forming process, inclusions of different types and their level in the final product should be controlled. For example, for ball bearing applications involving high contact load on the surface, brittle oxide or nitride inclusions in the steel are considered very harmful. Hence, ball bearing steels should be made by following a process where oxide and nitride inclusions could be minimised. Sulphide, being soft, helps in machining, but causes poor transverse property. Hence, the steelmaking process should try to modify the sulphide shape so that its beneficial effect for machining is retained but without much sacrifice to the transverse property.

Size-wise inclusions are most harmful when the size is large. In general, inclusions can be subdivided into various sizes—for example, macro- (>20 μm), micro- (1–20 μm), and sub-micro-inclusions (<1 μm). It is not only difficult to detect very fine size inclusions; it is also very difficult to avoid them in steels; because even a leftover of 1 ppm each of oxygen and sulphur in liquid steel can cause steel to contain 10^9–10^{12} non-metallic inclusion count per ton. Thus, theoretically, from the viewpoint of 'cleanness', steel is never free from inclusions.

Micro-inclusions which are <1 μm in size are considered harmless in use, and most harmful ones are in sizes above 20 μ. Larger inclusions of size above 20 μ are quite prevalent in air-melted steels. Hence, for fatigue sensitive parts, where inclusions play critical role in the fatigue crack initiation from the inclusion-steel matrix interface, steels are specified to be made through a route that can eliminate larger inclusions e.g. LF/VD process. Due to inert gas (e.g. argon) purging from the bottom in LF process (see Fig. 4-3), there is a good chance in general that most large size inclusions will float-off and make the steel reasonably clean for fatigue-sensitive end uses. Hence, steel specification where inclusion sizes of less than 10 μ can be tolerated, expensive VD might not be necessary from cost point of view. For such applications, only LF treatment, which is much more economical and commercially viable than expensive VD treatment can be prescribed. Vacuum treatment is called for when the applications require even finer inclusions and removal of dissolved gases from the steel—e.g. hydrogen gas, which can cause hydrogen flaking in steel on cooling. By VD process, inclusion sizes can be mostly controlled within the finer sizes of around 1–5 μ, which are acceptable for most applications.

Thus, inclusions in steel need not be completely removed, because that would entail higher cost without commensurate benefits, and may also limit availability—they should be removed to the extent to make the steel suitable for the intended applications. Tolerable limits of inclusions types and sizes are, therefore, specified in the customer specifications in addition to the process of acceptable steelmaking.

STEEL ROLLING

4.10. Hot Rolling Processes

Steels are rolled into semis (i.e. semi-finished product for the next stage of rolling or forging) or to a shape (e.g. beam, rail, bar, plate, etc.) from the ingot or slab or the section that has been cast. The size and weight of the input ingot or slab etc. is decided by the final shape and size of the finish-rolled product to be obtained. There are many downstream processing stages to obtain the final steel product, which can be listed as:

Primary rolling/forging: Primary rolling—or primary forging generally applies to heavy ingots used for large forgings—is carried out in a heavy mill like blooming mill, slab mill, etc. These products may be used as semis for further processing at users' end or can be further processed by steel mills, to give shape and properties as required by customers, specific or at large. For example, hot rolled slabs are the primary material for rolling into hot rolled strips. Similarly, hot rolled strips would be the primary material for cold rolling of strips.

Secondary rolling: This is the rolling given to obtain final shape of the product for sale to customers—e.g. bars, sheets, strips, sections, forms etc. This kind of rolling requires a more precise control of shape, size, profile, surface quality, etc. that is required for end uses and applications. For example, it refers to plates, sheets, sections, bars, etc. that are directly sold to the consumers.

Cold rolling/cold forging: This is the processing of steel by rolling or drawing or forging in the cold stage for obtaining bright surface, as well as certain characteristic properties, profiles, shapes etc. Examples of cold rolling include: bright bars, cold-rolled sheets and strips, wires, nails etc. Cold rolled products are either directly used for end applications—e.g. cold-rolled sheets for forming automobile parts—or for high-speed machining of components—e.g. automobile small parts, nuts, screws, bolts etc.

Rolling is generally a multi-strand operation, where 'rolling trains' consisting of number of strands are used to give required shape and profile—this is in contrast to forging where generally single station of forging is used to give shape. Also, cold rolled products (e.g. CR sheets) or cold processed parts (e.g. nails, bolts, small parts etc.) are often treated on the surface (e.g. galvanizing, phosphating, tinning etc.) for improved corrosion, weather or service environment protection. Cold rolling of steel is a specialised subject, hence only a brief outline of this process will be presented in this book.

Hot Rolling is an expensive and very energy consuming operation—hence an attempt is made to keep rolling sequences and schedules to as few steps as possible for giving shape and dimension to the rolled steel. *However, hot rolling is not only for giving shape, but it is an important step in steel technology for ensuring correct dimensions, homogeneity of the mass, and quality of the rolled product.* Efficiency of rolling operation is influenced by how effective is the designed roll-pass sequences for:

* Properly cogging down the as-cast structure,
* Giving perfect shape within the tolerance limits,
* Llimiting the number of reheating required for rolling to bare minimum, and
* Ensuring high quality surface.

Many customers of quality steel for engineering applications also call for minimum percentage reduction (reduction ratio) of the rolled product from the initial ingot, bloom or billet—in order to ensure proper homogeneity and 'kneading' of the mass for facilitating further operations. *Ingot steels do require higher reduction ratio than CC steel due to coarser macro-structure and higher segregation.*

However, there is no fixed rule for reduction ratio, which is dependent on the criticality of end use properties. For dynamic load bearing parts—i.e. parts subject to fatigue stresses—high reduction ratio is preferred. For standard applications, a reduction of 4:1 is generally called for—meaning that cross-sectional area of the rolled product should not be more than 1/4th of the starting material cross-sectional area—even when the quality of the cast section is of very high order. Such a requirement ensures that right size and quality of start material is used for rolling or forging for final applications—otherwise final usages of the steel may suffer from inadequacy of certain properties. For instance, in the application of steel for leaf springs manufactured for machines and automobiles, which are subjected to high dynamic load in service, a minimum reduction during hot deformation is required—otherwise consistent toughness of the structure (required for fatigue load bearing) at that high strength level as required for springs may not be achievable. Fatigue tests on spring steel flats show that if total hot-reduction of the flat from the starting billet is less than 20:1, the fatigue strength ratio with tensile strength may drop. Higher the degree of hot reduction in the rolling or forging—better is the homogeneity of internal structure and consistency in response to subsequent heat treatment. In this respect, type of rolling mill being used, roll pass design, furnace facility for proper soaking of the start-material at high temperature before rolling, and sequence of rolling play critical role in breaking down the coarse dendritic as-cast structure, for proper kneading of the material during rolling, and for producing homogeneity in the mass.

Primary mill size and type for rolling is determined by the choice of size and shape to be given in the rolling. Rolling mills can be generally grouped as:

A. Ingot rolling (long product)

- Roughing mill or blooming mill, where the ingot or slab is reduced to a semi-finished shape and size for next operation;
- Bar and billet mill, where the bloom or the billet is rolled to narrower sizes of billet or bar, respectively, for final usages (i.e. for the market); and
- Merchant mill or wire-rod mill, where the semis are rolled for a special section (e.g. T-section, channel, angles etc.) or to lower gauge wire rods, respectively, for industrial usages.

B. Slab rolling (flat product) mostly through continuous casting route:

- Plate mill—for rolling plates of various thickness;
- Sheet mill and strip mill—for rolling single length sheets and strips; and
- Hot strip mill—for rolling thinner gauges of sheets and strips in coil form. Coils can then be cut to individual sheets or rolled further either by hot process or by cold rolling.

For economic and productivity reasons, hot strip mill route of slab rolling is preferred now-a-days, and there have been many developments in this area, whereby wide range of strip thickness with high quality surface can be rolled at very high speed in these mills with coil weight varying between 5 tons to 50 tons. As a result, the coil quality and coil weight work as excellent in-put materials for *cold rolling* of automobile grade sheets with exotic forming properties. Surface quality of hot-rolled coils needs to be very good for consistent response to pickling and cold rolling at high speed. Metallurgical steps for hot-rolling—starting from ingot or slab—involve number of important steps for assuring the product quality, as shown in Table 4-3.

Cold Rolling of steel, whenever is called for, constitutes the most critical part of processing, because this must meet the final quality specifications—in terms of surface, profile and as well as metallurgical properties. Starting material for *cold rolling* includes the hot rolled products like the billets, bars, coils etc. Amongst the cold rolled products, cold rolling of sheets and strips is drawing most attention and technological development because of their extensive uses for manufacturing automobile body parts.

Table 4-3. General steps in hot-rolling process of ingot and slab.

Ingot	Slab
• Inspection for surface defects and conditioning	• Inspection for surface defects and conditioning
• Soaking in rolling mill furnace—avoid excessive scaling	• Soaking in rolling mill furnace—avoid excessive scaling
• Descaling before start of rolling (on-line)	• Descaling before start of rolling (on-line)
• Roughing/cogging	• Roughing to scalps
• Descaling (on-line)	• Descaling (on-line)
• Finish rolling to billets and bars	• Coiling of scalps
• Inspection and testing	• Hot strip rolling in continuous mill
• Marking and despatching	• Coil testing, marking and despatching

Most domestic applications of steels are in cold rolled forms—like the wires, nails, pins, cartularies, drums, etc. Engineering applications of cold rolled products are automobile body making, corrugated galvanised sheets, manufacturing of white-goods bodies etc. Cold rolled products need to be accurate in dimensions and quality—as being the final step in the line of making and using steels. Hence, cold rolling mills are more precise in controls than hot rolling mills. The general *process steps for cold rolling* involve the following sequences of operations:

- Pickling of HR product;
- Cold drawing, if wire rods/cold rolling, if sheets and strips, in coils;
- Intermediate annealing, if required for further cold rolling to thinner gauges;
- Surface coating, if required;
- Cutting to lengths, if required; and
- Marking and packing for despatch.

The critical part of cold rolling is to control the gauge and profile of the product, so that it conforms to specifications and application requirements. Deformation caused by cold rolling also causes *work-hardening*—a phenomenon whereby steel gets further hardened due to cold worked strain. Work-hardening makes further cold deformation difficult without intermediate annealing i.e. softening. This *annealing* must be carried out in most controlled environment and manner—often carried out under hydrogen atmosphere—so that surface is not damaged, by unfavourable environmental reactions and pitting, for further cold rolling to thinner gauges. More about cold rolling has been discussed later in this chapter, in Section 4.11.

4.11. Tasks of Steel Rolling

In general, rolling includes number of tasks related to:

- Attaining physical dimensions of the finished rolled product,
- Ensuring surface quality that is free from scabs, cracks, seams etc.,
- Internal soundness and homogeneity of structure,
- Control of hardness and strength, and
- Improvement or development of special characteristics or properties by extra care in rolling.

The latter refers to such factors as grain size control, by controlling finish rolling temperature in hot rolling or promoting 'pancake' structure by controlled rolling during cold rolling operations. Uses of cold rolled steels are largely in areas of flat products e.g. sheets and strips for press forming. In contrast, hot rolled products are used largely in long product areas e.g. billets and bars for forging or machining. Dimensional limits of hot rolled products are, therefore, wider compared to cold rolled products—as HR products are generally further forged or machined to generate new dimensions.

However, with the advances of CNC machining, dimensional tolerance limits of forgings are getting tighter because of consistency and closeness of geometry required for automatic CNC machining process. This, in turn, is giving rise to more demand for 'press forging' for closer dimensions than 'drop forging' with more liberal or open tolerance. As a consequence of this quality demand on forged parts, demands for closer dimensions in hot-rolled billets and bars are rising. One of the special requirements of press forging is the 'rhomboidity' of the section of the billet i.e. dimensions are not only about the size but also the shape. Similarly, for cold rolled sheets and strips, it is not the thickness alone that should be controlled but also 'profile' of the strip from centre to the edge, which is required for present-day high speed press-tool forming.

Hence, rolling is just not for shaping the steel alone—it is much more than that. Purpose of rolling can be summed up as:

- Ensuring soundness of macro-structure,
- Developing homogeneity of micro-structure,
- Shaping to close dimensions,
- Making the surface of the rolled product free of defects for further processing, and
- Ensuring special profile and special properties necessary for successful end uses.

Bad rolling can ruin good steels howsoever care might have been taken in making that steel. Hence, steel plants must be backed by good 'quality system management' (QSM) practices for assuring ultimate product quality. Steelmaking and rolling are in the same manufacturing-chain of events that assure quality, economy and availability. Hence, judicious choice of steel grade and technical correctness of steelmaking and rolling processes are the keys to successful usages and application of steels.

4.11.1. Types of Rolling Mills

For most products—except in the case of bar drawing—multi-station rolling mills in trains are used to give final shape and quality required. Aim of such multi-station deformation is to ensure that thorough and uniform deformation is given to the steel stock by sharing the task at different points of rolling. There are many different types of 'rolling trains' in use, depending on the charging temperature of the stock to be rolled and the shape and section to be rolled. In general, they all fall into: hot-rolling mills and cold-rolling mills. Table 4-4 shows a general list of rolling trains (mills) for some common steel products–either semis or finished.

These mills are characterised by roughing mill, intermediate and finishing mills, with tasks of increasing order. Some of these mills work as individual rolling stands (e.g. Blooming mill and Slabbing mill etc.) and some are continuous mills—e.g. hot strip mill cold strip mill, etc. When the capacity

Table 4-4. List of the general type of rolling mills (trains) in vogue.

Rolling Mill Name	Purpose/Definition
Blooming mill	Hot rolling of ingots into bloom
Slabbing mill	Hot rolling of slab/ingot into rough slabs
Billet mill	Hot rolling of blooms into billet
Bar mill	Hot rolling of bloom*/billet* to bars
Wire-rod mill	Hot rolling of billets into wire rods
Section mill	Hot rolling of blooms/billets into sections, rails, T-beams, angles etc.
Plate mill (heavy)	Hot rolling of slabs/roughed slabs* into plates
Hot wide strip mill	Hot rolling of roughed/cast slabs* into wide strips
Cold wide strip mill	Cold rolling of hot-rolled wide strips
Hot or cold strip mill	Hot/cold rolling of slabs/bars into narrow strips

* Denotes input could be from ingot or CC route.

of these rolling mills are planned for high output (i.e. with faster rolling speed), care has to be taken for advanced mechanisation and automation. In fact, in modern high speed continuous mills (e.g. hot strip mill), there is very little scope of any manual control or adjustment. Such mills demand better input quality of stock and uniform reheating of stocks for defect free rolling without any disruption. All continuous mills rely on automation for control and correction during rolling.

Rolling mills can be classified by the number of rolls in the stand and/or the roll

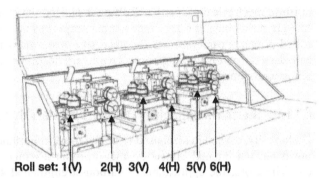

Roll set: 1(V)   2(H) 3(V) 4(H) 5(V) 6(H)

Figure 4-10. Illustrates the rolling arrangement in a six stand roughing mill block of a bar mill.

dimensions; but generally used criterion is the number of rolls and their arrangement. A typical six stand rolling mill, used for roughing billets for bar rolling, is shown in Fig. 4-10, where alternate vertical and horizontal positioning of stands could be noted. This arrangement is necessary for making the rolling stands more compact and of heavy duty for producing required shape with better size and tolerances.

Types of finish rolling mills are:

- Two-high mill,
- Three-high mill,
- Four-high mill,
- Cold-rolling cluster mill, and
- Planetary mill also known as *Sendzimir mill*.

As the layers of rolls increase, the work roll diameter gets decreased—except in the case of two- and three-high mills—in order to get better deformation and surface quality. Each basic arrangement can include additional aspects of rolling design for fulfilling the quality tasks. The four-high and cluster mills use special back-up rolls of larger diameter than the smaller work rolls for more contact pressure and better deformation. Cluster mills with 12–20-high stands are used for cold rolling of special grade strips where maximum dimensional accuracy is required. Where thickness reduction of over 80% is required with great dimensional accuracy, planetary mill is used.

However, with the increasing demand for better quality of rolled products along with higher dimensional accuracy and consistency at a competitive cost, many new developments in steel casting, especially continuous casting, and rolling technology are taking place—notably in the areas of flat steel rolling. Flat steels are now-a-days rolled with a process called 'thermo-mechanical (TM) rolling' or controlled rolling where—other than the deformation—the aim is to produce fine grain size and

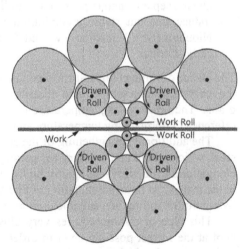

Figure 4-11. An illustrative 'cluster mill' for precision cold rolling of steel strips.

uniform mechanical properties. Rolling is carried out at a special temperature for producing uniformly fine grain size having higher mechanical properties and uniform microstructure. Thermo-mechanically rolled products can eliminate the necessity of normalizing or such heat treatment that is given to the rolled steel for developing some characteristic properties—like fine grain, uniform microstructure, along with dimensional accuracy.

4.11.2. Thermo-mechanical Rolling and Its Advantages

Historically, carbon in steel used to be the most important chemical element for strengthening the steel, but it has detrimental effects on many properties such as: elongation, poor formability, weldability etc. Therefore, the application of carbon-strengthened steels gets limited as against the requirement of strong and tough steel for many modern safety related construction activity. Conventional method of imparting higher strength and ductility/toughness to higher carbon steel for such applications has been to give—additional heat treatment like normalizing or hardening and tempering for obtaining the desired properties. But, these are additional and expensive operations. Hence, for wider application of cost-effective higher strength steels, the method of controlled rolling/thermo-mechanical rolling has been developed of late, which can (a) refine the grain size—a substitution of separate normalizing operation—and (b) promote some fine carbo-nitride precipitates, adding further strength. For the latter purpose, suitable micro-alloying elements are used in the steel, which on controlled rolling at an intermediate temperature leads to fine precipitation of carbides and nitrides. Thus, the process can lend to strengthening by both grain refinement as well as precipitation hardening. In industry, this process is called 'thermo-mechanical rolling' which is also referred sometime as 'controlled rolling'. Figure 4-12 illustrates the stages of thermo-mechanical rolling for improved properties.

The process can apply to the rolling of plate, strip, bars or section, and involves hot-rolling the steel at an intermediate temperature (below A_3) in a controlled manner, so that any or a combination of the following characteristics are developed in the steel:

- Fine and stable grain sizes are obtained in the steel, the formation of which would have otherwise taken a separate and expensive normalizing heat treatment, and
- Induce fine but stable precipitation in the steel by appropriate micro-alloying elements—like Niobium (Nb), Vanadium (V) and Titanium (Ti)—giving rise to fine precipitates of carbides and carbo-nitrides.

The choice of rolling temperature depends on the steel composition, and it has to be below or just around the recrystallisation temperature of austenite, so that rolling deformation can induce fine recrystallised grains. The finish rolling temperature is generally 650°C–750°C. Schematic stages of TM rolling with reference to normalizing temperature and recrystallisation temperature is shown in Fig. 4-12.

The additional strengthening of the steel by fine precipitation acts as substitution of higher carbon steel, which would be necessary for getting similar properties. Net gain in TM rolling is the increased elongation and ductility—as against higher carbon normalised steel due to finer grain and fine precipitates in the matrix. At time, combination of fine ferrite and tough bainite like structure can be also developed by controlled rolling of a suitable composition of steel.

This type of rolling requires very close control of slab reheating temperature, which should be kept at the lowest possible level in order to limit the grain growth due to reheating. Also, roughing deformation should be carried out in the range of recrystallisation temperature, and finishing just below the recrystallisation temperature. The process also requires very heavy deformation (80% +)

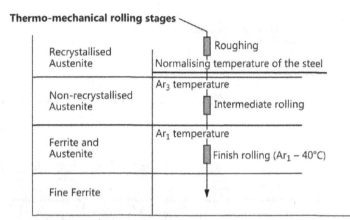

Figure 4-12. Schematic stages of 'thermo-mechanical' rolling for improved properties.

for producing very fine grain size. Because of the requirement of high deformation per pass and low processing temperature in the finishing, all rolling mills cannot carry out TM rolling. The mill has to be suitably upgraded so that high rolling force can be exerted and withstood by the rolling stands.

A popular example of TM rolling is the rolling of micro-alloyed HSLA steel. As the necessary total deformation in TM rolling has to be carried out without austenite recrystallisation, the recrystallisation stop temperature of the austenite has to be increased, and this is most effectively achieved—by adding niobium, vanadium etc. as micro-alloying in HSLA steel.

Thus, thermo-mechanically rolled HSLA steel is characterised by fine grain size along with fine carbide (or carbo-nitride) precipitates, which provide special strength to these micro-alloyed categories of steels. Because of fine grain size and precipitates, the steel is high on strength and ductility. As such, HSLA steels are being extensively used now-a-days for automobile frame parts, ship construction, line pipes and high strength off-shore structure construction projects. Both fine grain size and fine carbide precipitates considerably contribute to the increase in 'yield strength' of the steel and reduce ductile-brittle transition temperature—thereby increasing the load-bearing capacity of the formed parts along with other beneficial effects of finer grain size in steel.

4.12. Cold Rolling of Steel

Whenever close dimensions, good surface finish and close control of properties are required in steel products—like sheets, bars, wires, tubes etc.—the hot rolled stock is cold worked either by rolling or by drawing. However, amongst the cold working processes, cold rolling of steel sheets and strips occupy the bulk of activities and draw attention for technological innovation.

Cold rolling of steel sheets and strips are mainly focused to rolling for applications such as deep drawing, galvanising, tinning etc. Use of deep drawing and extra-deep drawing steel sheets is the backbone of modern automobile technology. And the field of application of cold rolled sheets has been the topic of extensive research and technological development in the cold rolling technology. To begin

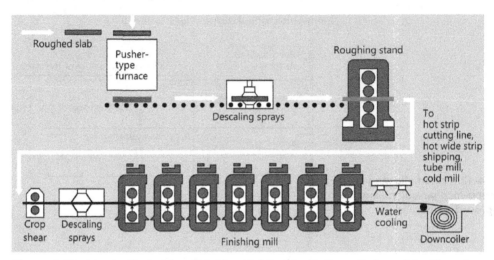

Figure 4-13. A typical hot-strip mill (wide strip) that feeds the coils for cold-rolling of steel trips

with, cold rolling of steel demands high quality hot-rolled stocks for producing better surface quality and dimensional accuracy in high-speed continuous cold-rolling mill. Hence, this technology is also the reason for many developments in hot rolling of steel strips and sheets as well.

For large scale production and economy of rolling, the input material for cold rolling is in the form of good quality *hot rolled coils*—coil weight can vary from 5 to 50 tons or more. Figure 4-13 depicts a typical hot-strip mill (wide strip) that produces and feeds the coils for cold-rolling of steel trips.

Hot strip mill coils have adherent thin scale on the surface; also it can have thicker scale patches at places. Presence of scales on the surface causes increase in frictional force during cold rolling, giving rise to patchy surfaces, which is not acceptable in the cold rolled products. Hence, the first step of cold rolling is 'pickling'—a chemical descaling process—of the HR stock before feeding to cold rolling mill. Some mills also adopt mechanical descaling such as by shot blasting, but pickling is more popular and economical for large scale production.

With the advent of automobile manufacturing and essentiality of cold rolled strips for automobile body parts, much attention is paid now-a-days to both hot strip rolling and cold rolling quality. Cold rolling of steel strips requires very high deformation load, which necessitates use of small work rolls with a number of large back-up rolls. This can be carried out by using two-high, four-high or cluster mills like the Cluster/Sendzimir mill (see Fig. 4-11). Along with high deformation load, the speed of rolling has to be such that the strip being rolled is always under tension between the rolling stands—allowing required deformation with high dimensional accuracy. High cold deformation causes metals to 'work-harden', making further cold rolling difficult without edge cracking. Hence, if necessary, provision for intermediate softening of the strip—either on-line or separately in the cold rolling mill—is provided to facilitate rolling to thinner gauges without any defect. Also, cold rolling being a very fast process requires precise control of strip thickness and shape. This is achieved by highly instrumented and automatic control of strip tension during rolling as well as with the help of other on-line measuring equipment for determining strip thickness and profile, and correcting on-line. Hence, cold rolling mills

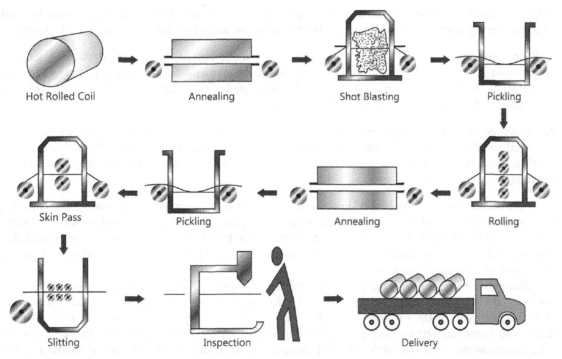

Figure 4-14. A schematic depiction of the essential steps in cold rolling of steel.

are mostly highly automated mills, provided with all facilities for continuous production e.g. on-line shear, strip joining stations, and strip coating line. A simple material flow diagram of cold rolling mill is shown in Fig. 4-14, showing all the elementary steps in cold rolling of steel.

The cold rolling mill features are briefly described as follows:

1. Cold rolling mill (CRM): This could be 4-high, 6-high or a cluster mill—depending on the degree of cold reduction required. These mills can be in series to perform even higher deformation tasks. For example, pickled coils can be rolled upto 0.25 mm thickness in a 4-high mill, upto 0.15 mm in a 6-high mill, and for lower thickness a cluster mill (Sendzimir mill) is used.

2. Annealing furnace: Many a time, cold rolled strips are to be annealed to obtain desired mechanical and forming properties as well as bright surface. Often, this annealing process is off-line and carried out in batch furnace. Important point is that the material has to be heated to a pre-determined temperature in a protective atmosphere and soaked for a specified time for developing bright surface and uniform annealed microstructure. Annealing is carried out towards the end of the CR rolling process—only skin pass is applied to the annealed material to improve flatness and suppress yield point phenomenon that may give rise to wavy surface (considered a defect) after forming due to discontinuous yielding. Quality of annealing towards the end of the cold rolling process largely determines the quality of surface and properties of the final product. Hence, optimum condition for annealing with good atmosphere control and sealing of the furnace to avoid air ingress is essential for bright surface with uniform properties.

Often, electrolytic cleaning is also used for removing any traces of oil from the strip surface that gets collected during cold rolling and may get burnt inside the annealing furnace causing dark patches.

3. Skin pass mill: This is the mill for very light reduction of the annealed strip to generate bright surface, better flatness and shape, and uniform properties—beside removing the 'yield point phenomenon' as mentioned earlier. Anti-rust oil can be used at this stage of skin rolling for protecting the cold rolled surface from rust formation before use.

4. CR slitter and cut to length unit: This is the final step in the cold rolling of strip where a slitter is placed on the line for de-coiling and slitting length wise and also trimming the sides for better width control. A separate cut to length unit can be installed for customer specific length cutting and packaging.

Cold rolled strips are widely used in automobile industries for forming critical body parts, requiring: (a) very good formability by stretching and deep drawing and (b) very good corrosion and weathering resistance of the car body. Hence, special measures have to be taken in cold rolling to impart extra deep drawing and other surface related properties, and the sheet metal also requires special surface coating for improved corrosion resistance.

There are a number of surface quality related tasks for cold rolling for ensuring successful forming and acceptable surface after forming the parts. Amongst them, 'stretcher strain' and 'orange peel effect' are critical for the surface finish. *Stretcher strain* marks (typical surface relief marks) appear due to non-homogeneous yielding during forming or pressing, and show up in areas that undergo very little deformation—less than 4%–5% cold elongation—in forming. The remedy consists in giving a slight 'pinch pass' or 'roller levelling' to remove sharp yield point phenomenon in the steel. The phenomenon is very similar to 'strain-ageing' of cold rolled steels due to the presence of dissolved unfixed nitrogen in the steel. Hence, free nitrogen in steel needs to be fixed by adequate Al content to make the steel of non-ageing type. Ageing of cold rolled steel causes change in hardness and other properties in the steel with time and condition of storing—higher the temperature and longer the time before use more is the chance of 'ageing' of the steel. Non-ageing steel—i.e. steel free from dissolved nitrogen content—is immune from such defects after forming as stretcher strain, interrupted yielding and loss of ductility. *Orange peel* effect is due to coarse grains in the steel, which may come due to faulty annealing in cold rolling.

Another important quality of cold-rolled sheets for extra deep drawing is the degree of anisotropy in properties. *Anisotropy in sheet steel is expressed in terms of $\bar{r}$* which is the ratio of strain in the width direction to strain in the thickness direction, and can be taken as the difference in strength—i.e. strain absorbing capacity without thinning—between through-thickness directions compared with strength measured in uniaxial tests in the plane of the sheet. In actual practice, measurement of $\bar{r}$ is carried out in three main directions, namely $0°$, $45°$ and $90°$. (For further details of $\bar{r}$ measurement, see Section 5.6 in Chapter 5). Anisotropy in cold rolled sheet influences the deep drawing operation of sheet steel—higher the $\bar{r}$ value better is the cupping value—this has been further illustrated in the next chapter. For low carbon sheet steel, $\bar{r}$ value should be in excess of 1.0, preferably 1.4–1.6. High $\bar{r}$ values are induced in steel sheets by Al-killing, which is believed to give a 'pancake' type grain structure during cold rolling due to interaction of Al and N in the steel during recrystallisation process and promoting favourable texture—a preferred grain orientation with maximum {111} crystal planes parallel to the sheet surface. In practice, for obtaining high $\bar{r}$ value in cold rolling, the aluminium content in the steel should be closely controlled within 0.02%–0.05% and avoid high coiling temperature during hot rolling.

4.12.1. Surface Coating of Cold Rolled Steel Strips

Steel surface, as it is, not weather or corrosion resistant; it may rapidly form oxides, rust, etc. on the surface which is exposed to environment. The surface can also get corroded when exposed to acidic atmosphere. Hence, to impart corrosion and oxidation resistance to cold rolled steel sheet surface, it requires some sort of coating or painting. Cold rolled strips, which are used in large tonnages for such critical parts as automotive body parts, are, therefore, often required in surface coated condition (e.g. galvanised or in galvannealing condition) for improved weather/corrosion resistance of its own. Additionally, cold rolled surface may also require good sub-surface quality for painting, if necessary or special decorative coating (e.g. colour coating) for special applications. Hence, most CR-lines are also equipped with coating units as well.

Coatings are provided to the steel for protecting the steel strip against corrosion, and could be also for providing decorative and protective surfaces. In this regard, coating process can be divided into: *metallic coating*—e.g. hot dip galvanizing, galvannealing, electrolytic galvanizing, tinning etc. and *non-metallic coating*—e.g. organic coating, enamelling, inorganic coating, plastic coating etc. However, for engineering applications and mass consumption, metallic coating like galvanizing (zinc-rich coating), tinning etc. are more in demand and widely practiced. If coating line is installed, it is generally before the slitting and length cutting unit so that coated product can be cut to required length and packed. Coating is also mostly carried out in the coil form for economies of scale in production in a continuous mill. Figure 4-15 illustrates a continuous wide strip galvanizing line in a Sendzimir mill set-up.

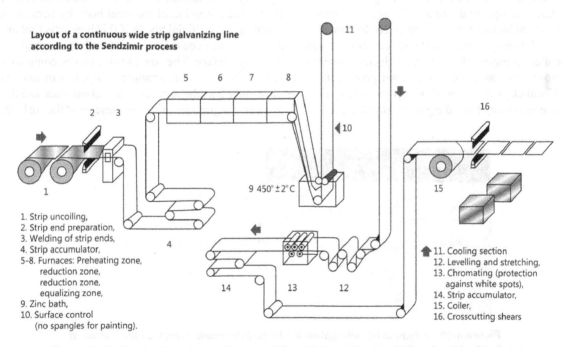

Layout of a continuous wide strip galvanizing line according to the Sendzimir process

1. Strip uncoiling,
2. Strip end preparation,
3. Welding of strip ends,
4. Strip accumulator,
5-8. Furnaces: Preheating zone,
   reduction zone,
   reduction zone,
   equalizing zone,
9. Zinc bath,
10. Surface control
   (no spangles for painting).

9 450°±2°C

11. Cooling section
12. Levelling and stretching,
13. Chromating (protection against white spots),
14. Strip accumulator,
15. Coiler,
16. Crosscutting shears

Figure 4-15. Illustration of a continuous wide strip hot-dip galvanising line in a Sendzimir mill set-up. *Source: Stahl-Eisen. Steel Manual, Verlag Stahliesen, Dusseldorf, 1992.*

Galvanized steels can provide protection from the atmospheric corrosion as well as from moisture and natural water—it is the most widely used coating process for steel—it can be used for cold-rolled strips as well as on formed steel parts like bolts, flanges, rods etc. Galvanizing can be done by 'hot-dip galvanizing' process or by electrolytic process called 'electro-galvanizing'. Hot galvanising process is most popular. The process deposits a thick robust layer of zinc/zinc alloy coating that protects the underlying steel surface. This process is okay for parts where no further painting would be carried out e.g. container body parts, vessels, frames, bolts, nuts, screws, pins etc. But for automobile or white goods body part manufacturing, where the surface is further coated with paints for protection as well as aesthetic look, thick coating layer is not desired. Hence, for automobile and white goods body parts manufacturing, electro-galvanized steel is preferred, because zinc layer deposited by this process could be controlled to very thin uniform layer, and such surface is suitable for further coating by paints etc. However, for many other automobile parts—e.g. flanges, hinges, joints, bolts etc.—hot dip galvanising is widely used for its economy and ability to provide necessary in-service protection.

Hot dip galvanising is carried out by applying a zinc coating to fabricated iron or steel parts by immersing the material in a hot bath (approx. 450°C) consisting primarily of molten zinc. The simplicity of the galvanizing process is a distinct advantage over other methods of providing corrosion protection. The automotive industry depends heavily on this process for corrosion protection of many components. Galvanizing forms a metallurgical bond between the zinc and the underlying steel or iron, creating a barrier that is part of the metal itself. Figure 4-16 depicts a typical microstructure that results from hot-dip galvanizing process. It should be noted that while the top surface is of pure zinc, subsequent layers contain progressively iron rich alloy of zinc with iron. Because of this alloying, the coating bond is very strong and protects the underlying steel surface better. Layer compositions and associated hardness is shown alongside the figure. In fact, zinc coating prevents the corrosion of the steel body by forming a physical barrier for the atmosphere to ingress and by acting as sacrificial anode if this barrier is broken.

A special feature of the galvanising process is that it can produce different sizes of zinc-crystals—called *spangle*—that can act as decorative parts on the body surface. The spangle size can be controlled from very fine crystal size—that gives uniform surface with no visible spangle—to grains of several centimetres wide that show up the spangle very prominently by adjusting the nucleation sites and the rate of cooling from the galvanizing temperature. However, for good painting appearance of the surface,

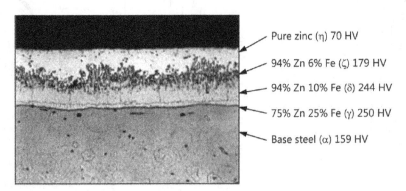

Pure zinc (η) 70 HV

94% Zn 6% Fe (ζ) 179 HV

94% Zn 10% Fe (δ) 244 HV

75% Zn 25% Fe (γ) 250 HV

Base steel (α) 159 HV

Figure 4-16. A typical hot-dip galvanised microstructural layers. *Source: Internet based American Galvanising Association (AGA) literature.*

spangles are not allowed for automotive body parts. Galvanising is widely used for protective coatings of steel parts and cold steel rolled sheets. The protection process relies upon the galvanic or sacrificial property of zinc relative to steel. Electro-galvanising process, employed for auto body parts, is adopted to give light, spangle-free, coating thickness which is smooth and has excellent forming properties with same mechanical property as the base metal. Electro-galvanised coating is thin, compact and strongly adherent to the mother steel surface, and can undergo press forming without damage to the coating. As such, this type of coating is preferred for automobile body parts or for press forming parts over hot-dip galvanising. Electro-galvanised steel is also suitable for spot welding or light welding.

Many a time, standard hot-dip galvanising process is modified with 'galvanneal' process, which involves post-processing by heating (between 540°C to 565°C) in an on-line oven to form an essentially zinc-iron (6%–15%Fe) alloy coating. The following is a brief summary of galvanneal process:

1. Steel strip is immersed into the zinc bath. In the bath, the alloy layer between zinc and the strip forms.
2. As the strip emerges, it drags excess zinc with it. The gas knives wipe off excess zinc to get the desired thickness.
3. The strip with the molten zinc coating is then passed through a heating furnace to heat it to about 540°C–560°C.
4. It then travels through a holding furnace to hold the strip at this desired temperature for a specified amount of time.
5. During the holding time, the molten zinc becomes fully alloyed with the iron to from a galvannealed coating—an alloy of zinc and about 10%Fe.

Galvanneal coating is harder and more adherent than simple hot-dip galvanised coating. Two primary reasons for considering use of galvannealed sheet rather than straight galvanised sheet are (1) improved coating adhesion and ease of painting and (2) improved spot welding, which is extensively used in auto-body forming.

The other common methods of coating steel surfaces are tinning and phosphating. Tinning is generally a hot dip process as the product is expected to give prolonged service life due to thicker coating. Tin coating is commonly used for preserving the foodstuffs such as milk etc. where the container can be sealed tightly to prevent air ingress. Phosphating on the other hand is a pre-treatment process and includes iron, zinc or manganese phosphate, as per choice. These are the three types of phosphating processes and mostly coated by immersion process in a phosphoric acid bath. Resultant bath reaction deposits metallic phosphate—as per bath composition—coating on the steel surface, which is grey in appearance. However, before immersion the material requires de-greasing for uniform coating and also post treatment for sealing the phosphate pores. Phosphating treatment is generally carried out on finished steel parts for corrosion resistance in storage and use. Some notable examples include: springs, bolts, nuts, bars etc. While manganese and zinc phosphate provide very good corrosion resistance, iron phosphate is not so good and it is used mainly for providing a good under paint bond that promotes better paintability and corrosion as well.

Though dipping is a very common method of coating, electrolytic coating is also popular. Electrolytic processes are cleaner and more controllable with regard to coating quality and thickness. Consequently, various energy efficient electrolytic processes are getting developed. Other coating methods are metal spraying and electro-deposition of passive/corrosion resistant metals or alloys like chromium, copper, nickel etc. as per service demands.

SUMMARY

1. The chapter highlights the major features and processes of steelmaking and rolling, emphasising that good steelmaking and rolling are at the route of quality steel necessary for economical and effective use and application of steels. Methods of clean steel making using secondary metallurgy or vacuum metallurgy, and defect free casting, using either conventional ingot casting or modern continuous casting have been explained. Possible steel defects and their sources and the constraints and limitations of inclusion content in steel have been described.

2. For clarity in the process of clean steel making, the chapter discusses with illustrations the steps of basic steelmaking processes as well as various secondary or vacuum metallurgical processes. Important processing steps for quality steelmaking—e.g. decarburisation, deoxidation, degassing, killing etc.—have been described and their role in steel quality mentioned. Since a bad casting can spoil all efforts of good steelmaking, care necessary for ensuring good quality casting free from cracks, segregations, and blow-holes has been highlighted.

3. Since quality steelmaking is at the route of successful end-use applications, the chapter specially discussed and compared the secondary metallurgy processes with that of vacuum degasing process for their merits, quality and cost implications. Benefits of LF and VD treated steels have been highlighted. Functions of secondary metallurgy have been especially listed both for economical and quality steelmaking practices.

4. Steel casting methods following conventional ingot casting and modern continuous casting have been discussed and illustrated. Various defects that can arise from such casting processes have been mentioned and their ill-effects elaborated.

5. For rolling of steels, salient features and tasks of hot-rolling and cold-rolling methods have been outlined, along with the illustration of different rolling mill features and their uses. It has been pointed out that the task of hot rolling is not only to give shape and dimension to the product, but also to ensure thorough working of the mass for refinement and compaction of structure.

6. Similarly, task of cold rolling is not only to control gauge and profile of the product, but also to take care of surface quality and anisotropy in property, influencing the r-bar value of CR sheets. The role of Al and N content in steel for developing a favourable texture has been pointed out. Details of steps and intermediate processes, including surface coating, involved in quality cold rolling of steel have been discussed and illustrated.

7. Considering the increasing demand of high strength sheet steel, the mechanism and advantage of thermo-mechanical rolling has been also highlighted and the salient features of the process have been pointed out.

FURTHER READINGS

1. *The making, Shaping and Treating of Steel.* United States Steel Corporation, Pittsburgh, 1971.
2. Price, David, ed. *Ironmaking and Steelmaking: Processes, Products and Applications.* Vol. 39. London: Institute of Materials, Minerals and Mining, 2012.
3. ASM Metals Handbook, Vol. 14. *Forming and Forging.* Ohio: ASM International, 2000.
4. Ghosh, Ahindra and Chatterjee, Amit. *Ironmaking and Steelmaking: Theory and Practice.* New Delhi: Prentice-Hall of India, 2011.
5. Williams, R.V. *Control and Analysis in Iron and Steelmaking.* New York: Butterworth-Heinemann, 1883.
6. Ginzburg, Vladimir B. *Steel-Rolling Technology: Theory and Practice.* New York: Marcel Dekker Inc., 1989.
7. Dieter, George E. *Mechanical Metallurgy.* New York: McGraw-Hill, 1961.

CHAPTER 5

Properties of Steel and Their Evaluation

Purpose of this chapter is to highlight different properties of steels and importance of their testing and evaluation for effective use and application. In the study of steel metallurgy, it is not enough to know about the properties of steel alone—Knowing the processes and purpose of steel testing and evaluation is also equally important. This is because the quality of steel has to be tested and confirmed before putting to use or application. The chapter, therefore, discusses the properties of steels, their relevance to different processes and applications of steel, and highlights the methods and purpose of testing of those properties. The chapter covers chemical, mechanical, metallographic and non-destructive testing of steels.

5.1. Introduction

Properties of steel are generally grouped under the categories—(1) physical properties, (2) chemical properties and (3) mechanical properties. Each of these properties influences the uses and applications of steels in its own way, but some are more important than the others and demand testing and evaluation before using them. For example, physical properties—such as elastic modulus, Poisson's ratio etc.—are considered important for design of structural parts for ensuring adequate load bearing capacity, and their values are often taken from the published data, because, these physical properties are considered, more or less, constant for steels. However, chemical and mechanical properties—such as chemical composition, tensile strength, hardness, impact strength, etc.—which have direct implications on the processing and applications of steel, need to be tested and evaluated for every batch of steel production before being put to use. Batch-wise evaluation of mechanical and forming behaviour of steel is necessary because of the sensitivity of steel properties to the variation of chemistry and structure (refer Chapter 2 for an explanation of relationship between composition and structure). This is particularly important in view of the fact that steelmaking is carried out in batches—the size of which depends on the steelmaking furnace or ladle size.

Due to the complexity of steelmaking operation, involving high temperature reactions, there is a batch to batch variation of chemistry, impurities and structure due to variability of steelmaking and rolling processes. This necessitates evaluation of required properties directly by testing the steel as per production batch. Thus, while physical properties of steel—like elastic modulus, Poisson's ratio, etc.—need not be tested before use, composition and mechanical properties—like hardness, tensile strength, yield strength, impact strength, etc.—are required to be tested and evaluated before use. And, this evaluation should be carried out as per their batch of production in order to know what is being used/accepted against what had been planned. Therefore, apart from knowing about the properties that are required for a specific application, it is also necessary to know how those properties can be evaluated. The chapter, therefore, highlights different steel properties, their importance in the processing and application behaviour of steel, and methods of their testing and evaluation.

Since the purpose of testing is to decide about the suitability of the steel for intended application, the test methods need to follow some standardised methods for common understanding and acceptance. There are vast numbers of 'standards' for testing and evaluation of different properties or attributes in steel, issued by different national or international standardisation bodies—such as American Society for Testing of Metals (ASTM), International Standards Organisations (ISO), British Standards (BS) and Bureau of Indian Standards (BIS), etc. The scope of this chapter is not to list these standards for testing, but to highlight necessity for testing—and briefly deal with the methods of testing—in order to ensure that steel being used is of right quality. Generally, tests are conducted by following standardised procedures and cover the scope of evaluation of:

- Chemical and mechanical properties,
- Formability characteristics,
- Macro and micro-structure,
- Internal soundness of steel and
- Surface quality of the steel.

However, tests are not the end—they are the means to the end uses and application of steel. Hence, understanding their purpose along with the process is equally important.

5.2. Types of Steel Properties, Their Significance and Evaluation

As mentioned in Chapter 1, steels have basically three types of properties: physical (intrinsic), chemical and mechanical. An understanding about the role of these properties, factors that contribute or control these properties, and tasks in their evaluation is necessary for correct selection and application of steels.

Physical properties are important for the design of steel structures. Important physical properties include:

a. Young's Modulus (E),
b. Shear Modulus (G),
c. Bulk Modulus (K) and
d. Poisson's Ratio (v).

These properties are generally taken as constant for all steel grades. Young's modulus, E is the proportionality constant between strain and stress within the elastic limit of the steel, and is expressed by

$$E = \sigma/e,$$

where σ is stress and e is strain within the limit of proportionality constant between stress and strain. This property can be estimated from stress–strain diagram of the steel, but for general design purpose, the value of E is taken as 210 GPa—where 1 GPa = 10^9 N/m^2. Poisson's ratio, v is another value that is taken as constant for steels and the value is 0.29. From these values, Shear modulus, G and Bulk modulus, K can be calculated by using the formula:

$$\text{Shear modulus, } G = \frac{1}{2}\frac{E}{(1+v)} \quad \text{and} \quad \text{Bulk modulus, } K = \frac{1}{2}\frac{E}{(1-v)},$$

where E is the Young's modulus and v is the Poisson's ratio.

There are other groups of physical properties, such as:

- Density,
- Thermal conductivity,
- Specific heat,
- Electrical resistivity and
- Coefficient of thermal expansion.

These values are also more or less constant at the room-temperature and their values are taken from published 'data sheet' for steels when required for any design calculation. One such physical data sheet has been provided in the appendices. Some important physical property values of steels are as follows (broadly, these values vary within the given range as per type of the steel):

Density, $\rho = 7.7 - 8.1$ [kg/dm^3]
Elastic modulus, E $= 190-210$ [GPa]
Poisson's ratio, $\nu = 0.27-0.30$
Thermal conductivity, $\kappa = 11.2-48.3$ [W/mK]
Thermal expansion, $\alpha = 9 - 27$ [10^{-6}/K], being higher with increasing temperature of applications

Of these, *elastic modulus and Poisson's ratio* are required to be used for structural designs—others may be necessary to be considered occasionally for special design purpose or design constraint. It may be necessary to know coefficient of thermal expansion for some special application relating to severe thermal fluctuation, where structural joints could be stressed due to thermal expansion and contraction in service. These property data are generally available in literature, and seldom these are evaluated by physical testing, excluding elastic modulus, E, which is important for all structural applications due to dependence of structures on 'yield strength' (YS)—a phenomenon in steel related to elastic modulus.

Chemical properties relating to steel usage are oxidation resistance, corrosion resistance, pitting resistance and acid resistance. Of these, corrosion and oxidation resistance of steels are often required for many domestic and industrial applications of steel—e.g. a water tank, roof sheet, chemical tank, automotive car body parts, room door handles, locks and keys etc. Any steel part exposed to atmosphere or in any special working environment, would require evaluation of the steel for its oxidation and corrosion resistance properties. This testing can be done in the laboratory or in the field, but requires longer time for exposure and accuracy to simulate the working conditions. Hence, oxidation or corrosion property of steel is often based on the 'development data' generated during development of the steel and the corresponding chemistry of steel. This is an indirect way of estimating chemical properties, but reliable enough for many applications. This is because chemical properties exhibited by steels come from the change in 'electro-galvanic character' of the steel due to its composition and presence of special alloying elements. For example, chromium and nickel, when present above a certain level, can make the steel inert to corrosion or oxidation e.g. 18-8 stainless steel. Therefore, chemical analysis of carbon and alloying elements of the steel can give fair idea about the corrosion and oxidation properties of steels within the limits of acceptability. Therefore, testing of steel for chemical composition assumes primary significance for getting some initial idea about the chemical, welding and mechanical properties of steel (see Fig. 2-14 for the composition–strength relationship in wrought steel).

Mechanical properties are the focus of most commonly adopted testing methods of steel. Commonly measured *mechanical properties* of steel include—hardness, tensile properties, impact strength, fatigue strength, creep strength, fracture toughness etc. These properties depend on the chemical composition as well as on the microstructure that gets developed by heat treatment or cold working of the steel. Composition has pronounced influence on the steel strength, hardness and other mechanical properties as well on the processability of steel—such as heat-treatability, weldability, machinability, etc. Chemical composition influences the mechanical properties partly through solid solution strengthening and largely through its contribution to microstructure development in the steel. Refer the topics on strengthening processes in steels and structure–property relationship in Chapters 1 and 2.

Hence, the first test that has to be conducted on steel is the *chemical analysis for composition*, which can indicate chemical behaviour, mechanical properties (in as-rolled condition), machining behaviour, heat treatability, welding behaviour and such other parameters that depend on chemistry of the steel. Carbon is the most common influencing element in steel composition that contributes to mechanical properties, like strength, ductility and impact transition temperature. Often, due to lack of other testing facilities, steel is tested for composition. Other behavioural properties of steel are inferred from the composition by using some empirical relationship—e.g. calculation of hardenability by Grossman formula (see Chapter 8 for further details).

Other than the tensile test related properties, there are some special end-use specific properties of steel that might require evaluation. These include fatigue strength, notch-toughness, creep strength (high-temperature property), low-temperature properties, wear resistance etc., all of which falls under mechanical testing. Table 5-1 illustrates some application specific steel properties—barring commonly used hardness and tensile properties—*vis-à-vis* how they get influenced by certain metallurgical attributes in steel. Variation of these properties with steel quality and attributes necessitates their testing and evaluation for individual applications. ASTM A1058 provides some standard test methods for mechanical testing of steels.

Study of physical metallurgy of steel is dominated by two phenomena in steel technology—*transformation study* and *deformation study*. Transformation relates to phase changes and production of different microstructures in steel—e.g. ferrite-pearlite structure, martensitic structure, carbides and

Table 5-1. Influence of steel character on mechanical properties for special purpose applications.

Steel Character → / Properties ↓	Grain Size (Fine)	Inclusion (Low)	Strength (High)	Impact Toughness (High)	Ductile-Brittle Transition Temperature (Low)	*Influence of Special Alloying (Ni/Cr/Mo/V)
Fatigue Strength	↑	↑ ↑	↑ ↑	↑	—	↑
Creep Strength	↓ ↓	↑	High-temp. ↑Strength	—	—	↑ ↑ ↑
Low-Temperature Properties	↑ ↑	↑	With high ↑ductility	↑	↑ ↑ ↑	↑
Wear Resistance	—	—	↑	—	—	↑
Fracture Resistance	↑	↑	↑	↑	↑	↑

* Influence of special alloying is through structural improvement and modification.

carbide types, etc as discussed in Chapter 2. There are other microstructural features in steel, like the grain size and inclusions, which also take part in the deformation process of steel. While final grain size is the product of nucleation and growth (N&G) process in the steel, inclusions originate from the steelmaking process. However, both of them play a key role in the deformation and fracture behaviour of steel. Microstructure and fracture are closely correlated—the former rules the behaviour of latter. Hence, understanding how steels can deform and fracture with respect to microstructural features is important. While deformation and fracture behaviour of steel is evaluated by different mechanical testing methods, microstructure evaluation is done by metallographic techniques. In the study of steel metallurgy, both microstructural features and deformation behaviour of steel need to be correlated, which is the subject matter of detailed 'structure-property correlation' study. In short, microstructure is the cause and fracture is the effect—hence, accurate estimation of microstructures, including grain size and inclusions, can provide important information about behaviour of steel in services.

The sum total of test results obtained by chemical analysis, mechanical testing and metallographic examination can clearly indicate as to how the steel will behave, in general, during processing or in service. However, there are other application specific properties of steels—like fatigue, creep and low-temperature applications—which may require critical evaluation for many engineering and structural applications under special environment. Hence, test programmes for evaluation of steel might require planning of such special testing. Fatigue and creep testing involve the study and evaluation of fracture behaviour of steel under respective loading type i.e. dynamic pulsating load for fatigue testing and static load under higher temperature for creep testing. All these tests require series of tests for establishing the behaviour of steel under fatigue, creep or impact.

Impact property measures the capacity of the steel to absorb energy ahead of an advancing crack tip; thereby blunting the crack and stopping the crack from further propagation. In other words, it measures the toughness of the steel under a given state of stress and temperature. Temperature is important, because normal ferritic steels, having Body Centered Cubic (BCC) crystal structure and can undergo ductile-brittle transition (discussed later in this chapter), rendering the steel brittle below a specific temperature. At low temperatures, many of the metallurgical basics of deformation and fracture theory at room temperature do not hold good due to freezing of atomic movements and dislocation immobility. Low temperature applications can lead to brittle or sudden fracture of steel parts and can lead to catastrophic accidents, if suitable steel is not chosen. Therefore, testing of fracture resistance properties of steel at low temperature becomes important for applications like—aircraft components, marine components and chemical processing plants—where equipment is required to work at sub-zero temperatures under high loads. Low temperature fracture resistance can be evaluated by 'Impact tests'—similar to the 'Charpy notch-toughness test' conducted over a range of temperature, covering sub-zero to above the room temperature tests. It is not a single test of impact value at room temperature; it calls for series of tests at different temperature levels ranging from above to below the room temperature for establishing the impact-transition behaviour of the steel.

However, testing for determination of chemical composition and determination of mechanical properties, along with the examination of macro and microstructure of the steel, forms the bulk of testing and evaluation tasks. While complete determination of chemical composition is always necessary, tests for mechanical properties or microstructure can be selective as per the requirements of the steel specification and service conditions. Tests for fatigue and creep or low temperature properties should be performed based on the intended application of the steel. This type of testing will be briefly discussed here to highlight their main features.

5.3. Mechanical and Forming Properties of Steels: Scope of Testing and Evaluation

For application of steels, two sets of property are required—one for the processing the steel—e.g. bending, stretching, cupping, machining, welding etc and the other for end-use application—e.g. strength, toughness, impact strength, corrosion, atmospheric oxidation etc. Testing is required for evaluation of both types of properties, though there could be some correlation between some properties. For example, hardness and strength are correlated, similarly as are ductility and toughness, and carbon in the composition and weldability are also correlated.

Processing of steel is influenced by a set of mechanical properties, as indicated in Table 5-2, which illustrates the qualitative relationship between forming behaviour and different mechanical properties of steel. The steel processing method, which is commonly termed as 'forming and fabrication', not only calls for evaluation of mechanical properties, but also the microstructures, like grain size, inclusion types and size, and few other structure related properties. For smooth forming, processing and fabrication, it is necessary to know some metallurgical features of steel. For example:

- If the microstructure is free from carbide or ferrite banding for trouble free pressing and forming;
- If the steel is sufficiently clean—i.e. free from harmful inclusions—for forming, machining or heat treatment;
- If the grain size is correct for machining or heat treatment;
- If the macro of the steel cross section is free from cracks and voids for forging, etc.

Table 5-2. Illustration of tentative influence of standard mechanical properties on various forming operations.

Mechanical Properties ➡ As they go up . . .

Forming Technique ↓	Tensile Strength ↑	Yield Strength ↑	%Elongation ↑	%Reduction (RA) ↑	Hardness ↑
Cold bending	↓	↓	↑	↑	↓
Cold drawing	↓	↓	↑	↑	↓
Stretching	↓	↓	↑	↑	↓
Machining	(Indirect) ↓	—	↓	↓	↑
Welding*	↓	—	(Indirect) ↑	(Indirect) ↑	↓
Cold forging	↓	↓↓	↑	↑↑	↓
Deep drawing (e.g. Bullet case)	↓↓	↓↓	↑↑	↑	↓↓
Cupping (biaxial stretching)	↓↓	↓↓↓	#Uniform El. ↑↑↑	↑	↓↓↓

* Ease of welding of steel depends on the 'carbon equivalent' (CE)—lower CE is better for welding. Since higher CE leads to higher tensile strength or the hardness, influence of these properties on welding comes from this effect.
Note: Arrow indicates the direction of increase or decrease.

Hence, determination of metallurgical character of steel becomes another important part of steel evaluation—involving macro-structure examination, microstructure examination, evaluation of inclusion content and grain size, hardenability tests, etc.

Steel being a very structure sensitive material, metallurgical character and structure can also be used as being indicative of mechanical properties and suitability of the steel for a given application. Thus, test of one kind may lend help to estimate the property of another kind, but that is an indirect estimate. For many applications, direct determination of some important mechanical properties as well as microstructural feature like grain size and inclusion testing are necessary. For example, for cold forming of steel by deep drawing, stretching or forming a component by deep machining—apart from the mechanical properties—the actual evaluation of grain size and inclusion types and level of presence is also necessary. Thus, there can not be any generalisation of testing programme for steel—it has to be either application or processing specific, or both.

To illustrate further, a flanged axle for automobile will require cold or hot forging of the flange, machining and heat treating. Therefore, steel has to be so selected that it can withstand the forging and upsetting for flange formation that can produce good surface on machining and can withstand the rigours of heat treatment to develop required properties. This requires strict control on composition, grain size and inclusion content in the steel. If the steel contains excessive hard inclusions, forging property of steel suffers. Similarly, another automobile part like the wheel-rim will call for cold forming, bending and welding of the steel, ensuring that the steel has sufficient strength to resist buckling and deformation under cyclical vehicle load. Steel for such components has to be tested for strength, elongation and bendability in one hand, and also for good weldability by controlling the carbon level in the steel composition, on the other.

For good drawing and cupping operation, softest steels with good surface quality and very low oxide inclusion content are generally sought after. Cupping involves biaxial stretching—hence the cold rolled steel sheet used for deep drawing and cupping should be so specified as to ensure removal of bi-axiality in the sheet by special temper rolling. For many critical automobile body parts, requiring *extra deep drawing and cupping*, special grades of extra-low carbon cold rolled, vacuum-degassed clean steel with controlled grain size and crystallographic texture are used. Testing and evaluation of these steels for the required cupping and stretching values are, therefore, necessary for ensuring trouble-free processing. These properties could not be accurately estimated from the standard mechanical properties like the hardness, elongation etc. They require special testing for evaluation of suitability of the steel.

Machining is a common fabricating technology for engineering steel parts, and this is affected by the presence of hard inclusion, fine grain size and harder matrix. If the steel is too soft, it can cause long and continuous chip at the tip of the cutting tool, leading to excessive tool wear. Hence, grain size, inclusion and hardness requirement for good machining is a matter of adjustment as per shop practice. To overcome such problem involving multiple testing, often special test—like the 'machinability' testing is carried out for direct evaluation of machining property. Otherwise, steels used for machining can be tested for hardness, strength, and inclusion content and grain size for estimating their possible response to machining operations, which are indirect but widely used methods. Too fine a grain size poses problem in machining by forming continuous chip at the tip of cutting tools, and, thereby, causes tool tip heating due to friction, leading to tool blunting. Hence, for god machining, coarser grains are preferred.

Welding is another common fabrication technology, and this is influenced by the *carbon equivalent* (CE) of the steel, which governs the tendency of martensite formation and cracking of steel weld joints. The CE factor is measured by considering all elements present in the steel in the order of their influence on martensite formation in the weld metal. During arc welding, very high heat input is used to get the electrode material melted and deposited on the joint being welded. This, in turn, raises the temperature of weld adjacent area to a level above the critical temperature (A_3) and, thereby, creates conditions where martensite can form in the weld adjacent area. Simultaneously, the molten weld pool can absorb hydrogen from the open atmosphere. Both these phenomena can lead to cracking of the steel—one by martensitic crack and the other is hydrogen crack. Thus, for sound welding, care is necessary to select steel composition with low CE value, which is easily weldable without martensite formation or take precautions like pre-heating of the weld area to avoid crack due to martensite formation, along with steps for protecting the molten weld pool from atmosphere. Widely adopted CE formula (as per BS: 4360) is as:

$$CE = \%C + Mn/6 + (Cr + Mo + V)/5 + (Ni = Cu)/15$$

General guidelines are:

for CE ≤ 0.40 good weldability; no special precaution is required,

≥ 0.41 to 0.45 use special low-hydrogen electrode or preheat,

≥ 0.46 and above use low hydrogen electrode + preheat.

Preheating causes slower cooling of weld region, allowing—(a) diffusion of some hydrogen atom out and (b) does not favour martensite formation. Thus, chances of cracking due to hydrogen absorption and martensite formation get reduced. More about welding and welding precautions will be discussed later in Chapter 8. But, potential problem in welding of steel calls for checking and using steel with appropriate composition to avoid weld crack.

Steels are also required to be evaluated for their chemical properties such as, *corrosion resistance, oxidation resistance* etc. which are encountered in actual applications involving industrial or sea water environment. Corrosion and oxidation resistance testing are generally time taking and require true simulation of environmental conditions. However, these properties are not structure sensitive like the mechanical properties—they are primarily controlled by the chemical composition of steel. Because, corrosion and oxidation properties of steels are a function of 'electro-galvanic' character of the steels, which is controlled by the composition—especially the presence of certain alloying elements like copper, chromium, nickel etc. Hence, for all practical purposes, corrosion and oxidation resistance can be deduced from the chemical composition of the steel concerned, and users might not have to go for corrosion testing in their own laboratory. If necessary, such tests can be referred to specialised laboratories in the country.

Thus, steel testing and evaluation is an important part of steel metallurgy and steel application. It provides insight into the steel properties and characteristics for their processing or application behaviour. There is no set formula for what tests are to be carried out—it is need-based, depending on the processing route and the severity of end application. Most steel standards prescribe tests are relevant testing procedures, including number of samples to be tested, as part of tests for the acceptance of steel.

5.4. Common Testing Methods for Steel

From the foregoing discussions, it is apparent that steel properties which are required to be tested and evaluated include:

- Chemical composition;
- Tensile strength; Yield strength; Proof stress; %Elongation (%El); %Reduction (%RA) in Area;

- Impact strength; Fracture toughness test; Fracture transition test; Nil Ductility transition test; Creep test; Fatigue test; and Corrosion test;
- Bend test; Cupping test; Deep Drawing test; and
- Inclusion test; Weld test; and Non-destructive test for integrity evaluation.

There could be other special purpose testing methods—like the corrosion test, corrosion fatigue test, hydrogen cracking test etc.—that are called upon due to specific nature of applications. All these tests when performed in a standard manner and by following certain established standardised procedures, yield information about test specific properties and behaviour of steel that could be used for either buying the steel or for successful application of the steel. It is, however, not necessary to conduct all these tests for all applications. The generally carried out tests are:

- Chemical analysis to conform the steel grade;
- Inclusion test to determine cleanliness of the steel;
- Hardness test to determine hardness for bending, cutting and machining etc.;
- Tensile test to determine YS, UTS, %El, and %RA;
- Jominy Hardenability test for determining hardenability of the steel; and
- Impact testing for understanding the impact toughness behaviour.

For conducting these tests, relevant Indian or foreign standards—like the IS, ASTM, BS etc.—should be used, because the purpose of any kind of testing is not only assessing properties and behaviour of steel but also establishing correctness of supplies as per order and specification. Specification and ordering as per relevant national or international standards not only refers to properties, but also the standard of testing to verify those properties. Hence, an acceptable reference standard for a test is necessary. There are also standardised testing machines and equipment that should be used for testing—e.g. Brinell Hardness Testing machine, Rockwell Hardness Testing machine, Universal Tensile Testing machine etc. Details of testing could be the subject of a complete book (see references 3 and 6 in the Further Readings section). Therefore only some salient features of mechanical testing would be mentioned here.

5.4.1. Mechanical Testing of Steels

Bulk of mechanical testing of steels is concerned with hardness testing, tensile testing and impact strength testing.

Hardness tests are carried out by causing an indentation on the smooth and flat surface by a hardened steel ball or a pre-shaped indenter under a specific load. In Brinell test, Hardness = load/area of impression, which is measured by the diameter of impression as shown in the following figure—larger the diameter of impression, softer is the steel.

$$BHN = \frac{F}{\frac{\pi}{2} D \cdot (D - \sqrt{D^2 - D^2_i})}$$

Rockwell test—which is used for harder steels and expressed as HR_A, HR_B or HR_C scale, depending on load used (60, 100 and 150 kg respectively) for testing—the indenter could be a steel ball (for A and B scale) or a 'diamond shaped cone' (for C-scale) having a cone angle of $120°$ for penetrating hard surface. Hardness is measured by reference to the 'hardness number' table or graduated scale in the respective machine of testing. Vickers Pyramid Number (VPN) can also be used for hard surface. It uses a $136°$ angular pyramid indenter, which can give impression of similar geometry under varying load that generally ranges from 5 to 120 Kg.

Hardness testing is the most common test and the hardness value can indicate properties—like softness, resistance to deformation (i.e. strength), wear, scratch resistance, bendability, machinability and ease of cutting. To a certain degree of accuracy, hardness data can be used as indicative tensile strength of the steel by conversion and there are standardised conversion tables available for this purpose (see appendices section at the end of the book). However, hardness testing cannot be the substitute for tensile testing, which provides a lot of other additional information necessary to gauge the steel behaviour—such as yield strength, elongation, reduction in area before fracture etc.

Tensile test indicates strength and ductility of the steel. Standard tensile testing involves applying increasing stress (or load) on the standard test specimen gripped between two screw- driven devices and then measuring the load versus extension of the marked up 'gauge length' of the specimen—either automatically or manually—for calculating the test result. A typical tensile test curve is shown in Fig. 5-1.

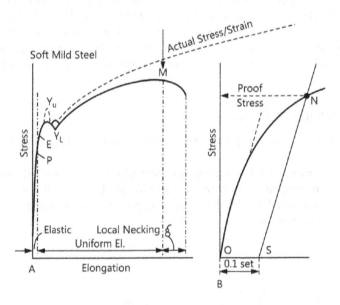

Figure 5-1. An illustration of the stress–strain (elongation) curve of low carbon mild steel. **A:** represents tensile test curve for normalised steel and **B:** shows how to arrive at a design stress under 'Proof Stress' (PS) when no sharp yield point is observed due to cold working of the steel.

Tensile test results must refer to the 'gauge length' used for testing as this influences the observed result and the same will not hold for another gauge length of the same steel. Results of tensile testing are derived by using the following relationship:

Stress = Load/Cross-sectional area of the specimen, expressed in N/mm^2

Strain = Extension of gauge length of the specimen/Original gauge length (expressed as percentage of original gauge length)

These data then allow calculation of Young's Modulus, E by applying Hook's Law, which states that—stress–strain is a constant so long as the stress–strain increment shows a linear relationship—as shown in point 'P' in Fig. 5-1A.

Actual stress–strain increase is shown in Fig. 5-1A is due to increase of working stress with change in cross section of test piece due to plastic deformation, leading to elongation. Tensile test curve when plotted gives 'lower yield point' (Y_L), 'upper yield point' (Y_U) and 'maximum load' (M) from which 'ultimate tensile strength' (UTS) is calculated. Formulae for calculating tensile test results are given in the following:

YP = Yield load (Y_L) /Original Cross-sectional area

UTS = Maximum load/Original Cross-sectional area: often expressed as tensile strength (TS) in Kgf/mm^2

Elongation (i.e. % on a gauge length) = Extension/Original gauge length × 100

Reduction of Area (%RA) = (Original area – Final area) /(Original area) × 100

Elongation obtained in tensile testing is dependent on the actual gauge length (GL) used—it decreases with increasing gauge length. Hence, for standard and comparable result, GL has to be a standard one. Also, the tensile test is strain rate sensitive i.e. sensitive to the rate of application of load, especially in softer materials like low-carbon sheet steel. Tensile testing is carried out till fracture, which takes place following 'necking' of the specimen—a phenomenon that shows thinning of the section or reduction of area, indicating the ductility of the steel. The more ductile the material more is the necking and %RA of the specimen before fracture.

Besides manually controlled tensile testing machines, there are computer controlled testing machines with electronic 'stress–strain' measuring devices. For example, 'Instron' or similar testing machines—which are very versatile and accurate—can be used for tensile testing, fracture testing and a host of other parameters like strain-rate sensitivity and formability measurement.

Tensile testing in such sophisticated electronically controlled machines can provide data about all commonly used properties of steels—such as YS, TS, %Elongation, Ductility, Toughness, Work-hardening rate as well as Stretchability, Strain-rate sensitivity, Springiness character etc.

Figure 5-2 illustrates a plot of stress versus strain in such instrumented tensile testing, indicating the character of steels at different carbon levels and their strength, ductility and toughness properties. Area under the curve determines the toughness in the steel, while ductility is measured by the total elongation of the specimen.

Figure 5-2 illustrates that low carbon steel has the most ductility, but medium carbon hardened steel shows highest toughness—as measured by the combination of strength and associated ductility in the steel. Toughness of steel implies that the steel is strong enough to resist early deformation and ductile enough at that strength level to resist any crack or fracture propagation. Resistance to fracture propagation in the steel comes from the capacity of the structure to absorb energy ahead of a crack tip. However, there are several variables that have profound influence on the toughness of a material—namely strain rate (rate of loading), temperature and notch sensitivity.

As indicated in Fig. 5-2, high carbon steel is the strongest, with high yield strength (note the shape of stress–strain curve for this steel). Area of the curve under high carbon steel is relatively smaller than the others, indicating that such steel lacks sufficient toughness. As such, high carbon steels can be used for springs and applications requiring springiness property in service, but not for any high load bearing application requiring toughness.

Rate of application of stress in tensile testing is rather slow and gradual, which is not always the case in all practical application of steels. Many a time, rate of loading could be faster or stressing the steel parts could be sudden, creating an impact load on the steel part. Behaviour of steel and its fracture mechanism under impact loading is different from the tensile loading. *Impact loading generally makes the steel behave in a relatively brittle manner compared to tensile loading.* Hence, evaluation of steel behaviour under impact load constitutes an important part of steel testing.

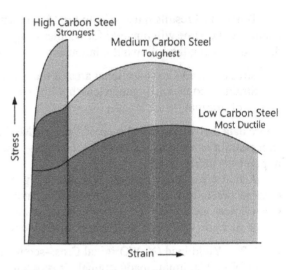

Figure 5-2. An illustration of the stress–strain curves of different carbon steels and their indicative strength and ductility property.

Impact testing is designed to assess toughness of steels under sharp notch and high rate of loading—i.e. under accelerated strain rate condition by applying an impact load onto a notched specimen. This is commonly carried out by using a standard 'V' or 'U' type notch cut on to a square test specimen of standard size. Test is carried out by using a test machine called 'Impact Testing Machine'—either Izod type or Charpy type—where a high impact load is released through the striking lever. The striking lever is made to hit opposite the notch of the test specimen for horizontal holding or just above the notch for vertically held specimen, which is kept firmly gripped on an anvil. This sudden application of impact blow on to the notch causes it to fracture under impact. Under the impact, fracture originates from the tip of the notch and progresses through the cross-section at high speed. If the steel has some toughness at the level of testing temperature and conditions, a part of the applied impact energy will get absorbed in the matrix, indicating the notch impact toughness of the steel.

Impact testing machines are fitted with a pendulum to strike the specimen from a pre-designed height and a graduated scale to indicate the swing, which in turn indicates the energy absorbed in propagating the fracture. If no energy is absorbed, the steel is fully brittle and *vice-versa*. Figure 5-3 shows the test set up of a 'Charpy Impact' testing process. The absorbed energy (generally indicated by a calibrated scale attached to the machine) is termed the impact energy—expressed in Joules. The machine can be: 'Izod Impact' testing machine' or 'Charpy Impact' testing machine' as per the principles of operation.

Impact toughness property of steel is temperature dependent—lower the temperature of testing more brittle could be the steel behaviour. Thus, it may not be always adequate to test the impact toughness value at room temperature alone—it must be carried out over a range of temperature above and below the room temperature, to know exactly how the steel will behave under notch and impact stresses with changes of temperature.

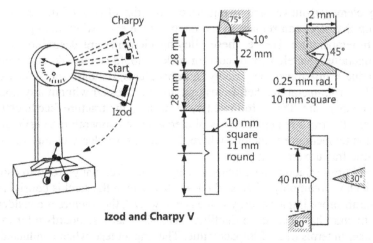

Figure 5-3. An Impact testing set up using 'Charpy notch test'. Mark the position of the notch and its configuration. *Source: Rollason, E.C. Metallurgy for Engineers. London: Edward Arnold, 1973.*

Impact test carried out over a range of temperature allows plotting of the complete curve for obtaining the 'ductile-brittle' fracture transition temperature of steels (see Fig. 5-4). Figure 5-4 shows—(a) how the impact energy to fracture changes with temperature, and (b) how impurities or residual elements present in steel can influence the impact property. Steel has many impurities in it—namely sulphur, phosphorus, oxygen, etc., which raise the transition temperature of the steel—i.e. where the steel changes from ductile behaviour to brittle behaviour. Of these, phosphorus has significant influence of raising the transition temperature; hence, care should be taken to limit the phosphorus content in the steel for such applications. Therefore, in the selection of steel care is taken to limit the presence of residuals.

Figure 5-4. Impact-transition curves for steels with low and high residuals (S + P + O + N) in steel plate. Steel with low residuals sharply increases the absorbed energy compared to steel with higher residuals. Higher the absorbed energy at a given temperature better is the toughness of the steel; note the level of change at −40°C.

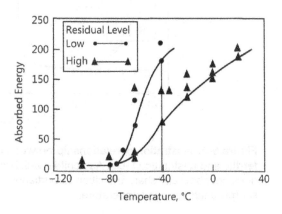

Of the alloying elements, nickel and manganese have the effects of depressing the transition temperature, implying that steel can remain in the ductile zone even when the operating temperature goes below the room-temperature. Hence, these alloying elements are desirable for controlling the impact transition temperature, but element of cost for nickel should be considered for final selection.

To establish 'ductile-brittle transition' energy in impact tests, standard specimens are subjected to soaking at series of lower temperatures before testing, and then tested without any time loss to retain the effect of soaking temperature. This is followed by examining the fracture face to estimate the brittle fracture percentage, and the fracture percentage is plotted against temperature to estimate the transition temperature. Transition temperature is the one where the fracture suddenly changes from having some ductility to 100% brittle fracture (see Fig. 5-5).

Since estimation of brittle fracture percentage is very subjective, many prefer to use a fixed energy absorption level—which is commonly taken as 27 J—below which the steel is considered fully brittle. This implies that if the absorbed impact energy value is below 27 J, the fracture is considered to be brittle and above 27 J, the fracture has acceptable ductility. As such, many standards refer to the 'transition temperature' of the steel in terms of 27 J impact value. The impact test, when conducted with care, can also provide data about 'nil ductility' of the material, which is critical for certain applications involving work at very low temperatures.

Figure 5-5 illustrates the typical fracture face appearances (schematically depicted) as the fracture progress from 'ductile' at higher temperature to 'brittle' at lower temperature. Taking 27 J level (indicated by arrow) as corresponding reference energy level for brittle transition, the transition temperature of this ferritic steel is about—20°C (as indicated by falling arrow). Steel with pearlite will have higher

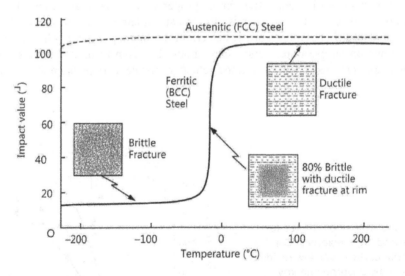

Figure 5-5. Illustration of relationship between brittle fracture and temperature for ferritic and austenitic steels. Austenite—which is a face centred cubic (FCC) structure—does not show any sharp transition to brittleness, whereas impact energy of ferritic steel sharply changes with temperature.

transition temperature than ferritic steel, moving the curves further to the right, which is not desirable. This is because if the transition temperature rises above $0°C$, the part may fail by brittle mode even when working in normal atmospheric conditions. The lower the transition temperature better is it for safety of the steel parts in services. Considering the importance of low-temperature toughness in steel, more about impact transition temperature and behaviour of steel will be discussed in Section 5.5 in this chapter.

Thus the preceding section discusses about some of the common mechanical property tests that are regularly carried out for conformance of steel quality. Acceptable values from the results of these tests are guided by the corresponding 'standards' and 'specifications' of the steel, which will be discussed in the next chapter. There are other specialised tests like fatigue tests, creep tests and formability tests (e.g. bending, stretching and cupping tests) which are also required to be performed on steels, if the application calls for. These tests will be illustrated in the subsequent section.

5.5. Fatigue and Creep Testing of Steels

Fatigue testing and fatigue fracture is an important area for applications of steel, because more than 80% of all engineering component failures are due to fatigue failure. Most engineering parts experience the so-called 'life load' i.e. dynamic load or fluctuating load in services. Dynamic or fluctuating stresses of relatively lower magnitude can cause 'fracture' in metallic materials, which otherwise could carry much higher static load. It is generally said that steel has become fatigued under those small but dynamic loads and gave way to failure by 'fatigue fracture'. Figs 5.6A&B, illustrate typical fatigue testing set up and a fatigue fracture face of an automobile shaft.

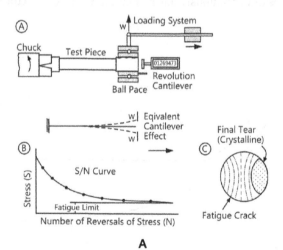
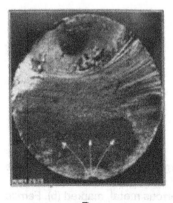

Figure 5.6A&B. A: Typical fatigue testing set up and B: a fatigue fracture face of an automobile shaft. Arrows indicate from where the fatigue crack had started, and the following serration marks are the foot-print of fatigue crack propagation. *Source of Fig. 5-6B: Rollason E.C., Metallurgy for Engineers, London: Edward Arnold, 1973.*

Fatigue is a type of failure of materials under *repeatedly applied stress*, though the stress may be much less than the corresponding yield stress of the material. *Fatigue failure* occurs under repetitive cyclical load working on the part in service, after large number of cycles. Fatigue accounts for over 80% of all industrial machine and component failures—where fluctuating loads under service could range from very small to very high level. For example, fatigue can cause failure of a *fastening bolt* where small vibrating load is at work. Similarly, fatigue is also active in large *landing gear of an aircraft* where the gear experiences high dynamic load, especially during landing. The dynamic fatigue load which a steel part can withstand infinitely without failure is termed as 'fatigue endurance' limit. This limit has been marked up in Fig. 5-6A in the region where stress versus number of fatigue cycles (N) are flattened off.

In actual testing, fatigue limit is taken as the stress level below which fatigue test sample does not fail when tested for a minimum reversible fatigue cycle of 10^6 or more. The nature of fatigue curve generated in testing clearly indicates that there is a minimum stress level below which sample would not fail, even if the test is continued indefinitely. This minimum stress level is taken as the $\sigma_{fatigue}$ of the steel—known as *fatigue limit* below which steel part does not fail. However, phenomenon of 'fatigue limit' stress—flattening the S-N curve—is limited to ferritic steels. Austenitic steels or non-ferrous alloys do not exhibit 'fatigue limit' (see Fig. 5-7).

Examination of fracture face of fatigue failure—as illustrated in Fig. 5-6A(c) and exhibited in Fig. 5-6B for an automobile shaft reveals typical foot-prints of applied stresses. These illustrations show that fatigue crack starts from the surface at the point of stress concentration or stress raiser, and then proceeds as if progressing in beach sand, leaving typical foot-prints. The beach-marks—resembling to sand waves in a beach—associated with fatigue fracture become finer when stress amplitude is lower and coarser when stress amplitude is higher. Any surface flaw or point of stress raiser, including radius or shoulder in the steel piece, can influence the initiation of fatigue crack.

Fatigue is a very common mode of failure of engineering parts—like the automobile parts, machine-tool parts, aircraft components, structural construction, ship halls etc. where fluctuating degree of stresses work intermittently at regular intervals or continuously during operations. Hence, extraordinary

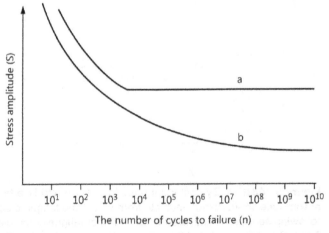

Figure 5-7. Typical fatigue curve of steel, marked (a), and non-ferrous metal, marked (b). Ferritic steel shows flattening off of fatigue stress level, whereas non-ferrous metals and austenitic steel show continually falling strength.

efforts are made to make steel parts even more stronger and tougher by special alloying—like addition of Cr, Mo and Ni or by providing additional *surface hardening processes* that produce some degree of *compressive residual stresses* on the surface. Compressive residual stresses on the surface counteract the applied tensile stresses under fatigue loading, and thereby minimise the effect of applied stress level. Such surface hardening processes include—case carburising, surface induction hardening, nitriding, carbo-nitriding, etc. Surface hardness produced by these processes could be very high (>750 VPN) and very effective, for resisting frictional wear and scuffing. More about these surface hardening processes will be discussed in the heat treatment section in Chapter 8.

There are two aspects of fatigue failure in ferritic steels—one, fatigue crack can start at stress level lower than the conventional yield strength of the steel (i.e. at very low stress level) and—the other is that once a fatigue damage (crack) has started, it cannot be eliminated from the steel structure even after withdrawal of the stresses. *Fatigue damage is cumulative in nature.* This implies that interruption in service or down time in maintenance does not improve fatigue life—damage once initiated in earlier operation will continue to grow with further services under fluctuating stress.

Fatigue load is a cyclic one—it gets applied on the test specimen in a certain frequency and can be conducted under different 'mean stress values'. Figure 5-8 shows different types of pulsating loads that can work in service or can be applied to the test specimen. Though most laboratory fatigue tests are carried out with mean stress equal to zero, most practical application of steels will involve higher mean stress and accordingly the fatigue limit value will change.

Fatigue resistance is quite sensitive to the surface finish, presence of radius in the design, smoothness of finish and quality of machining—in addition to the sensitivity of fatigue to quality of steel and its structure. Hence, for fatigue sensitive parts like the automobile and aircraft components, extreme care is taken for high surface finish and accurate machining along with the use of good quality clean steel. Despite care in maintaining surface finish in steel parts, fatigue crack always gets initiated from the surface, (see Fig. 5-6). Figure 5-6 shows that fatigue crack starts from surface—either from a surface flaw or discontinuity—making surface quality a critical factor for fatigue sensitive applications. In fact, harder the steel surface more is the propensity of notches for fatigue crack initiation. Hence, higher the tensile strength of steel, better should be the surface finish for reducing the chances of fatigue failure.

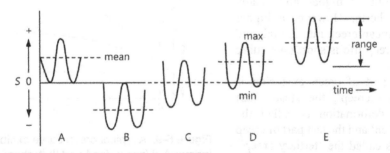

Figure 5-8. Illustration showing different pulsating stress pattern that can work in practice. In this figure only the type C is with mean stress = 0. With change in mean stress pattern, σ_f will change.

It is, therefore, no wonder that hardened steel—either bulk hardened or surface hardened are always specified for fatigue sensitive applications with superior surface finish generated by fine grinding. *Care has to be taken for finish grinding such jobs with softer grinding wheel and profuse coolant to avoid chances of generating fine grinding cracks or burning of the surface during grinding.*

Creep is the other failure mode of steel, but it occurs at higher temperatures. Creep failure takes place under static stress at higher temperature as opposed to fatigue failure that takes place under dynamic stress and at room temperature. Steel is said to creep with time under applied tensile load at temperature higher than the room temperature, and fails by fracture mode very similar to tensile fracture. Thus, creep deformation is a thermally activated process where rate of deformation called creep rate is dependent on the temperature of operation. Examples of creep related applications include: boilers and boiler parts, furnace parts, power plant equipment, nuclear plant equipment, etc. The characteristic difference between tensile fracture and creep fracture is that creep fracture could occur under load which is lower than normal tensile strength of the steel, because temperature brings down the hot tensile strength of the steel.

The process of creep involves number of stages of deformation with time. Figure 5-9 illustrates typical creep curves of steel, where strain generated in the creep test under a given load or stress is plotted against time. Exact shape of the creep curve depends on the temperature of testing, due to dependence of creep deformation on the thermally assisted deformation process i.e. temperature dependence. In general, creep involves three distinct stages of deformation:

- Instantaneous creep on loading the specimen-marked stage-I in the middle curve, called 'primary creep';
- Steady state creep, marked stage-II, where rate of creep slows down compared to stage-I, called 'secondary creep'; and
- Rapid creep stage where mean creep rate rapidly increases with time, called 'tertiary creep'.

At lower temperature (see curve A in Fig. 5-9), steady state creep is prolonged and mean creep rate is very low, whereas at intermediate temperature (curve B), mean creep rate is higher and steady state creep is shorter. However, at higher temperature, mean creep rate is high and steady state creep life is short (see curve C in Fig. 5-9).

The first part of creep extension is called 'primary creep', the steady part of the creep deformation is called the 'secondary creep' and the last part of creep deformation is called the 'tertiary creep'. Duration of these stages of creep depends on the temperature and loads in testing or service. Rate of creep at the steady state part of creep deformation is either steady

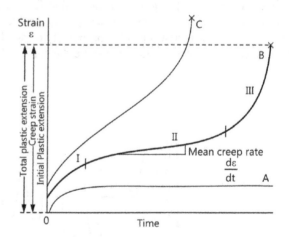

Figure 5-9. A typical creep curve exhibiting stage I (primary), II (secondary) and III (tertiary) of creep (curve B) at different temperature. Creep curve A pertains to lower temperature, curve C is for higher temperature and curve B is at an intermediate temperature between A and C.

or in near steady state due to dynamic recovery of stress acting on the steel at higher temperature. This 'time and temperature' dependent recovery process does not allow additional stress build-up in the matrix—which would have otherwise occurred due to work-hardening effect while the steel gets deformed under normal tensile load. Due to recovery effect—which is characteristic of the secondary creep stage, the steady state of creep gets prolonged and delays the onset of 'tertiary creep' which is responsible for the ultimate creep fracture. Tertiary creep progresses by coalescence of voids formed during secondary creep, leading to quick fracture.

Creep testing is done under standard 'creep testing machine', which is similar to tensile testing machine but with the provision of a furnace, wherein the specimen is gripped and tensile load is applied after heating and maintaining a constant temperature. The stress versus extension of the specimen under creep is recorded over a time and plotted as in Fig. 5-9. Creep test duration is generally high, ranging from a day to several months, depending on the conditions of testing.

The way to resist creep failure is to prolong the secondary creep stage as much as possible. This can be done by:

- Either controlling the operating temperature where deformation rate and recovery rate of the strain caused by the deformation can balance each other,
- By using suitable alloy steel where rate of creep deformation under the given load and temperature will be lower due to less softening of the alloy steel at higher temperature, and, thereby, prolonging the secondary creep stage, and
- Using coarser grain size steel which offers lesser scope for grain boundary deformation and void coalescence at lower and intermediate temperature and lower grain boundary sliding at higher temperature.

Creep data indicate as to exactly what should be the maximum possible design stress at the operating temperature for safe and useful service life. Creep strength is critical for applications like boilers, steam engines, turbine rotor blades and thermal plant parts. It should be noted from Table 5-1 that fine grained steel is not desirable for creep resistance, because it has been observed that at higher temperatures, deformation can progress by a process called *grain boundary sliding* where fine grain means higher scope of grain boundary area and sliding.

5.6. Low Temperature Properties of Steel and Their Evaluation

As fatigue fracture is a room temperature phenomenon and creep is a high temperature phenomenon. There is another mode of fracture which may occur at lower temperature i.e. at subzero temperatures which is termed as 'low-temperature fracture'. Low temperature fracture is brittle and fails without giving any sign of warning for repair or replacement of the component. It leads to sudden and catastrophic failure, and can cause huge damage e.g. the collapse of oil rig platform on high sea. Therefore, determination or evaluation of low temperature fracture properties of steel assumes great importance for some special applications concerning lower temperature of operations—especially at subzero temperatures down below to $-150°C$. Some of the examples of such low temperature operations are marine and oil rig structures operating at high sea cold water, submarines, industrial cold chambers, aircraft flying at high altitudes, cryogenic engines, etc. Here again, the steel part might be working in low temperature which is constant or in a cold environment where temperature can fluctuate over a range from low to extra low temperature.

Steel parts operating under fluctuating cold weather or with the possibility of experiencing cold temperature below $-130°C$ in operation are prone to catastrophic failure without warning. This is because of the following facts:

- At subzero temperatures, there is an increase of tensile and yield strength of steels with sharp decrease in ductility (this is probably because of the freezing of atoms in their respective lattice places in the absence of adequate thermal energy);
- Steels can have some impurities bringing up the impact transition temperature of the steel;
- Steel parts can have some surface defects that may act as sharp notch;
- The design of the part may have sharp bend or radius, giving rise to chance of higher stresses acting with tri-axiality from these points of stress concentration; and
- Sharp drop of elongation, ductility and reduction of area, %RA value of steels below $-130°C$.

However, poor low-temperature property is particularly a problem in BCC structure i.e. ferrite-based steels; face centred cubic (FCC) steel—like the austenitic steels—does not exhibit the aforesaid sharp drop of strength and toughness; hence does not exhibit sharp impact transition temperature(see Figure 5-5). Hence, evaluation of low temperature properties of ferritic steels—which includes all grades of steels excepting the austenitic stainless steels—assume importance for many critical applications.

Low temperature property of steel is generally evaluated from the 'ductile-brittle fracture transition' data of the steel, using notched specimen, which has been discussed under 'Impact testing'. Sudden transition of fracture behaviour in ferritic steels from ductile mode to brittle mode could be gauged from the fact that ferritic steels—that includes all ferritic, pearlitic, bainitic or martensitic steels—exhibit change from 'ductile-to-brittle' mode of fracture at subzero temperatures (see Fig. 5-5). Therefore, attempts are being made to push down this transition temperature as low as possible for safe working of the unit at lower temperature operations. The transition temperature below which brittle fracture can occur (this must be avoided for sudden and catastrophic failure of parts) are lowered by the following factors:

- Decrease in %carbon (%C) and phosphorus in the steel—lower carbon (≤0.15) is desirable;
- Increase in the radius of any designed notch e.g. 6 mm minimum;
- Decrease in notch depth, and freedom from deep surface defects, if any;
- Decrease in grain size and improvement in cleanliness of steels;
- Increase in Mn/C ratio to above 3 (preferably 8) by adjusting Mn content in the steel; and
- By increasing nickel (Ni) content in the steel for lowering of the transition temperature.

Hence, the prevention against brittle fracture at low-temperature applications requires—(a) careful design of the steel composition, and (b) careful design of the steel part to avoid stress concentration and tri-axiality. If the temperature of operation is much below the room temperature i.e. in the vicinity of $-130°C$, FCC material e.g. austenitic stainless steel should be used, which does not undergo ductile-brittle fracture transition. Figure 5-10 depicts the effect of low temperature on the mechanical properties of steels in plain and notched condition.

General trend is that un-notched tensile (and also fatigue) properties increases with lowering of temperature whereas notched test results show decrease of tensile strength with lowering of temperature. This is due to increase in notch-sensitivity factor and tri-axiality with lowering of temperature in notched sample. Hence, avoidance of notch in the design of steel parts for low-temperature applications is very important.

Brittle fracture at lower temperature is a big problem in welded structure, unless the weld joint is made free from brittle microstructure and also free from residual stresses. For this, critical weld structures for use in shipbuilding and marine vessels, which may encounter fluctuating temperatures and stresses, are given special heat treatment for stress relieving. Most submarines and nuclear vessels, therefore, make extensive use of high-nickel steels—e.g. 9% nickel steel—where the microstructure can be specially heat treated to make it tough with very low ductile-brittle transition temperature. In addition to these factors, corrosive and oxidising environment of operations can also contribute to further deterioration of the situation by damaging the surface areas from where cracks generally start. Hence, for avoiding brittle fracture under fluctuating type of stress, temperature and from corrosive atmosphere, ship structures and the hull are given special preventive protection.

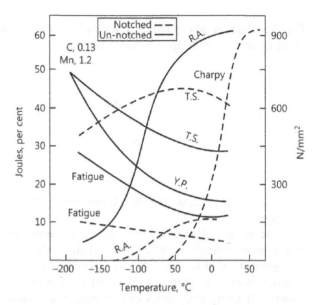

Figure 5-10. The effect of lowering of temperature on the mechanical properties of steels in notched and un-notched condition.

For most design purposes of steel parts for general structural and engineering applications, notch toughness data is adequate. But, for applications relating to some critical structural parts subjected to static stress, fracture toughness data is important. Notch toughness is closely related to *fracture toughness* property of steels, but they are not the same.

Fracture toughness (k_C) refers to the critical value of k_I, the *stress intensity factor* of a stress field, required for crack growth under a static load in elastic condition. Fracture toughness value in plain strain conditions is referred as k_{1C}, which is sensitive to the geometry of the plate. For testing fracture toughness property of the steel, cracks are intentionally introduced in experimental test piece to simulate the practical situation and measure the growth of the crack under plain strain condition. This is to simulate practical usage of steels, where some fine cracks or incongruity of the matrix are always present in the form of sharp planar discontinuities in the steel matrix. *Fracture toughness study* relates to controlled static stress under elastic condition required to cause crack growth in a given steel matrix. Thus, in fracture toughness experiment an attempt is made to evaluate how the crack grows under a controlled plain strain condition. This is in contrast to notch toughness test—where the impact of applied stress is high and the notch tip experiences a sudden and high intensity dynamic stress due to stress concentration factor (k_f), and fracture progresses very fast in an uncontrolled manner. Both notch toughness and fracture toughness parameters are extensively used for design of steel quality for end uses, but notch toughness data is easier to obtain and widely used in general practice. Figure 5-11 illustrates specimen configuration for fracture toughness study in plain strain conditions using edge crack sample under uniaxial stress.

For illustration, for a plate of dimensions $h \times b$ containing an edge crack of length α, if the dimensions of the plate are such that $h/b \geq 1$ and $a/b \leq 0.6$, the stress intensity factor at the crack tip under an uniaxial stress σ is:

$$K_I = s \sqrt{pa}\left[1.12 - 0.23\left(\frac{a}{b}\right) + 10.6\left(\frac{a}{b}\right)^2 - 21.7\left(\frac{a}{b}\right)^3 + 30.4\left(\frac{a}{b}\right)^4\right]$$

For the situation where $h/b \geq 1$ and $a/b \geq 0.3$, the stress intensity factor can be approximated by:

$$K_I = s \sqrt{p a} \left[\frac{1 + 3\dfrac{a}{b}}{2\sqrt{p\dfrac{a}{b}\left(1 - \dfrac{a}{b}\right)^{3/2}}} \right]$$

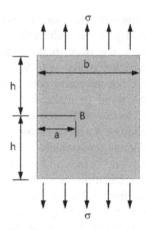

Figure 5-11. Specimen configuration for fracture toughness testing under uniaxial stress, using edge cracked plate sample.

where K_I is the stress intensity factor under plain strain conditions.

The stress intensity factor, K_I, is a parameter that amplifies the magnitude of the applied stress that includes the geometrical parameter. Stress intensity in any mode situation is directly proportional to the applied load on the material. If a very sharp crack can be made in a material, the minimum value of K_I can be empirically determined, which is the critical value of stress intensity required to propagate the crack. This critical value determined for mode I loading in plane strain is referred to as the *critical fracture toughness* (K_{Ic}) of the material. K_{Ic} has unit of stress times the root of a distance. The units of K_{Ic} imply that the fracture stress of the material must be reached over some critical distance for crack propagation to occur.

5.7. Formability Testing of Steels

Forming properties of steels are measured by the ease of forming a steel part without any crack and introducing surface defects (e.g. wrinkles and thinning). Forming properties of steel can be tested and evaluated with high degree of reliability of results. Examples of some formability tests are: bend tests, 'hole expansion tests', cupping tests, etc. However, many a time, data from hardness test and tensile test are used for gauging simple forming behaviour of the steel. For example, very often bendability or machinability of steel is gauged from its elongation or hardness value, respectively. Forming properties of steel are also microstructure sensitive, irrespective of its chemical composition. Thus, vast information about formability of the steel can be obtained from standard mechanical tests for hardness and tensile data, and from the microstructure examination of the steel.

Forming processes could be *hot forming*—e.g. forging and extrusion; or *cold forming*—e.g. press forming, die forming, stretch forming etc. At hot forming temperatures, strength or flow stress of steel becomes considerably lower, making hot forming of steel easier. Hence, the role of original microstructure in hot forming is less significant. Some of the examples of hot forming processes are forging, swaging, hot drawing etc. But, cold forming of steel involves working with high flow stress of the steel, which can increase with increasing deformation due to cold-working i.e. strain hardening of the structure. Cold forming, therefore, requires very favourable microstructure and softness of the steel with good uniform elongation. Examples of cold forming processes are—press forming, bending under dies, stretching, cupping, etc.—using press-tools and forming dies. Even machining of steels to give engineering shape can also be considered as cold forming process, and this process is also highly influenced by the microstructure.

In general, cold formability is influenced by: (a) *hardness*, which is a measure of resistance to deformation, (b) *microstructure*, which influences the strength and ductility of the steel and (c) *cleanliness of the steel*—because, presence of brittle inclusions in the steel can lead to cracking during forming by allowing the interface between the inclusions and the steel to act as discontinuities of matrix and sources of crack initiation. The word 'microstructure' includes all physical and chemical phases present in the steel, including inclusions, grain size and chemical compounds like carbides, nitrides etc. Influence of microstructure on mechanical and forming properties has been indicated in Tables 1-2 to 1-4 in Chapter 1.

Softness of steel, though a helpful parameter, does not necessarily ensure good forming if the steel does not have high uniform elongation—i.e. elongation before necking in tensile test (Figs. 5-1 and 5-2). Fine grain steel, having higher elongation, is helpful in this respect. Hence, in selecting steels for cold forming, attention must be paid to the hardness, ductility/elongation, cleanliness and grain size; and the combined effect of these parameters often decides the success. For good cold formability, steel should be softer, cleaner and with fine grain size that exhibits high uniform elongation. However, for machining, coarser grain size and some degree of strength/hardness in the steel are preferred, because of the necessity of 'chip breaking' during machining which prevents machine tool wear and tip burning.

Forming or shaping of steels at room temperature is carried out under dies and punches in a press. This requires not only proper elongation but also proper stretchability of the steel. As such, currently efforts are being made to develop these types of steels having good stretchability and formability by special design of chemistry and rolling process, including cold rolling. Demand for this type of steel is rapidly increasing due to critical shape of cold formed parts for automobile body application where good surface after forming is additionally required for good paintability. Generally, low-carbon thinner gauge lower strength steels with fine grain size, low inclusion content and specially rolled for improved forming are used for such stretch forming applications. For excellent surfaces after cold forming—as for automobile body parts or similar applications—steels are cold rolled to remove the yield point and/or to introduce special crystallographic texture in the steel microstructure for uniform elongation and proper stretchability. Texture is an important parameter in cold rolled steels as it induces plastic anisotropy in crystal structure favouring improved drawability of steels. The favourable texture for good drawability is grain orientation with {111} planes parallel to the plane of sheet metal being formed. Hence, cold rolling of steels for critical level of drawing is often carried out with the objective of producing as much of {111} grain orientation as possible.

However, softer grade cold forming steel is not the only sheet metal requirement of automobile industry. Auto industry also requires steels of higher strength for load bearing applications, e.g. body frame parts. Therefore, special category of steels called micro-alloyed 'high strength low-alloy steels' (HSLA)—have been designed where the combination of high elongation, good formability and higher strength are achieved by producing very fine grain size. These high strength sheet steels are produced by micro-alloying and controlled rolling/hot rolling to induce very fine precipitates of carbide and nitride for strength in the matrix (see controlled rolling of steel in Chapter 4).

Thus, whatever could be the demand on properties of steels either for forming or end-use applications, microstructure plays a significant role, and what is obvious is that no one type of test might be sufficient for a given application. Tests for detailed chemical analysis, relevant mechanical properties and microstructure examination are very common for general evaluation of steels. Fabrication property of steel not only implies forming and bending property—it also implies how well the steel can be fabricated by other manufacturing processes like machining, welding etc. Factors controlling these two processes have been already mentioned in Section 5.2.

Important among the forming properties of steels are cold forming behavior of steels such as bendability, formability, stretchability, cupping value, etc. These tests are very critical for sheet metals with steel thickness below 2.0 mm. Steels in thinner gauges are generally produced by cold rolling— though it is possible to hot roll steel sheet down to 1.6 mm thickness or so—which offers the scope of some rolling improvement for inducing better forming properties. Hence, automobile industry, which uses very large tonnage of cold-rolled sheets for forming different body parts, drives the quality needs of such steels, demanding good bendability, drwability, stretchability and cupping value.

Cold rolled steels of thinner gauge (<1.6 mm thick) are more widely used for different body panels and door parts in automobiles. Forming of these parts requires high drawability and stretchability, plus good cupping property. These properties can be tested by special testing, as illustrated in Fig. 5-12. Aim of this special testing is to simulate conditions that allow prediction of performance in severe bending, cupping and deep drawing operations involving friction and stretching due to contacts between press tools, dies and the steel surface. Various forces acting during press forming of sheet metal parts are shown in Fig. 5-12.

While the n value of the steel will depend on the composition and the grain size, $\bar{r}$ values will depend on the rolling temperature and degree of finish rolling deformation for producing the favourable

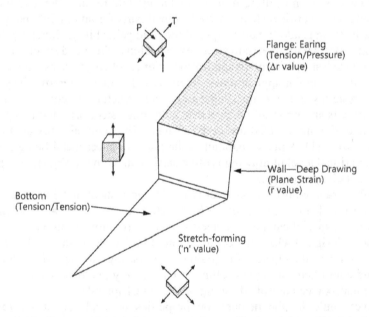

Figure 5-12. Stresses in drawing process in practice and the critical material properties. Notations: n = the work hardening exponent of material, $\bar{r}$ = normal anisotropy value and Δr = planar isotropy value.

texture (see Section 4.11 for definition of $\bar{r}$). Figure 5-12 indicates that drawing process involves—(a) stretch forming requiring adequate 'n' value, which is the strain hardening exponent of the steel and depends on the grain size and softness, (b) deep drawing involving high $\bar{r}$ value and (c) high planar anisotropy (Δr) for flanging and earing.

The work or strain hardening exponent, n, is characteristic of the steel and related to the stress-strain diagram by the relationship:

$$\sigma = K\varepsilon^n,$$

where σ represents the applied stress, ε is the resultant strain and K is the strength coefficient, a material constant. In other words, it is the stress–strain behaviour during work hardening of the steel. The value of work hardening exponent may lie between 0 and 1.0—a value of 0 means that the material is perfectly plastic while a value of 1.0 represents a perfectly elastic solid. Generally, n-value ranges in solids in between 0.10 to 0.50. For cold drawing steel the value should be minimum 0.20.

Sheet metal with textures exhibits anisotropy in properties, which is measured by $\bar{r}$ test. The $\bar{r}$ value expresses different contractile strain ratio between different directions of rolling and is used to denote 'index of anisotropy' in the steel. The test for $\bar{r}$ measures the 'true strain' in uniaxial tensile test specimen along the rolling direction, 45° and 90°. The results are recorded as R_0, R_{45} and R_{90} by using the relationship:

$$R = \varepsilon_w \div \varepsilon_t,$$

where ε_w and ε_t is strain in the width direction and strain in thickness direction, respectively. The $\bar{r}$ value is taken as sum of

$$(R_0 + 2R_{45} + R_{90})/4,$$

and Δr is taken as

$$(R_0 + 2R_{45} + R_{90})/2.$$

Fully isotropic steel has normal anisotropy value of 1.0, but for extra deep drawing operation, steel with higher $\bar{r}$ value (in the region of 1.40 and above) is preferred. The $\bar{r}$ value is a very sensitive measure—a small error in the measurement of strain can cause large error in the value. However, the accuracy of measurement can be improved by taking the strain to higher limit, but below the necking strain of the test specimen.

Bending is a prominent form of shaping plates or sheet steel of thickness 2.5 mm and above. Bending requires steel with good surface quality, softness with good elongation, and freedom from harmful inclusions. Since, surface area is mostly strained in a bending operation, quality of surface and freedom of surface area from scratch, pits, dents etc. are important. Elongation percentage, especially uniform elongation, is a good indicator of ductility and formability by bending.

Bend test is a simple test of bending a steel strip of standard dimension with smooth edges along the radius of a tool-bar by 90°. If there is no cracking of the strip on outer skin, then the result is expressed in terms of the tool-bar diameter e.g. 1T or 2T etc., where T = thickness of the strip. 2T means diameter of the tool bar—around which the steel could be bent—is two times the thickness of the plate. Flat bend test, often required for critical forming parts, means that the strip can be bent flat without any reference to any tool bar diameter. Most structural parts and automobile frame parts require various degrees of cold bending without cracking for giving shape to the component under press forming. Hence, bend test is an important test for these applications.

For very critical cold forming parts involving stretching of hot-rolled plates, a test known as *hole-expansion test* is also carried out. Hole-expansion test involves pressing a pre-designed rounded punch diameter through a pre-punched hole of 25 mm diameter at the middle of a square plate held rigidly under two dies. The result is expressed in terms of percentage (%) of expansion of the original hole without cracking.

Most critical assessment of forming properties of cold rolled sheets involves *cupping test*, where a specimen of the cold-rolled sheet is held firmly under two sets of dies and a punch of given diameter is pushed through the gap between holding dies till a crack or split appears at the bottom of the pressed dome. This test is known as 'deep drawing test' as well. Figure 5-13 illustrates a schematic deep drawing set up.

The regular cupping test is done by using standard machine like *Erichsen Test* with 20 mm diameter ball. In this test, friction is the main drag force that makes the

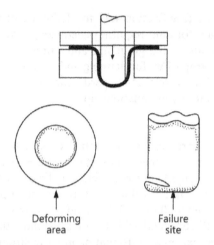

Figure 5-13. A simple illustrative 'deep drawing' test set up, indicating the probable failure sites.

sheet stretch and takes cup shape under the ball. The depth of the cup is the index of cupping, appearance of the dome gives an indication of uniformity of stretching or other defects like, 'orange peel effect' with uneven dome surface or cracks near the dome. The rough appearance of the dome indicates coarse grain size and if cracks occur in one direction, it indicates variation in ductility in different directions. Cupping test involves bi-axial stretching that simulates deformation in stretch forming operations as encountered in most automobile body panel parts. This leads to the fact that there is no even flow of material in all directions when forming the sheet metal. This situation calls for proper lubrication of the forming dies and tools on the one hand and optimisation of properties like n, $\bar{r}$ and Δr on the other, depending on the criticality of parts being formed—$\bar{r}$ values ranging between 1.3 to 1.8 are considered good for cold forming by stretching and drawing.

5.8. Metallographic Testing of Steels

Testing of steel can take three broad lines—*chemical tests* for determination of chemical composition of the steel, *mechanical tests* for determination of relevant mechanical properties like hardness, tensile strength, elongation, impact toughness etc. and *metallographic tests* for determination of macro- and micro-structures of the steel. These two metallographic tests (i.e. macro- and micro-examination) are often complimentary—one, to know if there are any internal defects in the steel like the internal crack, corner cracks, blow-holes, pin-holes, segregation etc.—and the other, for identifying micro features of the steel like the microstructure, grain sizes, inclusions etc.

Macro-examination: as the name suggests, is done on larger areas e.g. the entire cross section of the steel part. It is aimed at to reveal if any manufacturing defects are present in the steel, namely:

• Presence of centre cracks, pipes, blow-holes, pin-holes etc.,
• Segregation of chemical elements,

- Presence of any large size inclusion or slag patch and
- Pattern of grain-flow fibers arising from rolling or forging.

Macro-examination is generally carried out by either taking *sulphur print* (also called contact print) or by 'acid etching' of the cross section. Sulphur print involves allowing a flat steel surface to react with dilute acid, when hydrogen from the acid reacts with the sulphur spots in the steel and causes it to evolve hydrogen sulphide gas. When a photographic bromide paper is firmly placed on such reacting steel surface, the gases evolved due to reaction stain the bromide paper and give rise to a pattern of the macro-structure of the steel, exhibiting the segregation spots where sulphur has concentrated. Acid etching of a flat surface also involves exposure of the steel surface to mild acid, but by swabbing or immersing. The test requires well flattened surface, which is exposed to dilute 1:1 hydrochloric acid in water, either hot or cold. The acid attack causes some chemical reactions to occur with the surface areas where impurities have concentrated e.g. flow-lines arising from forging or rolling, and reveals them. Macro-etching also reveals the cracks, blow-holes and other macro-defects, if the etching process is well controlled. Figure 5-14 shows a typical macro-etched transverse cross section of steel billet where dark spot at the centre indicates centre-segregation, typical of continuous casting.

Macro-examination is very helpful for checking the internal defects in steel sections after forging or rolling. Defects revealed by macro-examination include—centre defects, subsurface defects, cracks, disrupted flow lines, ruptures, bursts, slag patches, large inclusions, segregation etc. Hence, it is an integral part of steel testing and evaluation.

Figure 5-15 illustrates the contour of forging flow lines in a forged flange look, which is revealed by deep macro-etching. Macro-etching of forged parts for examining the flow lines is widely practiced, because load-bearing strength of forged parts is largely controlled by the forged-fibers i.e. the flow

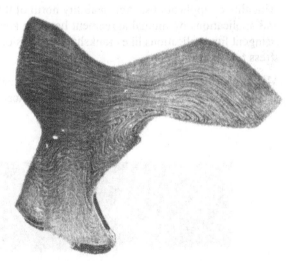

Figure 5-14. Macro-etched face of steel billet showing some centre segregation and inclusion pits caused by etching out of the inclusions. *Note:* Reminiscent of old dendritic pattern in the steel can also be seen.

Figure 5-15. Flow lines in a flange forging revealed by deep acid etching.

lines. There is one more macro test method by fracturing the steel section, called *Fracture Test* or *Blue Fracture Test*. It is aimed at revealing inclusion stringers in steels, which are harmful for fatigue sensitive applications. These inclusions are mostly oxide inclusions (e.g. silicate, alumina etc.) coming from steelmaking process, which, being solid at steel casting temperature, get segregated and concentrated in between the dendritic arms of steel during solidification. During subsequent rolling and forging, these oxides get fragmented and elongated along the rolling or forging fibers, giving rise to the presence of small and long oxide inclusion stringers. Inclusions or oxide stringers of macro-size are extremely harmful for fatigue strength. Hence, when necessary, fracture tests are carried out to check if the steel is free from such harmful inclusion stringers.

The test method involves cutting representative discs or slices of the steel section of about 12–20 mm thick, heating the pieces to about 300°C–350°C, held for few minutes for uniformity of temperature, and then fracturing hot under anvil or a press. The purpose of heating the steel pieces is to take advantage of 'blue brittleness' of ferritic steels at around that temperature and to make the steel section easy to fracture under press. Once the fracture has occurred, the hot fracture face gets exposed to the air and turns blue, except for the oxide stringers which do not change the colour. Specialty of the test is that due to brittleness in steel, fracture path travels through the interface of fibers, which is the weak zone, and this opens up any oxide stringer entrapped between the fibers. Figure 5-16 depicts one such blue fractured face, exposing the presence of few oxide stringers. In this test, the contrast between the blue matrix and grey inclusion stringers gets enhanced and makes it easy to identify presence of large inclusions. An alternative method is to cut notches at 90° to each other to facilitate easy fracture, harden by suitable quenching, fracture the sample under hammer, and blue the fracture pieces on a hot sand plate or in laboratory furnace. The fracture faces then turn blue, leaving the inclusion stringers light grey.

Blue fracture test requires a number of samples to be taken at random from a 'heat of steel' and tested. Each fracture face is examined for the presence of inclusion stringers, their length is measured or mapped and a decision about the quality of the steel is taken. Presence of such inclusion stringers is characteristic of steelmaking process and precautions adopted by the steel manufacturer. Blue fracture test is generally carried out as per internal standard of the customers of steels depending on the criticality of applications. Acceptability norm of this test is also arrived at as per criticality of end uses and applications by mutual agreement between supplier and users. The acceptability standard is most stringent for applications like crankshaft and torsional bars for automobiles which are subjected to high stress torsional fatigue.

Micro-examination: Micro-examination of steel involves examining a small area of the steel surface under 'microscope', where the magnification and resolution of the image can be controlled

Figure 5-16. A fracture face appearance after 'blue fracture', indicating the presence of light grey inclusion stringers across the fracture.

by respective lens fitted to the microscope. The surface under examination has to be absolutely flat and finely polished so that after light etching, the minutest of all features on the surface are clearly visible under the microscope lenses. Hence, an important part of metallographic examination for microstructure study is the surface preparation.

Surface preparation starts with cutting the sample from steel parts with care while cutting under an abrasive wheel, so that the cut surface is not burnt under cutting heat. The sample is then carefully polished in 'sand papers', starting from coarse and then gradually moving to finer grades of sand paper, often changing the direction of grinding, for obtaining finely polished surface. The sample is then gently washed to remove any polishing debris and etched in dilute 2% NITAL acid—2% mixture of 'nitric acid in alcohol' (NITAL) for short. Care should be taken to avoid over etching or under etching by repeated checking under microscope, and accepted only when the structure shown under magnification ×100 looks real and clear. There are number of etchant or special reagents available for examination of steel microstructure, but NITAL etching is the most popular and common. The degree of etching will also depend on the purpose of micro-examination; namely for grain size, inclusion or general microstructure. The polished and etched surface is then examined under microscope, starting at lower magnification but gradually going higher, as necessary to reveal more detailed structure. Figure 5-17 shows typical microstructures revealed under microscope after polishing and etching.

Micro sample preparation and microscopic examination of steel is a vast subject and received considerable attention of metallurgists and researchers for the very basic reason that steel is an extremely structure sensitive material. Grain sizes, inclusions and microstructure put together can closely predict the properties and behavior of steels in different applications. Various microstructures described and displayed in Chapters 1 and 2 are the results of such careful sample preparation and microscopic examination.

Microscopic examination can be done under a wide range of magnification, starting from ×50 to ×2000 under modern optical microscope. For higher magnification, electron microscope can be used which can clearly resolve the positions between few atoms in a structure. Therefore, electron microscopy is used for higher than ×5000 times magnification and can go upto one million time or more. Electron microscopy can take two roots—one 'Transmission electron microscopy' and the other

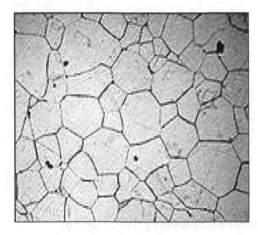 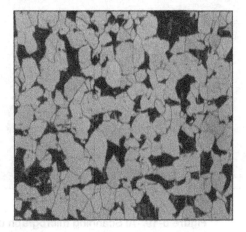

Figure 5-17. Light etching reveals 'grain boundaries' and a bit deeper etching reveals ferrite-pearlite microstructure of similar steel (Magnification: ×100).

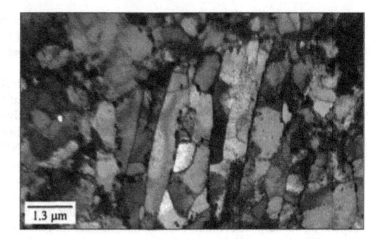

Figure 5-18. Transmission electron micrograph of martensitic structure in steel showing the platelets and dislocation tangles inside under high magnification (×12000).

'Scanning Electron microscopy'. While the former is extensively used for the study of structures of steels, the latter is mostly used for failure investigation and fracture face studies. Figure 5-18 depicts a high magnification transmission electron micrograph of martensitic structure, indicating the platelets of martensite and dislocation tangles inside the platelets—the latter is believed to be the source for high strength in martensitic steels. Figures 5-19A and B on the other hand show the scanning electron micrograph of steel fracture faces that failed by brittle fracture (i.e. cleavage mode) and ductile fracture by inter-granular mode, respectively.

While scanning electron microscopy requires very little surface preparation other than care in maintaining the purity of the fracture face to be examined, transmission electron study requires very elaborate specimen (thin-foil) preparation under electrolytic polishing and etching. Thin foil thickness

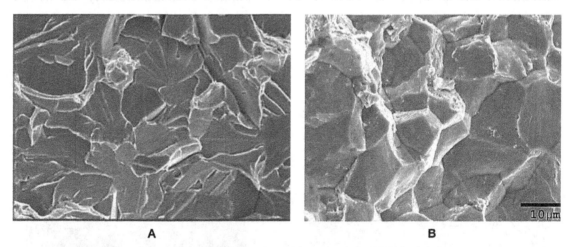

A   B

Figure 5-19. A: Scanning micrograph of steel fracture face failed by brittle cleavage fracture. B: Scanning micrograph of steel fracture face failed by ductile fracture mode. (Magnification: ×10,000).

needs to be few microns so that electron beam can transmit through the specimen and make the structural details inside the specimen visible, as if being viewed under light.

There is one more important laboratory test for steel which is generally conducted in metallurgical laboratories. This is 'Jominy Hardenability Test', and is designed to determine the depth of hardening of steel under a standard end-quenching condition. Since this test relates to identifying heat treatability of the steel, the subject will be discussed under heat treatment in Chapter 8.

5.9. Non-Destructive Testing of Steels

Steel parts cannot always be destroyed for testing. Economy and lack of opportunity for proper sampling or the necessity of 100% checking may be the deciding factors for 'non-destructive tests'. Non-destructive testing (NDT) is a wide group of techniques used in examining metals and materials, including steel parts, for such purposes as checking for internal defects, surface cracks, microstructure, etc. Since NDT does not require destroying, damaging or cutting the parts to test, it is a highly-valuable technique that can save both money and time in product evaluation, troubleshooting, and research. Popular NDT tests attempt to ascertain freedom of the parts or steel from certain defects, like the cracks, voids, laps, lamination etc. Non-destructive tests can either examine the surfaces of the steel or the mass of it, depending on the purpose of the tests. Examples of popular non-destructive testing include: Magnetic particle testing (MPT), Eddy current testing, Dye penetrant testing, Ultrasonic testing and Radiographic testing by use of x-rays. Like micro-examination, this area of testing can be very vast, especially with new developments being used for critical steel parts in nuclear and thermal plant applications. NDT is a subject of its own and there are many reference books for this fast emerging technology. Dealing with this subject in detail is beyond the scope of this book. Hence, few popular NDT methods will be very briefly highlighted here.

Magnetic particle testing (MPT): This method consists of examining the steel bar or sections by holding it between two electro-magnetic poles and pouring kerosene containing fine iron-oxide powder over it. Electro-magnetic induction creates flux lines across the body of the steel and if any surface crack lies along the bar, hence falls across the flux-lines, each side of the crack becomes a magnetic pole and attracts the iron-powder along it. If the steel article is then examined carefully under special light, the crack lines become visible. Sometimes, some fluorescent powder is added to the iron-oxide powder, which when collected by the crack edges and checked under fluorescent light makes the crack glow with clear visibility. An MPT test is widely used for cold drawn steel bars, pipes and welded parts where crack formation is a problem.

Eddy current testing: It involves placing a bar or pipe inside an AC coil when the coil induces a circulating (eddy) current on the exposed surface. The magnitude and phase of the eddy current will depend on the conductivity of the material and this can then be transmitted and read through an oscilloscope screen. Any flaw that interrupts eddy current flow will get recorded and exhibited on the screen. The screen reference can be suitably calibrated to identify only those defects that are harmful for the uses. Figure 4-20 illustrates an eddy current test set up for steel bars.

Similarly, eddy current can be induced on a flat surface when an energised probe coil is brought near the surface of a conductive material. Availability of various types of eddy current coils and probes makes it possible to apply this testing method to various parts of configuration and sections. Eddy current testing is ideally suited to steel pipe testing and weldments.

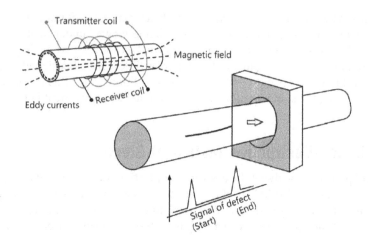

Figure 5-20. An illustration showing the test set up for eddy current testing of steel bars and pipes.

Dye penetrant test: It is a process that involves dipping or spraying the parts with a thin liquid mixed with fluorescence powder which can penetrate into the crack interface or into any surface discontinuities. The fluorescent powder in the liquid gets collected around the crack or open surface and shows up prominently. The surplus liquid is then carefully dried out and the dry surface is checked under fluorescent light for identifying the defects. This test is not limited to the magnetic material and can be applied to a large verity of steel parts and sections.

Ultrasonic tests (UST): In this test, pulses of high-frequency sound waves, in the frequency range of 0.1–15 MHz, are applied to the part under test by a 'piezo-electric crystal'. The transducer crystal is typically separated from the test piece by a couplant, such as oil or water (in case of immersion test method). In the intervals between pulses, the crystal detects echoes reflected either from the back-wall (far edge) of the test piece or from any flaws lying in the path of the sound beam. This reflection mode is called 'pulse-echo' system where the transducer performs both sending and receiving of pulsed waves back. The signals received are shown in a cathode-ray tube apparatus at regular time intervals. The reflected wave arising from defects is then examined for its position on the test piece from the time base connected to the screen. Modern diagnostic machines with frequency modulator or adjuster can display the results in the form of signal with amplitude representing the intensity of reflection and the distance, making it amenable to very accurate estimation of defects and defect locations.

Ultrasonic testing is very versatile and can be made fully automated and applied to large parts like the blades, drums or long plates. Since the technique is based on sound wave propagation and reflection, it can be used for all sorts of materials, such as steels, non-ferrous alloys, cement slabs and composites. The method of testing can be direct coupling or immersion in water. Figure 5-21 illustrates a UST set-up in immersion condition using single probe as pulser and receiver. An UST is widely used for checking long pipe lines and weldments of critical structural parts where parts being used should be free from harmful internal defects.

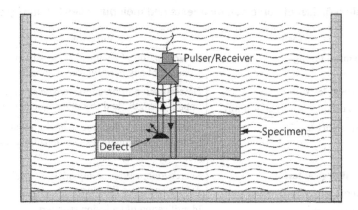

Figure 5-21. Illustration of an ultrasonic testing set-up where specimen is immersed in water.

X-ray radiography: This is another widely used non-destructive testing (NDT) method, especially for thick plates and welds. The test system involves passing a mixed beam of x-rays through the object and studying the shadow formed either on a fluorescence screen or on a photographic plate. In radiography, radiation is passed through the subject part or material, and a detector senses variations in the intensity of the radiation, exiting the sample and thereby enabling a profile of the internal defect distribution. Defects like blow holes, pipes, deep cracks, lamination, etc. absorb the rays less than the normal sound metal and give rise to darker shadows on the negative plate. Their positions are marked and types of defects are analysed from the x-ray plate or screen records. Figure 5-22 illustrates the general set up for radiographic examination.

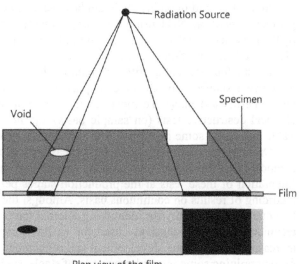

Figure 5-22. A schematic view of x-ray radiography testing set-up.

Table 5-3. List of some common tests and their purposes for quality checking and certifying

Tests	Purpose	Nature of Tests
Chemical testing	Determination of chemical composition and presence of any harmful trace elements	Destructive; sample basis
Physical tests (also called mechanical testing)	Determination of hardness, tensile strength, impact strength, and any other special tests	Destructive; sample basis
Formability tests	Determination of forming behavior of steels by bend test, cupping test, stretchability test etc.	Destructive; sample basis
Metallurgical tests Macro Fracture test Micro	Segregation, cracks or any other internal defects and grain flow etc. Identification of inclusion stringers. Microstructure, grain size, inclusions etc.	Destructive; sample basis Destructive; sample basis Destructive; sample basis
*Magnetic particle testing (MPT)	Detection of surface or sub-surface defects and cracks	Non-destructive; 100% or on sample basis
*Ultrasonic test (UST)	Detection of internal flaws and defects	Non-destructive; 100%
*Eddy current test	Detection of surface cracks, pits, scratch etc.	Non-destructive; 100%
*Dye penetrant test	Detection of surface flaws, cracks and discontinuity	Non-destructive; 100%

* Tests are not equally suitable for all profile of steel sections.

In practice, radiographic NDT can be classified according to the type of radiation employed: x-ray, gamma, or neutron. NDT using x-ray or gamma ray includes capturing and processing shadow created by the presence of defects on the photographic plates. The technique is very effective for testing thicker plates and heavy section weldments.

In sum, the extent of quality tests required for characterising steels depends on the recommendations of relevant standards or customer requirements; more stringent the quality demand, more elaborate would be the test plans. To meet quality requirements, steel mills and steel users generally employ classical destructive tests (on sample basis) and non-destructive tests (may be on sample/lot basis). Table 5-3 shows some important tests and their purposes.

On-line non-destructive tests are increasingly becoming important in steel mills due to changing technology in steelmaking, continuous casting and integrated hot-charging and hot rolling practices. Integration of these tests at the production stages permits application of on-line instrumentation and recording of results on continuous basis. Amongst the NDT tests, ultrasonic testing is widely used on-line due to its versatility for identifying internal flaws as well as it can be used as thickness measuring technique. UST is extensively used for on-line measuring of thickness in rolling mills. To further increase its applications, development and refinement are going on for making ultrasonic tests suitable for determining some internal properties of steels, such as inclusion and slag patches.

SUMMARY

1. The chapter discusses the types of steel properties, their characteristics and methods of evaluation for assessment of steel quality for applications. Properties under the classification—physical, chemical and mechanical properties have been considered for discussions. Relevance of each of these properties in the application of steels has been discussed.

2. Types of mechanical properties that commonly come into play for application of steels have been identified, their dependence on steel structure has been tabulated and importance of their evaluation has been emphasised. It has been pointed out that testing and determination of chemical composition and mechanical properties along with the examination of macro and microstructure of steel form the essential part of steel testing and evaluation.

3. For emphasising the importance of mechanical properties of steel, dependence of various types of forming behaviour of steel on mechanical properties have been highlighted and their evaluation emphasised. Since properties are closely correlated to structure of steels, examination and evaluation of metallurgical structures and their character have been advocated.

4. Common testing methods for steels have been described with reference to standard testing methods. In addition to common methods of testing for hardness and tensile properties, importance of impact testing and evaluation of low-temperature impact toughness have been specially illustrated due to their critical significance in many outdoor applications of steels. Fatigue and creep testing and their special testing features and interpretation of results have been also pointed out.

5. The bulk of uses of steel are concerned with sheet metal forming, such as in automobile industry. Hence, the chapter discusses in some detail the forming behaviour of steels and their evaluation technique, which include bend test, cupping test and test for measuring stretchability of steels.

6. The chapter also describes and illustrates the metallurgical testing of steels like macro-examination and micro-examination for evaluating the level of macro-segregation and microstructure.

7. Finally, the chapter deals with the non-destructive testing of steels, like magnetic parcel testing, eddy current testing, dye penetrant test, ultrasonic tests and x-ray radiography. Methods of testing under each of them have been briefly mentioned and their importance in the evaluation of steel quality has been highlighted. Purposes of such testing have been comparatively studied.

FURTHER READINGS

1. Rollason, *Metallurgy for Engineers,* 4th ELBS edition. London: Edward Arnold (Publishers) Ltd., 1973.
2. Baldev *et al. Practical Non-Destructive Testing.* Woodhead Publishing, 2001
3. ASM Handbook, Vol. 8. *Mechanical Testing and Evaluation,* Ohio: ASM International, 2000.
4. ASM Handbook, Vol. 17. *Non-Destructive Evaluation and Quality Control,* Ohio: ASM International, 1989.
5. ASM Handbook, Vol. 19. *Fatigue and Fracture,* Ohio: ASM International, 1996.
6. Nayar, Alok. *Testing Metals.* New Delhi: Tata McGraw-Hill Education, 2005.

CHAPTER 6

Classification and Specification of Steels
Purpose, Practice and Role

The purpose of this chapter is to emphasise that understanding the system of steel specification—through classification, standards and grades of steels—is central to the exercise of steel selection and application. The chapter, therefore, attempts to develop an understanding about steel specifications by highlighting the methods and systems for steel classification, standardisation and gradation in order to help the readers to choose right steel for right purpose.

6.1. Introduction

It is not enough to know only about the properties of steels—steels are to be used in industries by appropriate selection and application. Central to the selection and application of steels are the specifications. Steel specification system attempts to group the steels into application or user specific 'standards' for helping the users to focus in areas where they should be concerned. For example:

- British Standard 970: Part-1: Wrought Steels for Mechanical and Allied Engineering Purposes;
- ASTM Standard A36: Structural Steels; and
- Indian Standard IS-4432: Case Hardening Steels.

Thus, the standards of steel we refer to relate to some specific purpose for uses and applications. Each standard then covers a series of steel grades under that group, which are given a designated nomenclature for identification. To illustrate further (few examples only and not total coverage):

Standard	Grade	Remarks
BS 970: Part-1	080M40 (old EN-8)	Example of plain carbon steel
	530M40 (old EN-18)	Example of alloy steel
ASTM A36	Plates or Bars (as per shapes)	Structural steel
	Followed by steel grades/compositions	
IS-4432	C14	Plain carbon carburising steel
	20MnCr1	Alloy carburising steel

*Prefixes stand for: BS: British Standards, ASTM: American Society for Testing of Metals and IS: Indian Standards.

However, specification may not cover only standard and the respective grade—it could also specify the details of different conditions that are required for best uses of the steel or for fitness of the steel for intended use. For instance, specification may include steelmaking route, important rolling features, testing and evaluation method, certification system etc., for ensuring the quality and fitness for ready use. Preparation of such detailed specifications (which are extensively used in specialised industries like automobile, engineering etc.) demands certain knowledge and understanding about properties, testing and steelmaking quality capability. This understanding is also necessary for effective utilisation of steel standards and information given therein. Chapters 1 to 5 have discussed these points in order to develop such a knowledge base for selection and application of steels.

The number of steel grades used world over is huge; so is the number of standards covering these steel grades. These are all a part of the steel specification system. Details of specifications of steels covering different standards drawn by different countries may differ in certain respect, but the common aim of them remains the same—i.e. to ensure that the steel quality is fit for the intended use. However, not all the information about the standards, standardisation and specifications can be fully covered in a book of this size and purpose—readers and users of steel have to refer to relevant national or international standards for their specific purpose. This chapter only aims to develop an understanding of the underlying principles of classification, specification and standardisation of steel products. The primary purpose of this chapter is to cover the salient features of *steel classification* and *specification* for guidance and to provide some *grade specific properties* of few important grades of steels for better selection and application.

Discussing steel specification and grades requires references to various steel standards of different countries, where their nomenclature differs. Hence, steel grades have been often referred to in this book by their popular names as prevalent in the industries due to their long use and popularity. Fortunately, there are equivalent lists of these grades, which are referred to whenever required. These equivalent grades have been provided in the annexure of the chapter, and in the appendices section of the book, for reference and comparison. Owing to this constraint, illustrations in this chapter have been cited, as far as possible, by referring to IS (Indian Standards) grade steels or popular BS (British Standards) or SAE (Society for Automobile Engineers—a premier standardisation body in the USA) grade steels.

6.2. Purpose of Steel Standards, Specifications and Grades

Steels are procured by the 'grade' names under specific 'standard' e.g. 45C8, IS-5517 or 20MnCr1, IS-4432. The grade name is 45C8 and IS-5517 is the relevant Indian standard that governs the quality specification for flame and induction hardening steel grades. Same steel (45C8) can also be procured with reference to another Indian standard IS-4368, which covers the 'forging grade steels' for engineering applications. However, there could be some difference in the coverage of quality of steel under different standards—namely the flame or induction hardening steel covered under IS-5517 will limit the sulphur (S) and phosphorus (P) content in the steel relative to IS-4368—because the presence of S and P elements can cause surface crack during faster rate of heating and cooling used for flame or induction hardening. Thus, one of the objectives of steel standards is to ensure fitness for use. *Since there could be multiple uses of similar steels for different purposes, there could be multiple coverage of such steel under different standards too.*

This apart, similar steel can be procured by referring to another standard of another country e.g. 45C8, IS-5517 can be procured by referring to CK-45, DIN-17222—the German specification or as EN-8 (new designation 080M46) under BS-970—i.e. the British standard. Similarly, there are US standards for this steel, covered under the specification SAE 1045 and Japanese standards as JIS-S45C, where JIS stands for 'Japanese Industrial Standards'. Many a time, concerned 'standardisation' body (generally a national body) prefers to cover the detailed quality guarantee norms in their main standards book, without encumbering steel grades with application specific standards and conditions of supply. In such standards, a numbering (designation of steel) system is used that includes 'suffix' or 'prefix' to denote the steel type. To illustrate for instance, JIS uses the prefix 'G' to denote 'structural steel' and 'S' for plain carbon engineering steel—engineering steels are also termed as 'constructional steel' by some countries—SCR to denote chromium bearing alloy steel, SCM to denote chromium and molybdenum bearing alloy steel etc.

Different nomenclature or steel designation systems followed by different countries do indeed create some confusion and difficulties in handling steel related data and information. Though there are lists of equivalent steels, bearing designation of different national standards, there are efforts of unifying them through 'uniform numbering system' (UNI) in the United States of America and Euro-Norm numbering—following the ISO (International Standards Organisation, based in Geneva, Switzerland) system.

The purpose of all these systems followed by different countries is to safeguard the users' interests and help them to 'zero-in' on the exact grade of steel and quality level that is required either for processing and use of the steel or for an end-application. Towards this effort, three terms come into the process of selection of steels—namely 'grade of the steel', 'standard' covering the steel, and the 'specification'. Purpose of each of these is described in the following sections.

Grade: This is the designation of the steel, based on nationally or internationally recognised systems—the aim is to identify the steel with a name which is recognisable by all concerned. Designation of steels is generally based on chemical composition, applications or strength characteristics under a 'standard'. For example, SAE 5140 (also graded as DIN 41Cr4, En-18 or 530M40, IS 40Cr1, etc.) is a composition based grade, whereas S304 stainless steel is an application based grade, and St.42 or E42 is a strength based grade of structural steel.

Standard: This is the mother of all the 'grades' and 'specifications'. *Standard* is a broad-based coverage of various *grades of steels* with focus on specific uses, purpose or applications. Standards are generally application specific and describe the steel quality suitable for those applications—e.g. 'standards' for flame and induction hardening steels, 'standards' for case-hardening grade steels, 'standards' for forging quality steels, etc. To illustrate further, En-18 or 530M40 (mentioned above) is the grade name, and the standard which covers this grade is BS 970. Similarly, the same steel under grade name IS 40Cr1 is covered under the Indian Standards IS-5517 for 'Hardening and Tempering'. Reference standard for steel procurement can be a national standard (e.g. IS standard for India) or a foreign standard (e.g. BS standard of UK or SAE standard of USA or DIN standard of Germany).

Specification: *Specification* refers to the specific requirements of the steel for a given usage or application—it may be fully covered by a standard or may require drawing up of additional requirements by a user as per the specific requirements. Specification attempts to establish full requirements for material properties and quality for a given application, fulfilling the mandates of standards. Hence, specification could be part of the detailed standards or drawn out separately as per end-use requirements.

However, most standards cover sufficient details for the quality required for the end-uses as per that standard; namely chemical composition, physical, metallurgical and dimensional properties, conditions for steelmaking wherever applicable, important rolling conditions influencing the end-quality, methods of testing and certification, packing, marking and handling of steel lots etc. Hence, 'standards' could be sufficient to cover the detailed specifications of the steel. In sum, 'standards' specify the steel under a grade name along with properties and attributes required for the specified purpose for which the standards have been prepared.

As such, standards may specify number of steel *grades* under a given *standard*, mentioning chemical, physical, metallurgical and dimensional characteristics of all those grades for appropriate selection, supply and uses, for example, IS-2062—'Specification of Steel for General Structural Purposes'. This 'standard' refers to three different 'grades' (namely A, B and C grades) and covers what should be the chemical composition and relevant mechanical properties as reproduced in Table 6-1.

The standards, for example IS-2062, also provide the supply conditions and certification norm of the steels in order to ensure that the procured grade under the standard is fit for the intended applications.

Steel standards are prepared with focus on relevant areas of applications and uses as envisaged by the standard and mentioned in the title. Standards are prepared by the concerned 'Standardisation' agency of a country wherein different grades of steels are included for choice as per speciality of applications. The name 'standard' is derived from the fact that the quality information provided in

Table 6-1. An illustrative coverage of steel 'standards' (IS-2062 in this case)

A. Chemical Composition					
Grade	%C	%Mn	%S	%P	%Si	%CE*
A	0.23 max	1.5 max	0.05 max	0.05 max	—	0.42 max
B	0.22	1.5	0.045	0.045	0.04 max	0.41
C	0.20	1.5	0.04	0.04	0.04	0.39

B. Mechanical Properties				
Grade	UTS in Mpa (min)	YS in Mpa (min.)	%Elongation (min) GL = 5.65 So	Bend
A	410	250 for size <20 mm 240 for size 20–40 mm 230 for size >40 mm	23	3T
B	410	250, 240 and 230 respectively as above	23	2T and 3T †
C	410	250, 240, 230 respectively as above	23	2T

* Carbon Equivalent (CE) is mentioned in specification for ensuring good weldability, as structural parts are often required to undergo welding for fabrication.

† 2T for plate thickness 25 mm and below and 3T for plate thickness above 25 mm.

the said standard is measurable, comparable and certifiable by following standardised procedures and processes, which all parties in the contract—namely supplier, buyer (user) or a third party inspection agency, can follow. How standards guide to better steel selection—with focus on steel applications—can be further appreciated from the division of Indian Standards for Mild Steels, which is used for many structural applications.

Indian Standards for Mild Steels	
IS-226	Bars and structurals
IS-1977/IS-2002	Structurals and plates
IS-2062	Specification for structural steel for fabrication
IS-1786	High strength deformed steel bars and wires for concrete reinforcement
IS-10748	Specification for hot rolled carbon steel
IS-6240	HR sheets for LPG cylinder

Likewise, there are many other standards, either Indian or foreign, where the standards relate to different types of steels—e.g. mild steel, plain carbon steel, high carbon steel, alloy steel, stainless steel etc.—for varieties of applications, requiring different shapes, sizes, chemical compositions, mechanical properties etc. Therefore, attempts to standardise steels as per shape, size, chemistry, properties, applications etc. led to the development of numerous standards by different standardisation agencies. However, the aim of all such standards is to help the customers and users for appropriate selection and application of steels.

Thus, in sum, for steel selection and ordering, it is necessary to know: (a) the grade of steel and (b) standard under which the said steel is covered. Thereafter, detailed specifications—including range of chemistry, mechanical properties, dimensions, testing methods for certification, etc.—are necessary to avoid any ambiguity in supplies and uses. If this is not covered adequately by the relevant standard, it can be covered by separate 'specification'—which could be a part of the 'standards' or could be drawn out as separate specification—governing the quality requirements of the users and clearly communicated to the suppliers. What this system implies is that while 'grade' and the relevant 'standard' is a part of steel nomenclature, 'specification' (relating to properties etc.) could be either as per the provision of the standard or as per agreed requirements of users—documented and communicated in the form of 'supply specification'. Often, the specifications also mention the 'steel making and rolling' route and any other special measures required for ensuring quality for the uses and applications. The total system is popularly known as *steel designation and specification system.*

Steel being one of the most widely used materials in the world, steel designation procedure requires a clearly laid down system. Various world standards use different methods for designating steels—though their uses are similar. Being the age-old system, this got into the practice without any hitch so far. However, with increasing world trade and commerce, need for uniform numbering system of steels is becoming important.

An important feature of steel is that similar or same steel can be applied for a number of applications or purposes. For example, steel composition similar to grade IS-2062 Gr. A, can be used as structural plates or as rounds for machining some engineering parts (e.g. screws, nuts, bolts etc.) or as rods or shapes for specific purpose. This implies that same or similar steel can be graded differently for different purposes, and they can be classified as per end-uses or applications. Thus, understanding the classification of steels becomes relevant for this purpose.

6.3. Classification of Steels

Systematic grouping of steels first starts with 'classification'. Classification of steel means assigning the steels to 'restricted categories' by following certain user-specific 'norms'—in order to help the producers, users and buyers in their respective areas of interests. All steel products have been grouped into 'classes'—called classification of steels, and provided with 'classified designations' which are also known as 'grade of steels'. Therefore, if classification of similar steel changes, the designated names of the steel grades also change. Hence, it is not surprising if we find that similar steel composition when classified differently is also designated differently. For instance, medium carbon steel (or mild steel) in the carbon range of about 0.20–0.27%C is classified as St.42, IS-226 and Fe410, IS-1875, depending on the uses. IS-226 is the specification for structural steels whereas IS-1875 is the specification for forging steels.

There are a number of ways steels could be classified, such as:

- Shape—e.g. Flat steel Product, Long steel product, Tubes, Pipes, Sections, Angles, Beams, Bright bars, etc.
- Size—e.g. Blooms, Billets, Bars, Wire rods, etc.
- Application/Utility—e.g. Structural steels, Engineering steels, Forging steels, Carburising steels, Nitriding steels, Stainless steel, Corrosion-resistance steel, Tool steel, etc.
- Composition—e.g. Plain carbon steel, Alloy steel, Interstitial-free steel, etc.
- Structure—e.g. Ferritic, Ferritic-Pearlitic, Austenitic, Martensitic, Bainitic, etc.
- Properties—e.g. Mild steel, High-strength steel, Spring steels, etc.

Each of these has its own purpose and helps the users to 'zero-in' on to the grade of steel required for uses and applications. Hence, steels covered under the class of shape or size can be simultaneously classified as per their composition. As a result, there could be multiple standards covering similar steels but of a different class.

Table 6-2 shows a typical classification chart of steels as per commercial names of steels or their applications and the corresponding structure. Significance of grouping steels as per structure arises because of structure sensitivity of steels for their properties. However, for general commercial purpose, steels are popularly classified and grouped according to their chemical composition (i.e. chemistry-based names) and applications or utility with due consideration to properties for manufacturing and end-use applications. From such a stand point, popular steel classification systems can be said to be based on:

- Chemical composition of the steel—e.g. Plain carbon steel, Low-Alloy steel, High-alloy steel, Micro-alloyed steel, etc.
- Application or utility of the steel—e.g. Structural steel, Engineering steel, Stainless steel, Dye and Tool steel, etc.
- Properties—e.g. Mild steel, High-strength steel, HSLA steel, Stainless steel, Wear resistance steel, Heat resistance steel, Creep steel, etc.
- Processing route of the steel—e.g. Machining steel, Forging steel, Carburising steel, Nitriding steel, Deep Drawing steel, Through-hardening steel, Induction hardening steel etc.

The main points of such systems are described in the following.

Table 6-2. Classification chart of steels as per commercial names/applications and corresponding structures.

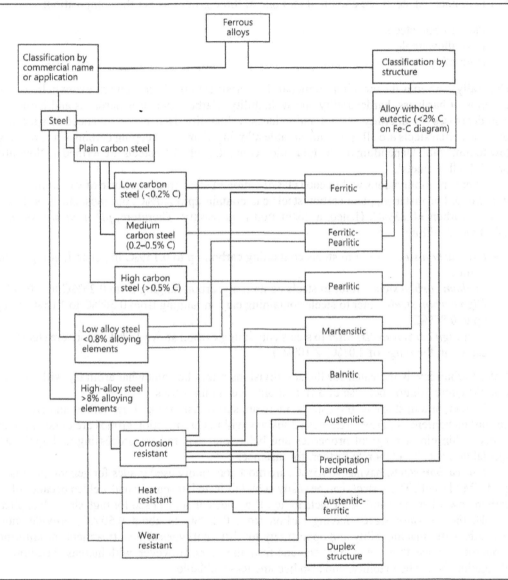

6.3.1. Steel Classification as per Composition

Most commercial steels are classified into one of three groups as per the composition:

Plain carbon steels
Low-alloy steels
High-alloy steels

Generally, carbon is the most important part for classification of steels due its overwhelming importance in terms of hardness, hardenability and weldability. Carbon increases hardness and hardenability, but also decreases the ductility in steel and reduces weldability. Hence, for enhanced or special properties of steels, often steels are alloyed with suitable alloying elements, keeping carbon at a level ranging from low to moderate. Depending on the total alloy content, such alloy steels are termed as 'low-alloy steels' or 'high-alloy steels'.

There are subdivisions within each classification of steel group, because of variation of composition in a group. For instance, plain carbon steels can contain upto 2.0%C, but most steels within this group contain carbon ±0.85%C. Hence, a distinction is necessary. Therefore, *plain carbon steels* are again sub-divided within:

- *Low carbon steels:* refer to steels containing carbon up to 0.15%C in general, going up to 0.25%C at time.
- *Medium carbon steels:* refer to steels containing carbon ranging from 0.25%C to ≤ 0.50%C.
- *High carbon steels:* refer to steels containing carbon ranging from 0.50%C to 1.0%C, but generally up to 0.85%C.
- *Very high carbon steels:* refer to steels containing carbon above 1.0%C (there are steels that contain carbon in the range of 2.0%C–2.10%C).

Note: Carbon levels indicated for the sub-divisions might be somewhat arbitrary with minor variation fromthe quoted figure, because of different practices in industries.

Further, amongst low-carbon steels, there are sub-divisions viz. Low carbon and Extra-low carbon i.e. carbon content <0.03%C. All these divisions within classification groups are done in order to group steels within closer range of properties and behaviour required for processing and applications for a special purpose. To elaborate further, for instance:

Low carbon steels, having <0.25%C, are most commonly used grades for general purpose under the grade 'Mild steel'. These steels can be easily bent, fabricated, machined and welded because of their lower carbon, lower hardness and better weldability. But, they cannot be used for high strength applications.

Medium carbon steels, having carbon from 0.25%C to about 0.50%C, provide much higher strength, better machinability and opportunity for different types of heat treatment for tailoring end-use properties. Hence, they are in great demand from the users. However, with increased carbon, steel loses elongation, hence bendability, and also become less weldable.

High carbon steels contain carbon between 0.50%C to 0.85%C. These steels are required for some special properties like higher hardness, higher strength, and 'springiness', etc. and widely used for shafts, springs, bolts etc. after heat-treatment. But, the welding of these steels is difficult, requiring much of precaution and preparation.

Very high carbon steels, having carbon content in the range of 1.0%C–2.0%C, produce structure which is very hard with presence of lots of carbide after heat-treatment, and popularly used for wear and abrasion resistance applications. These steels are used—instead of high-alloy steels—for wear and abrasion resistance applications more due to economy rather than superiority in properties over high-alloy carbide forming steels. However, high carbon steels are very difficult to weld due to high carbon content and need very elaborate precautions, if the plates have to be welded.

A closer examination may however indicate that though the steels might have been classified as per composition, but, in reality, they are also based on property.

Low-alloy steels typically contain alloying elements like chromium, nickel, manganese, molybdenum, silicon etc. to the extent of 5.0% in total, in general, but can go up to 10.0% in some individual cases (e.g. 9.0% nickel steel). Purpose of such alloying additions is to impart special strength and toughness properties to the steel (after heat-treatment) by keeping carbon levels lower. Carbon in low-alloy steels can range from 0.12%C (e.g. grade 15Cr3) to 0.70%C (e.g. grade 65Si7—i.e. the popular spring steel).

Alloy steels are meant to develop specific properties after heat-treatment in order to cater to wide ranging needs of engineering and construction industries. For example, molybdenum is added to the steel to reduce temper embrittlement in addition to its effect on hardenability and its role as carbide former. Similarly, nickel is used to improve low-temperature toughness—carbon, manganese, chromium etc. are added in balancing quantities for hardenability and strength after heat treatment. As carbon content influences hardness and hardenability of the steel, each alloying element has its own characteristic influence on some specific properties, like strength and toughness, depth of hardening after heat-treatment, etc. Hence, alloy steels can be further sub-divided as per alloy contents e.g. chrome-nickel steel, high-manganese steel, chrome-moly steel etc. Such a classification of steels by chemical composition under SAE classification system is shown in Table 6-3.

High-alloy steels contain 5.0% or more total alloying elements. Common examples of high-alloy steels are 'stainless group of steels' which contain at least 12% chromium and high nickel or its equivalent elements. Stainless steels are known for their corrosion resistance properties. There are other groups of high alloy steels e.g. tool steel, heat-resistance steels, wear-resistance steel, etc. High-alloy steels are also grouped as per their composition, but stainless steels can be also grouped according to their microstructural features—e.g. austenitic stainless steels, martensitic stainless steel, and ferritic stainless steel—because of distinctive properties imparted by such microstructures. Austenitic stainless steel of 18-8 grade (SAE 304 grade) is one such popular example. These microstructures impart their characteristic properties with regard to corrosion resistance, oxidation resistance, strength, ductility etc. in the stainless steel group of materials, and, thereby, justify this type of sub-division.

Table 6-3 provides the classification system and steel designation system of SAE grades of steels (SAE stands for Society of Automobile Engineers—a strong standardisation agency in the USA). This system is essentially based on chemical composition and prompted by the need for guiding common users of steel for better selection and application. There are further sub-groups of this classification and designation system (see Appendix to Chapter 6, Table 6A-1).

As per this SAE classification for designating steels, all carbon and alloy steels are designated by a four digit number,

Table 6-3. Primary SAE grading and classification system of steels based on chemical composition.

SAE Designation	Type/Classification
1XXX	Carbon steels
2XXX	Nickel steels
3XXX	Nickel-chromium steels
4XXX	Molybdenum steels
5XXX	Chromium steels
6XXX	Chromium-vanadium steels
7XXX	Tungsten steels
8XXX	Nickel-chromium-vanadium steels
9XXX	Silicon-manganese steels

where—the first digit indicates the main alloying element(s); the second digit indicates the secondary alloying element(s); and the last two digits indicate the amount of carbon, in hundredths of a percent by weight. For example, plain carbon steel containing 0.60 wt%C will be designated as SAE 1060 steel. The system also allows using a suffix "H" to any designation to denote, that hardenability is a major requirement for that grade of steel. There can be further sub-grouping of this steel grading system where, within each class, steels are further divided and grouped as per concentration of alloying elements (see Appendix to Chapter 6, Table 6A-1).

6.3.2. Steel Classification as per Applications

Steels are a very large family of materials. Some steels are designed as per composition and others as per application or for a purpose. Composition based classification allows some adjustment of chemistry and other quality features to suit specific needs of users. This is because of the uniqueness of steels which can be heat treated for developing exact properties. This situation allows the users to choose narrower composition limits within the permissible limits of a chemistry-based grade in order to tailor-make the steel properties. But, there are other types of steels which are required to give some special in-built property into the steel for given purposes, e.g. stainless steels, structural steels, tool or die steels, creep resistance steel etc. These steels are classified on the basis of applications or purpose. These steels are especially designed to give a 'specific application related property' that gets built-into the material by a combination of chemical and physical properties, where one grade of steel cannot exactly substitute the other grade in an application without compromising the end-use property. However, there could be different types of gradation within these classifications for bringing out their special application related properties—such as corrosion resistance, wear resistance, creep resistance etc. Hence, application-wise classification is predominant in these types of alloy steels.

   Table 6-4 illustrates how AISI system of classification takes care of classification of steels as per their application or purpose. Prominent among these grades of steels are stainless steels of all types and tool and die steels. Chemical composition of these application based steels—such as AISI 304, AISI 404, AISI 429, etc. or W1, W2, S1, S2, etc.—are then described in the AISI or corresponding standards, along with other quality requirements. Table 6-4 is not exhaustive—it is only an example of stainless steel and special steels prescribed for different utility or applications. Structural and engineering steels can also be classified on the basis of these applications, but because of their wide applications, they will be discussed separately.

6.3.3. Steel Classification as per Strength: Structural Steels

Amongst the application based steels, **structural steels** are a popular category where the steel is designated by the 'strength'. Structural steels are a group of construction materials produced in required forms and shapes with certain chemical composition and mechanical properties, the mechanical properties however are the critical criteria for selection and application. Structural parts have to bear mechanical load of varying dimensions, which vary not only from structure to structure but also in terms of the sections of the same structure. Shapes are also important for structural applications for rigidity, stability and effective load bearing capacity of a section. I-beams, Rails, Channels and angles are common examples of structural steels of different shapes. Since load bearing capacity is the key to structural design, this group of steels is generally classified or designated by its 'yield strength' which is a part of their mechanical properties—composition of such steels is secondary as long as the strength

Table 6-4. AISI classification of steels by application or purpose.

Stainless Steels		
AISI has established three-digit system for the stainless steel grades.		
AISI Designation	**Type/Classification**	**Examples**
2XX series	Chromium-nickel-manganese austenitic stainless steels	AISI 201, AISI 202 etc.
3XX series	Chromium-nickel austenitic stainless steels	AISI 301, AISI 303, AISI 304, etc.
4XX series	Chromium martensitic stainless steels or ferritic stainless steels	AISI 403, AISI 404, AISI 405, AISI 429, etc.
5XX series	Low chromium martensitic stainless steels	AISI 501, AISI 502
Tool and Die Steels		
Designation system of one-letter in combination with a number is accepted for tool steels.		
W	Water hardened plain carbon tool steels	W1, W2, W3, etc.
O	Oil hardening cold work alloy steels	O1, O2, O3, etc.
A	Air hardening cold work alloy steels	A2, A3, A4, etc.
D	Diffused hardening cold work alloy steels	D2, D3, D4, etc.
S	Shock resistant low carbon tool steels	S1, S2, S3, etc.
T	High speed tungsten tool steels	T1, T2, T3, etc.
M	High speed molybdenum tool steels	M1, M2, M3, M4, etc.
H	Hot work tool steels: Sub-divided as per Cr, Mo, W, alloyed, etc.	Cr-base: H10, H11, H12, etc. Mo-base: H41, H42, etc. W-base: H21, H22, H23, etc.
P	Plastic mould tool steels	P2, P3, P4, etc.

and bending properties could be achieved. In cases where extensive welding is involved for fabrication, limits on carbon composition are imposed for ensuring good weldability.

American Society for Testing of Metals (ASTM) standards—the most popular and internationally recognised standard for structural applications—uses a prefix 'A' to denote structural grade followed by digits to indicate yield strength. For example, structural steels for carbon and alloy steel structural shapes and plates are covered by A-36/A-36M, where A indicates structural type and 36 indicates 'yield strength' in Ksi unit. Similarly, there are grades with higher or lower yield strength also under the ASTM standards (e.g. A529 or A270), which are again designated as per yield strength—sometime with an appropriate suffix letter to denote any special property e.g. 'W' for weather resistance. Structural grade steels under Indian Standards (e.g. IS-226, IS-961, IS-1977, IS-1786, IS-2062 etc.) are also designated with 'yield strength' in Mpa scale. Table below shows some typical Indian soft structural grade steel with designation as per their 'yield strength'.

Grade Designation	Tensile Strength (Mpa)	Yield Strength (Mpa)	% Elongation	Internal Bend (Diam.)
E-165	290	165	23	2T
E-170	330	170	23	3T
E-215	370	215	23	3T

Note: T is the thickness of the plate. 1 Mpa = 1 N/mm^2 = 0.102 Kgf/mm^2.

Indian standards also have other grades with higher yield strength e.g. IS-961 or as IS-2062 grade structural steel, cited earlier in the chapter. Indian standards cover structural steels in grade E-165 to E-450 under different standards, in order to cater to the wide range of strength required for different structural applications. Such coverage is also common in other international standards, e.g. in ASTM or BS standards.

For imparting high yield strength with required toughness in structural steels, sometime alloying or micro-alloying of the steel composition is required; yet the designation of the steel is done by 'yield strength'. In fact, all national standards try to cover the structural steel quality as per strength and applications. Some important structural steel standards in India include: IS-226, IS-2830, IS-2831, IS-2062, IS-961, IS-1977 and IS-8500—the latter is for micro-alloyed high strength low-alloy structural steel. More of structural steel standards and material properties have been provided in the annexure to this chapter.

Purpose of classification is to map out the steels as per the shape or size of the material or their chemistry, uses, purposes and applications. These classified steels are then taken up for standardisation by following some established procedures and steel nomenclature or grade names that are assigned for their universal recognition and identification.

6.4. Steel Standards and Grades

The purpose of steel standards is to guide the selectors and users of steels for choosing the right steel grade with appropriate properties and conditions of supply. Classification of steels—as per shape and size, chemistry or purpose and application—lay the foundation of most standards. For example:

1. **Structural steel (standard quality) bars, sections, plates, etc.** are covered by:
   - American Standard: ASTM A-75
   - British Standard: BS-15
   - German Standard: DIN 17100
   - Indian Standard: IS-226

2. **High tensile structural steel** is covered by:
   - American Standard: ASTM A-242 and others like ASTM 440, 441
   - British Standard: BS-968, BS-4360
   - German Standard: DIN 17100
   - Indian Standard: IS-961

3. **Forging quality steel—alloy constructional steels** are covered by:
   - American Standard: SAE/AISI Alloy steels: Blooms, Billets, Slabs, Bars, etc.
   - British Standard: BS-970
   - German Standard: DIN 17210
   - Indian Standard: IS-4368, 4432, and 5517

Table 6-5. Illustrative examples of coverage of 'Steel Standards'.

Standard	Grade	Chemical Composition	Mechanical Properties				Remarks
			TS (Ksi)	YS (Ksi)	%El	Impact	
ASTM A-242	Type-1 and 2	C: 0.15–0.20%, Mn: 1.0–1.35%, S&P: 0.050 max	70	50	18	—	Further grading as per size and section
ASTM A-36 (Structural steel)	Shapes: Plates Bars	C 0.25% Si: 0.15%., Mn: 0.80–1.20%, P: 0.040, S: 0.050; Others: Cu 0.20% min C: 0.26%, Mn: 0.60–0.90%, P: 0.040%, S: 0.050%, CU: 0.20% min	58–80 58–80	36 36	23 23		
BS-970	080M40 (Old: En-8)	C: 0.36–0.44, Si: 0.10–0.40%, Mn: 0.60–1.0%, S&P: 0.050%	Mechanical properties as per different supply conditions or 'ruling section' for heat-treatment; (see the Standard)				
DIN-17210	42CrMo4	C: 0.36–0.44%, Si: 0.35% max Mn: 0.70–1.0%, Cr: 0.90–1.20%, Mo: 0.15–0.25%	Mechanical properties as per ruling section (see the DIN standards)				Heat treatable steel
IS: 4432	20MnCr1 (Old: 20MnCr5)	C: 0.17–0.23%, Si: 0.15–0.35%, Mn: 1.10–1.40%, S&P: 0.025% max	Hardness (as-rolled): 250 BHN max				Case-carburising Grade
IS-5517	40Cr1	C: 0.35–0.45%, Si: 0.10–0.35%, Mn: 0.60–0.90%, Cr: 0.90–1.20%, S&P: 0.050% max	Hardness (as-rolled or normalised): 250 BHN max				Hardening and Tempering steel

1 Ksi = 0.70 N/mm^2 = 0.70 MPa.

These are few examples of coverage of standards, wherein grades of steel along with mechanical properties (wherever applicable) are mentioned. A list of some useful standards has been given in the annexure for further reference. Few examples of how steel grades are covered under different standards are given in Table 6-5.

Thus, standards cover steel grades that are classified according to their purpose, use, application or composition. They may also cover one or more grades of steel which are appropriate for the purpose and uses. Standards provide the details of chemical composition and relevant properties, which may include limits of impurity (trace elements), inclusion, grain size etc. as well as supply conditions that specify testing, acceptance and certification norms. Therefore, standards can be the basis for ordering and procurement of steels. Role of steel specification, in this regard, is to further build certain specific requirements of the steel users such as—requirements of any specific steelmaking route, rolling conditions for ensuring soundness of the product, requirement of special testing and certification, and any other relevant information necessary for further building quality for best use of the steel for a given purpose. Thus, classification effectively groups the steels as per use or purpose and standards build on it by providing the nomenclature (designation) of the steels under coverage, their chemical composition, relevant properties, and testing and certification requirements.

A major purpose of standard is to facilitate standardised steel quality and availability within a country and across the countries. Unfortunately, various grades of steels covered under different applications and specific standards, have different nomenclature (grade names) in accordance with the system of naming and grading of steel adopted by different countries. Although ISO or UNS system has been in vogue of late for uniform nomenclature across the countries, it has still not become popular amongst many users, due to wide acquaintances with the older and traditional systems.

Hence, steel grades are more often than not called by old designation, like the C-45 of Indian standard, EN-8 of British standard, AISI 1045 of American standard, S45C of Japanese standard etc. Many of these grade names have undergone a number of changes from time to time—e.g. EN-8, a very popular plain carbon steel, is now named as 080M46 or EN-18 and a popular chrome-alloyed steel of BS is now named 530M40 in BS or EN-19 is now called 708M40. These new nomenclatures are based on carbon and alloy content. This change of names is due to change in numbering systems adopted by the country's standardisation body according to the times and trends. Table 6-6 illustrates few examples of steel nomenclature of same steel by different standards or countries. The table also illustrates the current system of designating engineering steels is predominantly based on chemical composition.

The Task of standardisation, thus, includes—classifying steels as per use, purpose or chemistry, designating steels as appropriately as possible for common understanding, and formulates respective standards as per class of steels. Table 6-7 shows some important product-wise (purpose specific) standards of different countries and grades of steel covered therein.

Table 6-6. Gradation of some important steels by different standardisation bodies of the major steel producing countries.

Germany DIN	USA AISI/SAE	Great Britain BS (New)	(Old)	Japan JIS	India IS
*St 14	C1008	14491CR*		—	IS-513D*
CK10	C1010	040A10	(EN-2)	S10C	C10
C45	1045	080M46	(EN-8)	S45C	45C8
C55	1055	070M55	(EN-9)	S55C	55C6
41Cr4	5140	530M40	(EN-18)	SCR440 (H)	40Cr1
16MnCr5	5115	527M20		SCR415	17Mn1Cr95
42CrMo4	4140	708M40	(En-19)	SCM 440 (H)	40Cr1Mo28
50CrV4	6150	735A50		SUP10	—
100Cr6	52100	534A99	(En-31)	SUJ2	100Cr3

* Cold rolled sheet steel specification with low yield strength.

Table 6-7. List of Standards formulated by different countries to cover similar use or purpose.

Product	Indian	British	German	American	Japanese
Structural steel (weldable quality) bars, sections, plates etc.	IS-2062	BS:2762 BS:4360	DIN:17100	ASTM:A-373, 36,283	JIS:G:31060 G:3113,3101
High tensile structural steel	IS:961	BS:548 BS:968 BS:4360	DIN:17100	ASTM: A-242 440,94,441	CORTEN-A JIS:A:5528
Carbon steel blooms, bars, billets and slabs for forging carbon constructional steels	IS:1875 IS:4432 IS:5517 IS:4369	BS: 970: EN3S,4,5,8,9, 32,43A,438	(DIN:17222, 7111,1613) C-10,C-15, C-25,C-45, CK-45,C-53	(ASTM A-73, 235,236,273) SAE1010 SAE1012 SAE1015 SAE1021 SAE1025 SAE1030 SAE1040 SAE1045 SAE1050 SAE1055	JIS:G:3251, 320,3505
Bars and rods for cold heading/cold extrusion	IS:11169 IS:2255	BS:3111	DIN:17111, DIN:1654	AISI:1008, 1010, 1012, 1015, 10B21, 15B25, 10B21, 15B25, 10B30, 10B38, 15B41, 1541, ASTM:A-545	JIS:G:3507
Hot rolled sheets/ strips Hot rolled sheets/strips for steel tubes and pipes	IS:1079 IS:10748 IS:11513	BS:1449	DIN:1623	ASTM:A-569	JIS:G:3202
Forging quality steel— Alloy constructional steels	IS:4368 IS:5517 IS:4432	BS: 970 EN-15,16,17, 18,19,19C,24, 207,29A,29B, 14A/B,206,12, 22,33,13,40, 41,34,35,30, 39,351,352,25, 26,100,110	DIN:17210 40Mn4 34Cr4 37Cr4 41Cr4 42CrMo4 50CrV4 20Mn6 28Mn6 36Mn5	SAE:4150, 4340,1524, 1536,1541, 5140,5115, 5120,4023, 4027,4032, 4042,4140, 4620,3120, 4340,8640	
Ball bearing steels including wires	IS:4398 103Cr1 103Cr2 102Cr2Mn70	BS 970: EN 31	DIN: 17220 100Cr6	SAE:52100 ASTM:A-295	

Note: Old steel designations have been mentioned for their popularity.

6.5. Steel Designation System

Whatever could be the steel classification or standard, primary requirement of steel specification is—its chemical composition and properties to which the steel should be made and certified. For example, if the steel part is a structural component, its mechanical properties are important. Hence, such steels are designated by 'strength'. But, if the steel is a dynamically loaded or experience fluctuating loads, (e.g. engineering steels), then the steel is designated by chemistry—so that the choice of steel allows development of properties by subsequent heat treatment as per end use requirements. The reason being that chemistry is the main source of properties of steels—as chemistry influences the development of microstructure (refer Chapters 1 and 2)—and microstructure denotes the properties. Hence, chemistry being at the route of various properties that steels can develop or exhibit, designating steels as per chemistry is the most popular method, except in the case of structural steels where designation is generally as per 'yield strength'—which determines the load bearing capacity of the component. Thus, designation of steel varies with its class.

Many a time, steels would require some special property or attribute to meet the processing or end-use requirements. For example, a structural steel might require 'weldability' guarantee (by ensuring carbon-equivalent, CE, value) in view of extensive welding required to fabricate the component. Similarly, engineering steel might require guarantee in impact properties as the part is to be used under shock load or at lower than room temperature operations. Steel designation system should be able to take care of such requirements in a way that the system is understandable by all concerned with the steel.

Amongst the national standards, American standards issued by ASTM, and SAE and AISI (American Iron and Steel Institute) are very popular. ASTM mostly deals with the structural materials using steel designation system based on strength. The SAE and AISI mostly deal with engineering and special grades of steels, using chemistry based designation of steels. Typical SAE/AISI steel designation system has been illustrated in Table 6-3, where type or class of steel is represented by first two digits and the carbon content is represented by the next two digits. To illustrate further:

SAE-1040: 1040 is the grade name of the steel, where:

- First digit 1 represents for Plain carbon steel (not alloy steel),
- Second digit 0 for Plain carbon, unmodified, and
- Last two digits: 40 for steel containing 0.40%C (average).

SAE-4140: 4140 is the grade name of the steel, where:

- First digit 4 represents for steel containing Cr & Mo as alloys,
- Second digit 1 for concentration of major alloying element, in this case Cr (1.0%), and
- Last two digits for average %C in the steel (0.40%).

Note: A prefix like 'E' can be included in the grade to indicate the steelmaking route, where 'E' stands for 'Electric Arc Furnace' route or a suffix like 'H' can be included to indicate 'hardening grade steel', which is optional.

Table 6-3 illustrates the steel designation system of plain carbon and alloy steels using the traditional 4-digit system of AISI classification. The system follows designating steel as per chemistry. The system of steel designation by chemistry again follows two routes: one to designate the steel by letters or codes (using letters and numerals) and the other is by directly referring to the chemistry. Example of former designation method—based on letters and codes—is the steel SAE 4140—a chrome-moly steel with carbon 0.40% and Cr of 1.0% and molybdenum of 0.20%, whereas the same steel is designated by DIN as 42CrMo4—making a direct reference to carbon, chromium and molybdenum content in the steel. The suffix after the elements (e.g. 4 after Mo in this case) represents the percent of main alloying element, which is here 1.0% (4/4 parts) chromium (average). Similarly, Indian standards also designate steel by direct reference to chemical composition e.g. 45C8, 20MnCr1 etc. It is important to note that both systems indicate the chemistry of the steels.

The SAE/AISI of USA and BS of UK seem to prefer designating steels as per some letters and numerals, and others like Germany, Japan or India prefer to go by direct reference to the steel chemistry. For instance: plain carbon steel containing 0.15%C (average) is designated by different national standardisation agencies as:

Euro-Norm*	SAE (USA)	BS (UK)	DIN (Germany)	JIS (Japan)	IS (India)
C15D	1015 or	040A15	C15 or Ck15	S15, S15CK	15C6
	1018	080M15			18C8
		(Old: EN3B)			

* Euro-Norm standards of steel refer to European Union standards.

Designation by code is shorter and looks simple compared to, what could be lengthy chemistry based designation for some grade of steels, especially alloy steels. For example, 18-8 Cr-Ni grade austenitic stainless steel is designated by AISI as 304 (see Table 6-2), but the same steel is designated as x5Cr Ni 18 10 by DIN, German Standard, or 03KH18N11 by GOST, the Russian standard.

To overcome such difficulties, there is move now to go for uniform steel numbering system. The ISO in Europe and ASM in USA have taken the lead for promoting *unified numbering system* (UNS) of steels. Aim of the system is to adequately describe the class, attributes and chemistry of the steels. The 'unified numbering system' of ASM is aimed to unify the steel grades existing under the SAE and AISI standards to a common nomenclature, and covered under the 'standards for steel nomenclature of SAE', (SAE J 1086) and (ASTM E 527). Table 6-8 outlines the UNS coding system of materials (steel related codes have been highlighted in bold). The system allows sub-grouping of steels as per their characteristics within the number range provided for each class (refer the full list in the Appendix to Chapter 6).

The UNS numbering system of putting a 'letter' as prefix to the designated 5-digit number of the grade is intended to make identification of the material or steel type (i.e. class) easier. *However, the unified numbering system is an alloy designation system; it is not a specification of properties or uses.* It consists of a prefix letter and five digits that designate a material composition. For example, a prefix of S indicates stainless steel alloys, C for copper, brass, or bronze alloys, T for tool steels, G for AISI/SAE carbon and alloy steels (except tool steels), H for AISI/SAE 'hardening' grade steels, etc. A UNS number alone does not constitute a full material specification because it establishes no requirements for material properties, heat treatment, form or quality. It is not a standard of steel. Nonetheless, using the UNS numbering system, existing AISI/SAE steel grades can be easily converted to universal numbers for reference. For example, AISI/SAE 1020 steel can be designated as 'G 10200' in UNS system, where 'G' stands for carbon and alloy steels of AISI/SAE grades under the UNS classification system and additional zero at the end of steel grade to comply with 5-digit numbering. Similarly, popular stainless steel grades like AISI 304, 316 or 410 can be designated as S30400, S31600 and S4100 as per the UNS system.

Table 6-8. An outline of the 'unique numbering system' (UNS) of steels and other materials.

A00001 to A99999	Aluminium and aluminium alloys
C00001 to C99999	Copper and copper alloys
D00001 to D99999	**Specified mechanical property steels**
E00001 to E99999	Rare earth and rare earth like metals and alloys
F00001 to F99999	Cast irons
G00001 to G99999	**AISI and SAE carbon and alloy steels (except tool steels)**
J00001 to J99999	**AISI and SAE—Hardening grade steels**
J00001 to J99999	Cast steels (except tool steels)
K00001 to K99999	**Miscellaneous steels and ferrous alloys**
L00001 to L99999	Low-melting metals and alloys
M00001 to M99999	Miscellaneous nonferrous metals and alloys
N00001 to N99999	Nickel and nickel alloys
P00001 to P99999	Precious metals and alloys
R00001 to R99999	Reactive and refractory metals and alloys
S00001 to S99999	**Heat and corrosion resistant (stainless) steels**
T00001 to T99999	**Tool steels, wrought and cast**

Note: Steel related numbering systems have been highlighted by bold.

Despite this effort of the ASM or ASTM, most steel users continue to call the steel grades by the older country specific designation—e.g. SAE 1045 or EN-8—due to familiarity and popularity of names being used for decades. There is no single answer as to which system is better. Every system has some limitations. For example, the UNS system designates steels specified with mechanical property with the prefix 'D', to be followed by another 5-digit grade description (see Table 6-11). If any steel—e.g. structural steel, which is designated by strength—requires inclusion of 'yield strength' into the grade name, then it has to be accommodated within the 5-digits allotted to the steel. Thus, this numbering system will have limitations in describing steel in terms of all propriety/character.

At times, prefix letter in application based classification of some standards might interfere with the UNS system of prefixing letters for denoting another group of steel. For example, H for 'Hot work tool steel' in application based classification of ASTM interferes with and H for 'Hardening grade steels' in UNS (see Table 6-11). Similarly, designation of structural steels with prefix 'S' as per Euro-Norm can cause confusion, because as per UNS system, 'S' denotes heat and corrosion resistant steels. Some standards attempt to overcome this problem by using 'St' in place of 'S' for strength or 'E' for yield strength; Indian standards are an example of the latter. But, this requires conscious efforts. More about UNS system will be discussed as we go along with relevant discussions.

Table 6-9 presents the ISO system for designating 'Engineering steels'. As per ISO system—adopted by Euro-Norm—the steels are first classified into various groups as per purpose or use (see Table 6-9A). Steels are then designated with a prefix letter indicating their class, followed by a numbering system to represent their characteristics (see Table 6-9B).

Table 6-9A. ISO/EN classification of steels.

Designation	Meaning	Examples
S	Steels for structural steel engineering	S235JR; S355J0
P	Steels for pressure vessel construction	P265GH; P355M
L	Steels for pipeline construction	L360A; L360QB
E	Engineering steels	E295; E360
B	Reinforcing steels	B500A; B500B
Y	Prestressing steels	Y1770C; Y1230H
R	Steel for rails	R350GHT
H	Cold rolled flat rolled steels with higher-strength drawing quality	H400LA
D	Flat products made of soft steels for cold reforming	DC04; DD14

Classification of Steels (Application-based) by ISO

Steels under each of these designation and class are then taken up for assigning appropriate grade name (or number) following the ISO approved system. Table 6-9B shows one such numbering system of steel for engineering applications–designated E in Table 6-9A.

The system works on 'coding system' for accommodating different steel characteristics and properties, instead of direct reference to steel composition as used in DIN or IS system. UNS system of ASM has also evolved to accommodate different characteristics of steel by sub-grouping, within a group of material, which is nothing but coding again. This would be evident from table in the annexure, giving the sub-grouping of UNS system within the groups originally conceived by ASM. The system advanced by ISO or ASM has the common aim of making steel grades recognisable, but in practice, there seems to be much inertia stopping their wide uses amongst steel users. Nonetheless, these systems are gradually getting into the process of standardisation.

ISO or Euro-Norm system can apparently get into more details of steel characters than the UNS system, especially for structural grade steels where quite a few attributes like yield strength, state of heat treatment, weldability, etc. are required. The European standards (e.g. Euro-Norm: EN10025 for structural steels) designate structural steels by following ISO/EN system with more property details. For instance: S275J2 or S355K2W where 275 or 355 represent 'yield strength' in N/mm, J2 or K2 denotes material toughness by reference to 'Chirpy Impact test' and W denotes 'weathering steel'.

Therefore, numbering system of any standard has to make an attempt to designate steels giving as much indication of application properties as possible. Traditional ASTM/SAE and few other standards also follow the system of providing suffix to denote any special quality of the structural steels—such as heat treatment, welding etc.—which are apparently absent in UNS system. For example, notations for Normalising (N), Quenched and Tempered (Q), Weldability (W) etc. as suffixes are common to many traditional steel standards, indicating the state of materials for end applications. In contrast to these established systems, UNS system calls for prefixing of letter 'D' for steel type followed by another five digit number for denoting steel within which all conditions of supply have to be denoted.

Table 6-9B. ISO/EN steel designation system for Engineering Steels

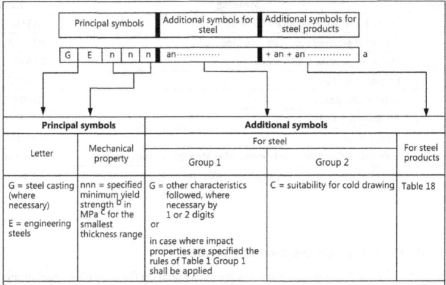

Principal symbols		Additional symbols		For steel products
		For steel		
Letter	Mechanical property	Group 1	Group 2	For steel products
G = steel casting (where necessary) E = engineering steels	nnn = specified minimum yield strength [b] in MPa [c] for the smallest thickness range	G = other characteristics followed, where necessary by 1 or 2 digits or in case where impact properties are specified the rules of Table 1 Group 1 shall be applied	C = suitability for cold drawing	Table 18

[a] n = numerical characters, a = alphacharacters, an = alphanumeric characters.

[b] The term "yield strength refers to upper or lower yield strength (ReH) or (ReL) or proof strength (Rp) or proof strength total extension (Rt) depending on the requirement specified in the relevant product standard.

[c] 1 MPa = 1 N/mm^2

Examples of steel names	
Standard	Steel name according to EN 10027-1
EN 10025-2	E205 E295GC E335 E360
EN 10293	GE240
EN 10296-1	E355K2

Note: There are similar tables for other classes of steels e.g. Structural steels, Reinforced steels, etc.(as per classification in Table 5-9A.

6.6. Equivalent Grade of Steels

Since steels are differently designated by different countries, despite their closeness in chemistry and properties, there is a need for steel charts showing comparative steel grades for easy reference. With increasing international trade and business, such need has become stronger. Table 6-10 gives comparative grades of different types of steels covered by various international standards This list is for some important grades only, and does not cover all types and grades. For further details or reference, one may consult Appendices at the end of the book and reference 8 from Further Readings section.

Table 6-10. Comparison of important grades of steels as designated by different standards bodies.

Euro-Norm Steel Number	EN Steel name	SAE Grade	DIN	BS 970	JIS
Carbon Steels					
1.1141 1.0401 1.0453	C15D C18D	1018	CK15 C15 C16	040A15 080M15 080A15 EN3B*	S15 S15CK S15C
1.0503 1.1191 1.1193 1.1194	C45	1045	C45 CK45 CF45 CQ45	060A47 080A46 080M46	S45C S48C
1.0726 1.0727	35S20 45S20	1140/1146	35S20 45S20	212M40 En8M*	
1.0715 1.0736	11SMn37	1215	9SMn28 9SMn36	230M07 En1A*	SUM25 SUM 22
1.0718 1.0737	11SMnPb30 11SMnPb37	12L14	9SMnPb28 9SMnPb36	230M07 Leaded En1A* Leaded	SUM22L SUM23L SUM 24L
Alloy Steels					
1.7218		4130	25CrMo4 GS-25CrMo4	708A30 CDS110	SCM420 SCM430 SCCrM1
1.7223 1.7225 1.7227 1.3563	42CrMo4	4140/4142	41CrMo4 42CrMo4 42CrMoS4 43CrMo4	708M40 708A42 709M40 En19* En19C*	SCM440 SCM440H SNB7 SCM4M SCM 4
1.6582 1.6562	34CrNiMo6	4340	34CrNiMo6 40NiCrMo8-4	817M40 En24*	SNCM447 SNB24-1-5
1.6543 1.6523	20NiCrMo2-2	8620	21NiCrMo22 21NiCrMo2	805A20 805M20	SNCM200(H)

Euro-Norm Steel Number	EN Steel name	SAE Grade	DIN	BS 970	JIS
Stainless Steels (UNS)					
1.4310	X10CrNi18-8	301			
1.4318	X2CrNiN18-7	301LN			
1.4305	X8CrNiS18-9	303	X10CrNiS18-9	202S21 En58M*	SUS 303
1.4301	X2CrNi19-11 X2CrNi18-10	304	X5CrNi18-9 X5CrNi18-10 XCrNi19-9	304S15 304S16 304S18 304S25 En58E*	SUS304 SUS304-CSP
1.4306	X2CrNi19-11	304L		304S 11	SUS304L
1.4311	X2CrNiN18-10	304LN			
1.4948	X6CrNi18-11	304H			
1.4303	X5CrNi18-12	305			
1.4401 1.4436	X5CrNi-Mo17-12-2 X5CrNi-Mo18-14-3	316	X5CrNiMo1712 2 X5CrNiMo17 13 3 X5CrNiMo19 11 X5CrNiMo18 11	316S 29 316S 31 316S 33 En58J*	SUS316 SUS316TP
1.4404	X2CrNi-Mo17-12-2	316L		316S 11	SUS316L
1.4406 1.4429	X2CrNi-MoN17-12-2 X2CrNi-MoN17-13-3	316LN			
1.4571		316Ti	X6CrNiMoTi1712	320S 33	
		410			
1.4016		430	X6Cr17	430S 17	SUS430
Tool Steels					
1.2363	X100CrMoV5	A-2	X100CrMoV51	BA 2	SKD 12
1.2379	X153CrMoV12	D-2	X153CrMoV12-1	BD 2	SKD 11
1.2510		O-1	100MnCrW4	BO 1	

Note: Comparison is only by chemistry. IS specification nomenclature for corresponding grades closely follows the DIN standard; and hence omitted here.

This international comparison chart is for illustration of typical numbering system followed by different standards—some number them by directly referring to the chemical composition and some designate them by codes. Both systems have merit, but do not make the system of identifying comparable grades any easier. The comparison table establishes the need for a simple but less complex system of steel designation. Uniform numbering system introduced by ISO and UNS of ASM are efforts in this direction, but as it stands now, the system is far from simple. The ISO system of designating steels for 'engineering' applications has been shown in Table 6-9. The system works on 'coding system' for accommodating different steel characteristics and properties, instead of direct reference to steel composition as found in DIN or IS system discussed. The UNS system of ASM has also evolved itself to accommodate different characteristics of steel by sub-grouping within a group of material, which is nothing but coding again. More detailed steel comparison charts of different countries have been attached at the end of this book.

The foregoing discussion on steel specifications and standards unambiguously illustrate—the importance of steel classification, designation and standardisation as means for effective selection and utilisation of steels. Steel standards are at the core of making right choice of steels for applications for industrial uses. Therefore, knowledge about how standards can help and information about what standards are available covering the different types of steel is important. In view of this knowledge, annexure to this chapter covers some more information about steel standards and international grades. A reference to these tables in the annexure will throw further light on the uses and utility of steel standards and specifications.

SUMMARY

1. The chapter discusses the methods and systems of steel classification, specification and standardisation, including the system of assigning the grade names to different types of steels. Understanding of different types of steel classification and standardisation systems is essential for choosing the right steel for right purpose. The purpose of such classification, specification and standards has been highlighted and illustrated with examples in this chapter.

2. Throughout the chapter, large number of working examples, illustrative grade tables, specifications and grading system and chart for equivalent grades has been provided.

3. Discussions point out with examples that one of the objectives of steel standards is to ensure that it is fit for use. Since there could be multiple uses of similar steels for different purposes, there would be multiple coverage of such steel under different standards in order to lead to the correct steel grade and quality. Discussions and illustrations in the chapter, therefore, have been focused on enabling the users of steel to zero-in to the right grade under right standard for a given application.

4. Different methods of classification of steels have been described and discussed; pointing out that classification by chemical composition, application or properties is by far the most popular method. Main points of this classification of systems have been highlighted with examples.

5. The purpose and methods of standardisation have been discussed and illustrated, emphasising that while classification of steels lays the foundation of most standards, the purpose of standards is to provide as much details of the steel quality as necessary for guiding the users of steel for choosing the right type for right purpose. Thus, steels of similar chemistry but for different applications would require coverage under different standards for guiding the users.

6. The task of standardisation is, therefore, classifying steels according to its use, purpose or chemistry, designating steels as appropriately as possible for common understanding, and formulating respective standards with enumeration of as much quality specifications as is a necessary for the right uses of steels.

7. However, whatever could be the steel classification or standard, primary requirement of steel specification is its chemical composition and properties according to which steel should be made and certified. This necessitates formulation of the system by which steel can be named or designated

for common understanding. This is done by grading the steels following a steel designated system based on chemistry or properties.

8. The chapter also discusses the efforts for unified numbering system of steels by AISI and ISO for the common understanding of all. However, the system is yet to become popular universally— instead the older designation of steels under the popular international standards like the AISI, BS or DIN rule the current practices by which steels are specified.

9. Finally, in view of different steel designating systems of different countries, the chapter provides some insight into how similar grades of steel compare with different standards of other countries.

FURTHER READINGS

1. UK Steel. *Steel Specification Handbook,* 13th edition. London, 2009.
2. Bringas, John E. *Handbook of Comparative World Steel Standards,* 3rd edition. ASTM (DS67B): Philadelphia, 2004.
3. ASTM, USA. *ASTM Standards for Steel,* Vol. 8. Philadelphia, 2012.
4. British Standard Institution, UK. *Summary of British and American Standard Specifications.* London, 1988.
5. American Society for Testing of Materials, USA. *Book of ASTM Standards.* Philadelphia, 2004.
6. British Steel Corporation. *Iron and Steel Specifications.* London, 1986.
7. Nayar, Alok. *The Steel Handbook.* New Delhi: McGraw-Hill, 2001.
8. ASM International, USA. *Worldwide Guide to Equivalent Irons and Steels,* 5th edition. 2006.

APPENDIX TO CHAPTER 6

SAE (American) and BS (British Standards) are the most popular standards world over; covering almost all the grades of steels used anywhere. Some details of steel designation system of these two standardisation bodies have been given here; corresponding designation of other standards for similar steel can be obtained from the 'Equivalent chart'.

Table 6A-1. Conventional SAE classification and designation system of steel grades and their sub-divisions.

SAE Designation	Type
	Carbon Steels
10xx	Plain carbon (Mn 1.00% max)
11xx	Resulfurized
12xx	Resulfurized and rephosphorized
15xx	Plain carbon (Mn 1.00% to 1.65%)
	Manganese Steels
13xx	Mn 1.75%
	Nickel Steels
23xx	Ni 3.50%
25xx	Ni 5.00%
	Nickel-Chromium Steels
31xx	Ni 1.25%, Cr 0.65% or 0.80%
32xx	Ni 1.25%, Cr 1.07%
33xx	Ni 3.50%, Cr 1.50% or 1.57%
34xx	Ni 3.00%, Cr 0.77%
	Molybdenum Steels
40xx	Mo 0.20% or 0.25% or 0.25% Mo & 0.042 S
44xx	Mo 0.40% or 0.52%
	Chromium-Molybdenum (Chrome-moly) Steels
41xx	Cr 0.50% or 0.80% or 0.95%, Mo 0.12% or 0.20% or 0.25% or 0.30%

(Continued)

SAE Designation	Type
Nickel-chromium-molybdenum Steels	
43xx	Ni 1.82%, Cr 0.50% to 0.80%, Mo 0.25%
43BVxx	Ni 1.82%, Cr 0.50%, Mo 0.12% or 0.35%, V 0.03% min
47xx	Ni 1.05%, Cr 0.45%, Mo 0.20% or 0.35%
81xx	Ni 0.30%, Cr 0.40%, Mo 0.12%
81Bxx	Ni 0.30%, Cr 0.45%, Mo 0.12%
86xx	Ni 0.55%, Cr 0.50%, Mo 0.20%
87xx	Ni 0.55%, Cr 0.50%, Mo 0.25%
88xx	Ni 0.55%, Cr 0.50%, Mo 0.35%
93xx	Ni 3.25%, Cr 1.20%, Mo 0.12%
94xx	Ni 0.45%, Cr 0.40%, Mo 0.12%
97xx	Ni 0.55%, Cr 0.20%, Mo 0.20%
98xx	Ni 1.00%, Cr 0.80%, Mo 0.25%
Nickel-molybdenum Steels	
46xx	Ni 0.85% or 1.82%, Mo 0.20% or 0.25%
48xx	Ni 3.50%, Mo 0.25%
Chromium Steels	
50xx	Cr 0.27% or 0.40% or 0.50% or 0.65%
50xxx	Cr 0.50%, C 1.00% min
50Bxx	Cr 0.28% or 0.50%
51xx	Cr 0.80% or 0.87% or 0.92% or 1.00% or 1.05%
51xxx	Cr 1.02%, C 1.00% min
51Bxx	Cr 0.80%
52xxx	Cr 1.45%, C 1.00% min
Chromium-Vanadium Steels	
61xx	Cr 0.60% or 0.80% or 0.95%, V 0.10% or 0.15% min
Tungsten-Chromium Steels	
72xx	W 1.75%, Cr 0.75%
Silicon-Manganese Steels	
92xx	Si 1.40% or 2.00%, Mn 0.65% or 0.82% or 0.85%, Cr 0.00% or 0.65%
High-Strength Low-alloy Steels	
9xx	Various SAE grades
xxBxx	Boron steels
xxLxx	Leaded steels

Table 6A-2. Corresponding British standards system of designating steels (replacing old En-designation).

Prefix	Type/Class	Example
01–09 series	Plain carbon steels without alloying	080M40 (old En8)
10 series	Carbon-manganese steels	150M28 (En14B)
20 series	Carbon Free-cutting steel/Alloy Free-cutting steels	212M36 (EN8M)
50 series	Carbon-chromium steels	530M40 (En18)
60 series	Manganese-moly steels	605A32 (En16B)
70 series	Chrome-moly steels	708M40 (En19A)
80 series	Chrome-nickel-moly steels	817M40 (En24)
90 series	Manganese-chrome-nickel-moly Or Aluminium alloyed steels	905M39 (En41B)

Table 6A-3. Comparative Indian and International standards (application/purpose-based standards).

S.No.	Product	Indian	British	German	American	Japanese	Russian
1	Steel billets, blooms and slabs for rerolling into structural steel (standard quality)	IS:2830					
2	Steel billets, blooms and slabs for rerolling into structural steel (ordinary quality)	IS:2831					
3	Structural Steel (standard quality) bars, sections, plates etc.	IS:226	BS:15	DIN:17100	ASTM:A-75	JIS:G:3101	GOST-380
4	Structural steel (Weldable Quality) Bars, Sections, plates etc.	IS:2062	BS:2762 BS:4360	DIN:17100	ASTM:A-373, 36,283	JIS:G:31060 G:3113,3101	GOST-4637, 915
5	High tensile Structural Steel	IS:961	BS:548 BS:968 BS:4360	DIN:17100	ASTM: A-242 440,94,441	CORTEN-A JIS:A:5528	GOST-5058
6	Medium and high Strength Weldable Structural steel	IS:8500	BS:4360	DIN:17100	ASTM:A- 242 A440,441,514	JIS:G:5528	GOST-5058
7	Structural steel—ordinary quality	IS:1977					
8	Hot rolled bars for production of bright bars	IS:7283					
9	High Strength deformed steel bars and wires for concrete reinforcement	IS:1786	BS:4461	DIN:1045, 488	ASTM:A-615	JIS:G:3112	
10	Mild steel wire rods for general engineering purposes	IS:7887	BS:1052	DIN:278	ASTM:A-510	JIS:G:3505 3506	
11	Carbon Steel wire rods	IS:7904	BS:2763	DIN:278	ASTM:A-603	JIS:G:3521 3525	

S.No.	Product	Indian	British	German	American	Japanese	Russian
12	Ship building quality structural steel	IS:3039 IS:2985	L LOYDS-GR A,B,C,D,E	DIN:1016	ASTM:A-131		GOST-5521
13	Carbon steel blooms, bars, billets and slabs for forging carbon constructional steels	IS:1875 IS:4432 IS:5517 IS:4369	BS970: EN3S,4, 5,8,9,32, 43A,438	C-10,C-15, C-25,C-45, CK-45,C-53 (DIN:17222, 7111,1613)	SAE1010 SAE1012 SAE1015 SAE1021 SAE1025 SAE1030 SAE1040 SAE1045 SAE1050 SAE1055 (ASTM A-73, 235,236,273)	JIS:G:3251, 320,3505	
14	Mild steel and medium tensile steel bars and hard drawn steel wires for concrete reinforcement	IS:432	BS:4449	DIN:488	ASTM: A61,A15	JIS:G:3112	
15	Rivet bars for structural purposes (upto 40 mm diam.)	IS:1148	BS:548	DIN:17111	ASTM:A31	JIS:G:3104	
16	High tensile steel rivet bars for structural purpose	IS:1149 BS:1502	BS:1113 DIN: 59130	ASTM: A-502	JIS:G:3104		
17	Carbon steel bars for production of mechined parts for general engineering purpose	IS:2073	BS:970 (EN-3,4, 5,6,8, 9, 14,15,41, 43,49)	DIN:17210, DIN:1623	AISI:1010, 1012,1015, 1017,1020, 1023,1025, 1030,1033, 1035,1038	JIS:G:3507	
18	Hot rolled mild steel, medium tensile steel and high yield strength steel, deformed bars for concentrate reinforcement	IS:1139					
19	Bars and rods for cold heading/cold extrusion	IS:11169 IS:2255	BS:3111	DIN:17111, DIN:1654	AISI:1008, 1010,1012, 1015,10B21, 15B25,10B21, 15B25,10B30, 10B38,15B41, 1541, ASTM:A-545	JIS:G:3507	
20	Billets,Bars and Sections for boilers	IS:2100					
21	Mildsteel for metal arc welding electrode core wire	IS:2879 IS:814				JIS:G:3523	
22	Chequered plates	IS:3502					
23	Hot rolled sheets/strips Hot rolled sheets/strips for steel tubes &pipes	IS:1079 IS:10748 IS:11513	BS:1449	DIN:1623	ASTM:A-569	JIS:G:3202	
24	Galvanised steel sheets (plain and corruged)	IS:277	BS:2989 BS:3083	DIN:59231	ASTM:A-163	JIS:G:3202	

S.No.	Product	Indian	British	German	American	Japanese	Russian
25	Hot rolled sheets for LPG cylinders	IS:6240 IS:10787	BS:1501		ASTM:A-621		
26	Black plates	IS:597	BS:2920		ASTM:A 625	JIS:G:3303	
27	Forging quality steel—Alloy constructional steel	IS:4368 IS:5517 IS:4432	EN-15,16, 17, 18,19, 19C,24, 207,29A,29B, 14A/B,206, 12,22,33,13, 40,41,34,35, 30,39,351, 352,25, 26, 100,110	DIN:40MN4 34CR4 37CR4 41CR4 42CRMO4 50CRV4 20MN6 28MN6 36MN5	SAE:4150, 4340,1524, 1536,1541, 5140,5115, 5120,4023, 4027,4032, 4042,4140, 4620,3120, 4340,8640		
28	Case hardening steels (cycle chain rivet wire)	IS:4432	EN 32B,36C, 353,354,355, 361,362	15CR3 16MnCR5 20MnCr5 15CrNi6 17CrNiMO6 20CRMO4 25CRMO4	SAE:8620, SAE:8615		
29	Seamless bars				ASTM:A106 A161,A181, A182, A200, A209,A213		
30	Ball bearing steels including wires	IS:4398 103CR1, 103CR2, 102CR2Mn70	EN 31	100CR6	SAE:52100 ASTM:A-295		
31	Spring Steel	IRSM:24 IS:3195 55Si7 60Si7 50Cr4 50Cr4V2 60Cr4V2 50Criv23 55siMn90	EN 45,45A, 42,44,47	C-75, MK-101 65Si7, 55Sc7 50CrV4	SAE:1075, 1085,1095, 9255,9260 6150		
32	Free cutting steel	IS 4431 13S25 40S18 10S11 10C8S10, 14C14S14 25C12S14 40C10S18 11C10S25 40C15S12	EN: 1A 8M 32M	DIN:1052	SAE:1111		
33	Low Alloy Steel Wire Rods		EN1A-pb		SAE:8620		
34	Electrical Steel Sheets	IS:648					
35	Forged Rounds	IS:2004					
36	Closed die forged blanks and rolled rings	IS:5517					

S.No.	Product	Indian	British	German	American	Japanese	Russian
37	Welded Tubes seamless tubes for general application	IS:1239 IS:4923 IS:1161	BS:1387 BS:806				
38	Precision tubes for boiler and super heater purpose	IS:1914 IS:2416	BS:3059/ 3606 BS:806	DIN:17175 DIN:17177	ASTM:A-178, A-179,A-210, A-214,A-192, A-216	JIS:G:3461	
39	Precision tubes for line pipe	IS:1978 IS:1979			API-5L,5CT		
40	Precision tubes for bicycles and allied purpose	IS:2039	BS:1717				
41	Precision tubes for automotive purpose	IS:3074	BS:980 BS:6323	DIN:2391pt2	SAE:1040		
42	Precision tubes for Mechanical and General Engineering Purpose	IS:3601	BS:1775 BS:6323				
43	Precision tubes for furniture	IS:7138					
44	Precision tubes for transformers	IS:8036					
45	Precision tubes for Heat Exchange and condenser tubes				ASTM:A-214, A-179,A-200		
46	Precision tubes for structural purpose					JIS:G-3445	
47	Seamless tubes for oil industry	IS:3589			API-5L,5CT		
48	Seamless tubes for rotary core drilling accessories		BS:4019 PTI				
49	Seamlesstubes for water wells	IS:4270					
50	Seamless tubes for roller conveyors	IS:9295					
51	Seamless tubes for water, gas, steam services	IS:1239	BS:1387		ASTM:A-53		
52	Seamless tubes for high temperature services		BS:3602		ASTM:A-106		
53	Seamless tubes for low temperature services				ASTM:A-333		
54	Cold Rolled sheets/strips	IS:4030	BS:1449	DIN:1623	ASTM:A-366	JIS:G:3141	GOST-8596
55	Agrico Products-powarah-picks and Beaters, point and tee end, shovels, crow bar	IS:1759 IS:273 IS:274 IS:704					
56	Rolling and cutting tolerances for hot rolled steel products	IS:1852 IS:1730 IS:1731 IS:1732 IS:808 IS:1864 IS:3954	BS:4	DIN:1014 DIN:1035 DIN:1013 DIN:1017 DIN:1025 DIN:1026 DIN:1028 DIN:1029	ASTM:A-709	JIS:G:3191 JIS:G:3192	

S.No.	Product	Indian	British	German	American	Japanese	Russian
57	Dimensional Tolerances for carbon and alloy constructional steel	IS:3739					
58	Axle for carriage wagon	IRS:R16					
59	Axle for locomotive and tender	IRS:R:18					
60	Axle for Diesel Electric Locomotive (DMU)	IRS:R43					
61	Tyre for carriages and wagons	IRS:R15					
62	Wheels and Axles for carriage and wagon	IRS:R19					
6 3	Pig Iron for steel making and foundry purpose	IS:13502			ASTM:A-43	JIS:G2202	
64	Corrosion resistant structural steel	IRS:M41					
65	Fish plates to be rolled from billets conforming to M37/64	IRS:T1					
66	Flat bottom rails	IRS:T12					
67	Hot rolled steel plates, strips and flats for flanging and forming operation	IS:5986					
68	Electrolytic Tin plates	IS:597 IS:1993	BS:2920	DIN:1541	ASTM:A-624 ASTM:A-626		
69	Steel plates for boilers	IS:2002 IS:2041 IS:2100	BS:1501	DIN:17155	ASTM:A-285	JIS:G:3103	GOST:5520
70	High tensile steel bars	IS:2090					
71	Steel for manufacture of volute and helical springs (for railway rolling stock)	IS:3195					
72	Steel for manufacture of laminated springs (for Railway rolling stock)	IS:3885					
73	Sponge Iron/ Direct reduced iron						
74	Stainless steel Austenetic Martenesetic Wrought Ferritic				AISI-201,202, 301,348, AISI-403,410, 414,440,501 AISI-405, 430,446		
75	Bright Bar wire and Industrial chain		EN51,EN52 En18D,EN24	16 MnCr5 4140,8620	SAE:52100,		
76	Flat/Round cable armour wire/tape	IS:3795 IS:3975	BS:1441 BS:1442		ASTM:A411		
77	ACSR (Aluminium cable steel reinforced) core wire single wire and standard wire	IS:398	BS:4565 BS:215				

S.No.	Product	Indian	British	German	American	Japanese	Russian
78	Galvanised Wire	IS:279			ASTM:A641	JIS:G:3432	
79	Galvanised steel ropes/strand	IS:1855 IS:1856 IS:2114	BS:183, 236,330	DIN:48200 DIN:48201 DIN:48202	ASTM:A475, 363	JIS:G:3536	
80	Pins/clips/staples-copper coated and galvanised	IS:4224					
81	Pre-stressed concrete wires and strands	IS:6003 IS:6008	BS:2691		ASTM:A421, 416	JIS:G:3536	
82	Wires for Auto/Cycle tyre bead	IS:4824					
83	Umbrella Rib wire	IS:4223				JIS:G:3521	
84	Cycle spoke wire	IS:6902	BS:2453				
85	High Tensile bolt wire		BS:3111	DIN:1654	ASTM:A546,547 AISI:4140, 1541,4137, 8620,8735		
86	Wires for Screws, rivets low tensile bolts etc.				AISI:1008, 1010,1051		
87	Nail wire	IS:280					
88	Wires for springs	IS:4454	BS:1420				

CHAPTER 7

Properties and Grades of Steels

Structural, Engineering and Stainless Steels

The purpose of this chapter is to first highlight the influence or effect of different alloying elements in steel, including carbon and other microstructural factors that contribute to the build-up of steel properties, and then to relate them to the properties of structural, engineering and stainless steel for industrial applications. Discussions about the properties of structural, engineering and stainless steels have been made with references to national and international standards, as far as possible, in order to re-emphasise the role of 'steel standards' for providing a common platform for steel selection and application.

7.1. Introduction

An understanding of the specifications and standards is necessary for focusing on the right group of steel for selection, but that alone does not ensure efficient and effective selection and application of steels. The ultimate choice of steel rests on how well the steel fits to the purpose and fulfils the required end-use properties. Properties can be of two types: (a) manufacturing properties i.e. how well the steel fits the manufacturing and processing needs and (b) end-application properties i.e. how well the steel meets the service related properties and can endure the service conditions. The former involves forming, welding, forging, machining, heat treating etc. and the latter involves requirement of hardness, strength (tensile and yield strength), elongation, ductility/toughness, impact toughness, oxidation and corrosion resistance, etc.

Steel classification and standards (discussed in the previous chapter) help in zeroing-in on the class and grade of steels for choice, but the ultimate choice should be made by considering how well the steel can develop relevant properties (of both types) and their cost and availability. While role of cost and quality in the selection and application of steel has been discussed in Chapter 9, this chapter discusses about the influence of different chemical constituents and related microstructural features that are important for the properties of steels for different classes of application, viz. structural, engineering and stainless steels. The chapter attempts to highlight the metallurgical approach and apply logic in the selection of steels for structural, engineering and stainless steel applications by analysing the character and properties required for functional performance in these grades. The chapter discusses the properties of steel grades in various categories with reference to their respective standards.

7.2. Influence of Carbon and Alloying Elements on Properties of Steels

Carbon is the primary constituent of steel, after iron. Steels of all description—barring interstitial free steel—contain carbon, which can normally range from 0.02% to 2.0%. Carbon up to 0.85% increases hardness (or strength) of steels, but any increase of carbon thereafter tends to lower the strength (see Fig. 2-14 in Chapter 2). Reason for the tendency of decreasing strength with increasing carbon beyond 0.85% in wrought steel is due to 'carbide' formation in the structure. Similar trend is also observed in hardened steel (see Fig. 2-12) which depicts the variation of hardness in martensitic and bainite structure with carbon. Figure 7-1 shows a similar plot of hardness versus %carbon (%C) in hardened steel—without reference to structure type—which also confirms tapering off of hardness with increasing carbon. Thus, the phenomenon of optimum hardness (or strength) in steel with increasing carbon upto about 0.80–0.85%C is well established—though the reason for such behaviour in wrought steel and hardened steel could be different.

However, while carbon (upto eutectoid carbon level of 0.80–0.85%C) increases the strength of steel, ductility/ductility related properties, and the weldability, sharply decreases with increasing carbon. Hence, there are few distinct trends of using carbon levels in steel for different purposes, properties and applications. They include:

1. Lowering of carbon, as in low or extra low carbon steel, where high ductility of the steel is the primary concern for critical forming and deep drawing applications.
2. Adjustment of carbon, which is necessary for improving structure and strength in low carbon steel where ductility is high but strength is low. This type of steel is often used for sheet metal and structural applications requiring optimum strength with ductility, but with care for ensuring good weldability.
3. Adjustment of carbon is also needed for improving strength and structure in medium carbon steel for optimum combination of strength and ductility/toughness, to be used with or without heat treatment. Steel is used for various machine parts for engineering applications, with or without heat treatment, is a common example of such steel—e.g. SAE 1035, 1040, 1045 grades of steel.
4. Carbon for improved hardness level and hardenability in steel for hardening to the martensitic structure; to be used along with or without other alloying metals for required level of strength and toughness—e.g. SAE 1045 plain carbon steel, SAE 5140 and 4140 grades of alloy steels. Steel used for most engineering components, such as shafts, gears, cams, levers, etc. are common examples of this type of steel. The higher the dynamic nature of the applied load, higher is the toughness sought for in such steel by adjustment of alloying elements with carbon.

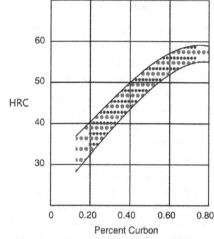

Figure 7-1. An illustration of how hardness in hardened steels increases with increase in carbon percent.

5. High carbon steel where the requirement is either high yield strength for spring type applications (see tensile curve for high carbon steel in Fig. 5-2) or high surface hardness with hard carbide phases for wear and abrasion resistance. This type of steel is used mostly after heat treatment of some kind or other.

These are a few main trends of how carbon level should be maintained in the selected steel, but there could be other special uses of carbon in steels—like the low-carbon high alloy stainless steel, high carbon high alloy heat resistance steel, creep steel, low-carbon carburising grade steel, etc. Thus, level of carbon in steel is used with discretion in order to produce structure and properties as appropriate for the end uses and applications, especially with care for optimising strength and toughness/ductility.

In general, contributions of carbon to steel properties come from three sources: (a) solid solution strengthening of steel, (b) structural changes in steel with increasing carbon under equilibrium or normal cooling (refer austenite decomposition in Chapter 2) and (c) contribution of carbon to martensite formation by quenching where carbon content is adjusted for increasing the level of hardness, on the one hand, and also for contributing to the depth of hardness i.e. hardenability, on the other (refer chapter 8 for heat treatability of steel). Evaluation of exact contribution of carbon for each of these factors in the development of properties of steel is difficult, especially due to variation of grain sizes in steel and the combined effect of several elements present in the steel. Nonetheless, there are empirical formulas to provide some ball-path figure for calculating response of steel to heat treatment under identical conditions or the arithmetical mean of tensile strength of rolled/wrought steel. Table 7-1 provides some factors for carbon and other alloying elements for calculating 'hardness factor' of steel grades containing the same elements—this is not the Grossman formula for calculation of hardenability of steel, as discussed in chapter 8. The values are relative index of hardness factor, and may not be very accurate when used for comparing steel of one type with the other. The table indicates that the effect of carbon is nearly 4 times that of Mn and 6 times that of Cr for developing hardness after heat treatment.

Carbon is one part of the steel; the other part is the presence of residuals—e.g. S, P, N, Al, Si, etc.—and different alloying elements. Table 7-1 indicates that not all elements present in steel contribute equally for hardness (or strength); in effect, each element has its own foot-print and influence. Combined effect of other alloying elements may change the course of different structure formations and the result. Hence, for all practical purpose of heat treatment, especially for hardening operation, heat treatability of steel and steel composition by following 'Jominy hardenability test' is preferred, which has been described in Chapter 8.

Nonetheless, understanding of individual contribution of alloying elements in the development of application specific properties of steels is necessary for choosing the right grade and composition of

Table 7-1. Effect of carbon and other alloying elements for increasing 'hardness factor' in heat treatable steels.

Carbon	0.01% = 30	Nickel	0.01% = 4
Manganese	0.01% = 8	Chromium	0.01% = 5
Phosphorus	0.001% = 4	Vanadium	0.01% = 20
Sulphur	0.001% = 1	Molybdenum	0.01% = 16
Silicon	0.01% = 5	Tungsten	0.01% = 4
		Copper	0.01% = 4

Source: Heat Treating Data Book, 7th edition. SECO/Warwick Corporation, Meadville, PA, USA.

steel for different applications. Table 7-2 presents the summary of principal effects of different alloying elements in steel for a more accurate guide to the design and selection of steels. The table includes elements like Al, Si, S, P, etc. which are generally present in steels as residuals. But, at times, some residuals like Al and Si can be intentionally alloyed with steel for developing special properties—such as 'nitriding steel containing Al' or 'electrical steels containing Si' (transformer grade steels).

Table 7-2 shows the basic effects of alloying elements on various properties of steel. The table also indicates the intensity of effects by multiple arrows. Influence of alloying elements on basic properties of steel is due to their role in producing different combinations of microstructures. For example, most allying elements, including carbon, tend to form carbides in addition to changes in main microstructure containing ferrite, pearlite, bainite or martensite as per treatment. The exceptions to this rule are nickel and manganese which when present in higher quantities promote austenite, which is the basis of getting austenitic stainless steel.

Table 7-2. Commonly observed effects of various alloying elements on different properties of steels. These effects can change when more than one element is present and have a combined synergistic effect.

The Effect of Alloying Elements on the Properties of Steels													
Alloying element	Mechanical properties							Hight temperature stability	Cooling rate	Carbide formation	Resistance to wear	Forgeability	Machinability	Resistance to corrosion
	Hardness	Strength	Yield point	Elongation	Reduction of area	Impact value	Elasticity							
Aluminium	—	—	—	—	↓	↓	—	—	—	—	—	↓↓	—	—
Chromium	↑↑	↑↑	↑↑	↑↑	↓	↓	↑	↑	↓↓↓	↑↑	↑	↓	—	↑↑↑
Cobalt	↑	↑	↑	↓	↓	↓	—	↑↑	↑↑	—	↑↑↑	↓	~	—
Copper	↑	↑	↑↑	~	~	~	—	↑	—	—	—	↓↓↓	~	↑
Manganese in pearlitic steels	↑	↑	↑	~	~	~	↑	~	↓	~	↓↓	↑	↓	—
Manganese in sustenitic steels	↓↓↓	↑	↓	↑↑↑	~	—	—	—	↓↓	—	—	↓↓↓	↓↓↓	—
Molybdenum	↑	↑	↑	↓	↓	↑	—	↑↑	↓↓	↑↑↑	↑↑	↓	↓	—
Nickel in pearlitic steels	↑	↑	↑	~	~	~	—	↑	↓↓	—	↓↓	↓	↓	—
Nickel in austenitic steels	↓↓	↑	↓	↑↑↑	↑↑	↑↑↑	—	↑↑↑	↓↓	—	—	↑↑↑	↓↓↓	↑↑
Phosphorous	↑	↑	↑	↓	↓	↓↓↓	—	—	—	—	—	↓	↑↑	
Silicon	↑	↑	↑↑	↓	~	↓	—	↑	↓	↓	↑↑↑	↓	↓	—
Sulphur	—	—	—	↓	↓	↓	—	—	—	—	—	↓↓↓	↑↑↑	↓
Tungsten	↑	↓	↑	↓	↓	~	—	↑↑↑	↓↓	↑↑	↑↑↑	↓↓	↓↓	—
Vanadium	↑	↑	↑	~	~	↑	↑	↑↑	↓↓	↑↑↑↑	↑↑	↑	—	↑

Symbols: Upward arrow = increase; downward arrow = decrease; (~) means constant; (–) means no influence or unknown; several arrows means intensified action.

Table 1-1 in chapter 1 illustrated how different structures can influence the properties of steel. What actually happens is that it is not the alloying elements but the structure they help to produce that causes or influences the steel properties, except for chemical properties—like corrosion and oxidation resistance, which are composition specific and independent of structure.

It could be observed from Table 7-2 that the effects of alloying elements generally follow the following lines:

1. All alloying elements increase the hardness and strength of steel; with some having more effect than others—e.g. Cr has more intense effect than other (not including the effect of carbon).
2. If the requirement is higher yield strength—as in structural steels—presence of Si, Cu and Cr are helpful.
3. For improved ductility or toughness, presence of Mn and Ni in steel is desirable.
4. Phosphorus drastically reduces impact strength value of steels; and hence needs to be carefully controlled.
5. Co, Mo, V and W increases high temperature strength of steels. These elements also form strong carbides, promoting wear and abrasion resistance properties in steel.
6. Sulphur has strong effect in improving machinability of steel, but reduces impact toughness. Hence, any extra addition of sulphur in steel for improved machinability needs to be taken care of by adding elements which preferentially combinewith sulphur to form unharmful sulphur compound e.g. Mn that forms MnS.
7. Resistance to corrosion is improved by Cr, Cu and Ni (This is a chemical property and it is due to electro-chemical nature of these elements).

These are the instances of direct influence of alloying elements on properties. But, there are other structural features and attributes like the grain size, inclusion, carbide types and size, etc. which work along with the effect of alloys and either favourably or unfavourably modify the final outcome. The role of grain size and inclusions in steel has been discussed in some detail in Chapter 3, and their influence will be considered during the current discussion as well.

Table 7-2 is a qualitative list of influence of alloying elements on steels. Table 7-3 lists the presence and percentage of various alloying elements in steels and their principal effects. These two tables are complimentary to each other for arriving at the right choice of alloying. Between Tables 7-2 and 7-3, influence of various added elements on steel, including carbon and some residual elements, can be worked out for the purpose of selecting the appropriate steel composition. Effects of various elements shown in Tables 7-2 and 7-3 are applicable to all grades of steels—be that structural, engineering, or stainless steel. From these tables, chemical composition required for a function can be worked out and taken as a reference for selecting steel from the available standards. For example, if the application involves a mild atmospheric corrosive environment—but not demanding use of expensive stainless steel—then some Cu can be added to the steel as per Table 7-3 for improving the atmospheric corrosion resistance. Similarly, if the application involves cold bending and forming—as in sheet-metal forming—not only is leaner chemistry with low carbon required, but also the steel should be free from nitrogen to avoid reduction of ductility by 'ageing' effect. If the steel is required to be hardened by quenching, Tables 7-2 and 7-3 show that the steel must have enough of carbon plus alloying elements like Cr, Mn, etc. which improve hardenability.

However, as mentioned earlier, properties of steels are also influenced by grain size and inclusions. Hence, their effects on the properties of steels should have to be also considered along with the effect of added elements. Effects of grain size and inclusions have been discussed in Chapter 3. Thus, while selecting the grades of steel from the specifications and standards, it is of utmost importance to examine

Table 7-3. Principal effects and function of some common alloying elements in steel, including the residual elements.

Element	Percentage	Primary Function
Aluminium alloying	0.95–1.30	Help in faster nitriding of steels when present as an alloy. When present as residual or in lower level (<0.05%) for degassing and killing of steels, Al helps in getting fine grained steel and also resists grain growth in heat-treatment.
Bismuth	In traces	Improves machinability
Boron	0.001–0.003	A powerful hardenability agent i.e. rapidly increases hardenability of steels
Chromium	0.5–2	Increases hardenability
	4–18	Increases corrosion resistance
Copper	0.1–0.4	Improves corrosion resistance; atmospheric corrosion resistance
Lead	—	Improves machinability of steels
Manganese	0.25–0.40	It combines with sulphur and with phosphorus to reduce the brittleness. It also helps to remove excess oxygen from molten steel.
	>1	It increases hardenability by lowering transformation points and causing transformations to be sluggish.
Molybdenum	0.2–5	It forms stable carbides, inhibits grain growth and counters temper embrittlement effect. It increases the toughness of steel, thus making molybdenum a very valuable alloying element for making the cutting parts of machine tools and also high temperature applications like turbine blades of turbojet engines.
Nickel	2–5	Increases toughness of steels
	12–20	Increases corrosion resistance
Silicon	0.2–0.7	It increases strength.
	2.0	It increases springiness of steel by increasing yield strength (a popular alloy for spring steel application).
	Higher percentages	It improves magnetic properties (used in electrical stampings for transformer).
Titanium	Small amount	Stable carbide former; it produces fine carbides and grain size (used for micro-alloying).
Tungsten	—	It forms hard carbides; and also increases the melting/softening point of steel (used for increasing hot hardness in steel).
Vanadium	0.15	Stable carbides; it increases strength while retaining ductility; promotes fine grain structure and increases the toughness at high temperatures.
Cobalt	4.0%	It increases hot-hardness in high grade tool steels.
Carbon	0.02–2.0%	It has a major effect on steel properties. Carbon is the primary hardening element in steel. Hardness and tensile strength increase as carbon content increases up to about 0.85% C, while ductility and weldability decrease with increasing carbon.

Sulphur	<0.50%	It decreases ductility and notch impact toughness especially in the transverse direction. Weldability decreases with increasing S content. Sulphur is found primarily in the form of sulphide inclusions, which help in improving machinability. Sulphur is controlled at lower level (less than 0.05%) for most steels except re-sulphurised machining grade steels.
Phosphorus	<0.50%	Phosphorus increases strength and hardness and decreases ductility and notch impact toughness of steel. The adverse effects on ductility and toughness are greater in quenched and tempered higher-carbon steels. Phosphorous levels are normally controlled low levels. Higher phosphorus is specified in low-carbon free-machining steels to improve machinability.
Nitrogen	<0.10%	It acts like carbon for strengthening, but reduces ductility more drastically due to 'ageing' effect.

if the steel—including its composition (the carbon and alloy level in the grade) and attributes (like grain size and inclusion)—would be good enough to produce the right type of properties necessary for: (a) manufacturing, and (b) end-use application. While the contributions of compositions have been discussed in this section, contribution of other attributes of steel would be discussed next.

7.3. Other Factors Influencing the Properties of Steels

In general, it can be summed up that properties of steels are basically dependent on:

1. The composition (i.e. grade) of the steel,
2. Internal quality of the steel (e.g. homogeneity, cleanliness etc.), and
3. Different treatments it gets during steelmaking, rolling and heat treatment.

Microstructure, which is an integral part of steel, is the product of steel composition and treatment. Carbide size and type, mentioned earlier, is considered as a part of microstructure, because of their dependence on the nature of composition. Other factors or attributes that influence the steel properties such as the cleanliness of steel (i.e. inclusion rating) and grain size have been discussed in chapter 3.

Out of these basic factors, influence of composition with regard to carbon and alloying elements has been discussed in Chapter 2 and also reviewed in Section 7.2 in this chapter. The internal quality—including cleanliness and grain size control—has been discussed in Chapters 3 and 4. The third factor relating to different steel treatments refers to various treatments during steelmaking, thermal treatments and controlled rolling, and mechanical working of steels. The purpose of these treatments is to further improve the properties or fine tuning the properties for a given manufacturing process or application. This can be done by additional treatment or by controlling some attributes of the steel during steelmaking or rolling. For example, if the chosen steel is to be machined for giving shape, there may be need for heat treating the steel and also controlling the grain size and inclusion content of the steel. Heat treatment like annealing or normalising will produce the right microstructure for machining and a relatively coarser grain size (ASTM 5–8) will facilitate the machining process. It is to be recalled that coarser grain size helps in better machining by way of facilitating easier chip breaking at the cutting tool tip. Inclusions, especially the oxide inclusion types, are not helpful for machining, because, inclusions tend to blunt the cutting tool tip by wear and abrasion.

Hence, if the steel structure is found difficult for machining after controlling the microstructure and grain-size, action might be necessary for controlling the inclusion type and size for better machining (refer Chapter 3). These parameters can be achieved by specifying the steelmaking route in the specification, such as LF treated fully-killed steel with grain size range of ASTM sizes 5–8 and also by limiting the level of oxide inclusions in the steel. This scope of choice has been provided by all user specific steel standards.

Similarly, grain size has considerable influence in the hardening of steel, which is at the centre of steel technology. Coarser grains have higher hardenability (refer Chapter 8). Hence, coarser grains with excessive inclusion content in the steel may give rise to cracking during hardening by quenching from higher temperature. Therefore, if the steel is to be first hardened by quenching and then machined, as in many engineering parts, very coarse grains with high inclusion content might lead to cracking of quenched parts. Hence, careful balancing of steel with control of inclusion level is necessary. Because of the deleterious effect of inclusions in machining, heat treating and for applications in fatigue, much attention is paid for controlling the inclusions and inclusion type in steel by various methods of steelmaking (refer Chapters 3 and 4). As such, most steel standards provide the acceptable range of inclusions in a grade and allow further reduction in inclusion level by agreement between user and producer of steels.

Treatment of steel can be also extended in the rolling process itself, by controlling the rolling deformation and temperature whereby the steel microstructure is altered to finer and stronger structure with higher ductility. Such microstructures produced by controlled rolling not only have high ductility, but also strong due to finer grains. The structure is very helpful for press forming (cold bending) of parts which are subjected to severe bending in forming. The HSLA steel for cold forming and bending is one such example, which is thermo-mechanically rolled at intermediate temperature for increased strength and ductility (refer Chapter 4). Control-rolled HSLA steels are widely used for structural applications carrying high dynamic load e.g. long-members of commercial vehicle frame and other load carrying engineering body parts.

For structural grade steel, which are required to be formed by severe bending, it would be necessary to control grain size and inclusions along with limiting the carbon level. If the steel has to be severely bent and drawn during forming, only drawing quality steel of low carbon, fine grain size and with low inclusion content has to be chosen and specified, because of the requirement of extra ductility in the steel. For less severe drawing, standard low-carbon steel can be used at times by giving additional softening treatment like annealing for increased ductility, but its use is very limited due to extra cost for additional operations. For extreme draw and stretch operations in forming, steels should be very soft and ductile, requiring extra low carbon residual free steel with cold-rolling and appropriate texture.

In the ultimate count, steel is a structure sensitive material—it develops its properties through the manifestation of different microstructures, including grain size and inclusions. This has been pointed out and discussed earlier, but in the context of understanding properties of different grades of steels and their applications, a summary of structure-composition-properties and applications has been presented in Table 7-4 for general guidance. The table presents an overview of the structure and properties of steels along with their general applications.

The purpose of Table 7-4 is to provide an overview of areas concerning applications and selection of steels with respect to structure and composition, but it is not the complete list. There are other factors that control ultimate steel properties. For example:

Character of the chosen alloying element: For example, while Ni could contribute to increased toughness, P and S in the same steel have to be controlled for retaining the toughness. P increases susceptibility to brittleness and S decreases through-thickness toughness. But, if Mn is used for increasing strength, no such control would be essential as Mn would take care of excess S by

Table 7-4. Examples of steel structure and their broad composition, important properties and general applications.

Steel Structure	Major Chemical Constituents	Important Properties	Applications
Austenitic	Low C and high alloy: Cr, Ni, Mo	Corrosion resistance and heat resistance	Stainless steel structurals, vessels, utensils and miscellaneous
Ferritic	Very low C, or Specially alloyed to turn ferritic in stainless steel by Cr, Si etc.	Excellent cold formability and corrosion resistance	Sheet metal, structural, stainless steel utensils and miscellaneous
Ferritic-Pearlitic	Medium C or Medium C + low alloy	Moderate strength and good ductility	General commercial applications for structural (sheet, plates etc.) and engineering parts (rods, bars, forgings etc.) with or without HT
Pearlitic	High C or Medium C + low alloy	Good strength and low ductility	Sheets and plates: structurals; rods: wire drawing and miscelleneous engineering parts with HT
Martensitic	Medium C + low alloy Low C + high alloy (to turn stainless steel)	High strength, high toughness and corrosion resistance if stainless steel	Engineering parts and components after HT
Bainitic	Medium C + low alloy (preferably with Ni and Mo)	High strength and high toughness	Engineering and structural parts and components requiring toughness

combining and forming MnS. MnS so formed might even add to machinability of the steel by providing lubrication to the cutting tool tip.

Grain size: Finer grain size obtained by specifying Al or Al-Si killed steel is most favoured for increased strength (especially yield strength), ductility, fatigue and fracture strength, impact strength etc. But, fine grain size might not be favourable for certain applications like heavy duty machining or for applications at elevated temperature (creep), where coarser grain size is better.

Inclusion nature and type: Inclusions in all forms are harmful, but some are more damaging than the others, especially the harder inclusions like the oxides and silicates. Hence, these are to be minimized in the steel by appropriately specifying the limits while ordering the steel. Also, coarser inclusions are more harmful than finer size of inclusions for certain applications—like fatigue, fracture, ball-bearing or anything that undergoes wear and tear. The chart below illustrates how inclusions and grain sizes influence various applications and uses.

	Fatigue Strength	Fracture Strength	Impact Strength	Machining	Cold-Forming	Hot-Forming	Corrosion and Pitting	Yield Strength	Creep Strength
Grain size (Finer)	↑↑	↑	↑↑	↓	↑	—	↑	↑↑↑	↓↓
Inclusions (Harder)	↓↓	↓↓	↓↓	↓*	↓	↓	↓	↓	↓

* Sulphide inclusions improve machinability.

It is hard to get the total properties of steels by fixing one parameter; it has to be built in the steel during ordering and procurement by considering the following parameters:

- Chemical composition, including the level of S and P;
- Nature and character of alloying element chosen for the steel;
- Process of steelmaking i.e. the killing and teeming practice, and any special treatment (e.g. LF/VD treatment etc.) given to the steel for controlling grain size and cleanliness;
- Control of any harmful residual or tramp elements (e.g. tin, antimony, bismuth etc. which are found in steelmaking process using recycled steel scraps);
- Reheating and rolling practices, influencing the internal structure, internal cracks, surface cracks, homogeneity etc.;
- Proper roll-pass design and good degree of deformation under each roll-pass during rolling is known to produce more homogeneous structure, fragmented and finer inclusions, and defect free interior; and
- Preciseness and conformance to chemical composition and other specified parameters such as inclusion level, grain size, hardenability, etc. and properties—e.g. as-rolled hardness, hardenability, heat-treated properties, impact strength etc.—as per the prescribed standards.

Most standards are designed to provide these provisions in the steel specifications. However, it is upto the users of steels to choose the right grade under right standards for an application. At time, quality specification of a user might have to consider additional measure or precautions in testing for ensuring the right quality of steel. Properties and application of structural, engineering and stainless grade of steels will be further discussed in the light of these observations on 'property-composition' relationship.

7.4. Structural Steels: Properties and Applications

Structural steels, as the name implies, are required to support loads acting on a structure. The load could be static load due to internal weight of the structure, external and imposed load in service, fluctuating load caused by winds, vibration, earth movement etc. Examples of structural steels uses are buildings and construction, plants, fabricated structures for storage and sheds, telecommunication towers, load-carrying members of vehicles and transportation systems, pressure vessels, containers and vessels, boilers etc. Common amongst these applications is the requirement of suitable strength of the steel that is being used. Since structure must not deform or yield in service, its 'yield strength' plays a dominant role in the choice of structural steels.

Yield strength of steels gets increased by the addition of 'interstitial alloying'—e.g. carbon and nitrogen—which have smaller atomic diameter and can form interstitial solid solutions. Therefore, increasing yield strength in steel by increasing carbon is a popular and economical method. Nitrogen is not popular because it promotes 'ageing' in steels, causing loss in ductility with time—an important parameter for forming and application of structural steel. Increasing carbon in structural steels for increasing strength has its limitations, because it decreases ductility and reduces weldability. Therefore, composition of structural steels (i.e. the grade) is carefully selected keeping in view of property requirements.

Since higher strength is always associated with decrease of ductility, often recourse is taken to support and 'reinforce' the main structural frame by jointing additional reinforcement plates of appropriate shape rather than increasing the steel strength. This is for increasing the load-bearing capacity of the

Table 7-5. Common types of steels as per ASTM standards for structural applications.

Carbon Steels (Plain Carbon Steel)	High Strength Low Alloy Steels (HSLA)[1]	Corrosion Resistant Low Alloy Steels[2]	Quenched and Tempered Alloy Steels[3]
ASTM A36: Structural shapes and plates	ASTM A441: Structural shapes and plates	ASTM A242: Structural shapes and plates	ASTM A514: Structural shapes and plates
ASTM A53: Structural pipes and tubing	ASTM A572: Structural shapes and plates	ASTM A588: Structural shapes and plates	ASTM A517: Boilers and pressure vessels
ASTM A500: Structural pipes and tubing	ASTM A618: Structural pipes and tubing		
ASTM A501: Structural pipes and tubing	ASTM A992: Wide flanged beams		
ASTM A529: Structural shapes and plates	ASTM A270: Structural shapes and plates		
Uses: Common and commercial structures; and general applications.	Uses: Towers, vehicle frames and members; and uses requiring higher YS.	Uses: Structures at sea cost and wet applications;transportation and containers etc.	Uses: Pressure vessels, boilers; and heavy-duty containers etc.

[1] Micro-alloyed with Ti/Nb/orV. [2] C-Mn +Cu varieties; and Corten Grades (Cr alloyed). [3] Alloyed with Cr/Mo/V etc.

structure without having to use higher strength steels with lower ductility or use more expensive HSLA or heat treated steel. As such, selection of structural steels offers varieties of options, depending on the strength, cost and availability. To keep cost down, large part of structural steels is generally made of lower carbon content or by manganese alloying for strength.

To fit to wide varieties of uses, structural steels have different shapes and forms—e.g. flats, plates, sheets, pipes, tubes, angles, channels etc.—but they are universally specified by their strength, especially the 'yield strength'. In addition to strength, the steels should have good ductility (for forming as well as for shock load absorption) and weldability, if required. Most structures work at outdoor environment; hence it might also require additional properties like atmospheric oxidation resistance and corrosion resistance, which should be built into the steel by composition control, such as by addition of copper and chromium. If the structure is operating at high sea or places, where temperature is low or fluctuates over and below the room temperature, impact toughness and low-temperature property of the steel is also required. Table 7-5 illustrates some common types of ASTM grade plain carbon and alloyed structural steels as per their recommended uses.

Table 7-5 shows that depending on the duty of the structure steel grades can range from simple plain carbon mild steel to HSLA and heat treated alloy steel. However, there is no bar for using any grade of steel for structural applications as long as the steel fulfils the criteria of shape, size, strength and fabrication properties required for the application. While detailed specification of structural steel and their properties can be obtained from the respective ASTM or national standards, Table 7-6 mentions the grade names of some common ASTM grade structural steels and their international equivalent. Table 7-6 furher shows that steel strength for structural steel can range from 18.5 kgf/mm^2 yield strength to about 34 kgf/mm^2 (see grade EN 10025) for general purpose cold formed sections and parts. However, for special requirements, steel yield strength can be increased by micro-alloying (HSLA) to about 42–45 kgf/mm^2 and further up to the level of 50–55 kgf/mm^2 by heat treatment.

Table 7-6. A list of some popular structural grade steels, designated as per their YS and other applicable conditions, and their international equivalent.

ASTM (ASTM)	Europe (EN 10025(93))	Germany (DIN 17100)	India (IS)	Japan (JIS) 3101	Japan (JIS) 3106	ISO (ISO 630)
A283: A B C D	S185	St.33		SS310		E185
	S235 (JR)	RST 37-2	IS226: Fe410S	SS400	SM400A	E235B
	S235J2G3	RST 37-3N		SM400C		E235D
A36	S235J2G4					
A529 Gr.50, 55	S275JR S275JO	St.44-2 St.44-3U	IS2062: Fe410WA			E275B E275C
A572 Gr.42, 50	S275GJ2G4	St.44-3N	Fe410WB			E275D
A633 Gr.58, 65,70	S355JR S355JO	St.52-3U St.52-3N	Fe570HT Fe540HT	SS490A SS490B SS490C		E355C E355D
A656 Gr.50	S355J2G4			SS490YA SS490YB		

Note: First 3 digit of Euro-Norm indicates the yield strength in N/mm^2 and first two digits of DIN standard refers to tensile strength.

Table 7-6 mentions not only the yield strength, but also other conditions of supply (e.g. N, U, HT, AR, etc.) and other critical properties—e.g. impact property represented by J (Joules)—for applications. For example, consider the case of Grade S235 steels, where:

- When designated S235JR: it denotes a structural grade steel (S) with minimum YS of 235 N/mm^2 with guaranteed longitudinal Charpy V-notch impact test value (JR) of 27 Joules at room temperature; and
- When designated S235AR: it denotes the same type of structural steel with minimum YS of 235 N/mm^2, supplied in 'as-rolled' (AR) condition.

Similarly, if the steel grade is S275, it implies minimum YS of 275 N/mm^2; meaning of other notation (JR or AR) remains same. If the notation is JO, it means longitudinal impact value of 27J at 0°C, and if J2, it would mean longitudinal impact value of 27J at –20°C, etc. Thus, steel standards and grades attempt to cover various combinations of properties that might be required for an application, and thereby facilitate right selection of steel.

Properties of structural steels may call for a combination of:

- Higher strength i.e. yield strength—which sets the stress level up to which the steel can withstand load without deformation;
- Adequate elongation (i.e. ductility) so that parts can be formed easily by bending, whenever necessary, and also allow certain elasticity without deformation under shock load or vibration;
- Good oxidation resistance or paintability to protect atmospheric oxidation;
- Good weldability (i.e. lower carbon equivalent, CE) for jointing and fabrication;

- Good toughness for resisting sudden fracture under high stresses at the temperature of operation, expressed as charpy impact value of min. 27J at that temperature;
- Good low-temperature property (i.e. low ductile-brittle transition temperature) to withstand cold service weather or fluctuating temperature in service; and
- Good corrosion resistance property for special applications in marine or saline water.

Other than good corrosion property—which is a chemical characteristic and might even require uses of stainless steels in extreme cases—all other properties can be met with by using carefully balanced carbon steel or HSLA or low-alloy steel in as-rolled or heat treated conditions. Generally, structural steels include steels with yield points ranging from about 21–50 kgf/mm^2, and these strength levels are obtainable by varying the chemical composition or by heat treatment of plain carbon or appropriate micro-alloy steels.

Increasing carbon for increasing strength in structural steels is not always recommended, because, as strength increases with carbon content in the steel, its ductility, toughness and ductile-brittle transition temperature decreases with increasing carbon. This limits the use of economical means of increasing strength by carbon all the time. Instead, where higher strength is required for application demanding good ductility either a combination of C-Mn alloying or micro-alloyed HSLA steel or heat treated steels can be used, depending on the exact requirements. Where the strength requirement is beyond the scope of normal HSLA steels, quenching and tempering (HT) of low-alloy structural steel are used, with controlled carbon content for higher toughness and better welding. Other factor that may adversely affect toughness related properties of structural steels include 'residual elements' in the steel. Some residual elements like tin, antimony etc. can induce brittleness in steel—hence such harmful residual elements are not desirable in structural steels. Therefore, for critical structural applications of higher strength requirement, steels are often required to be procured with limited residual content, especially elements like tin, antimony, etc. which are known to induce brittleness in steel. Therefore, major sources of supply of structural steels have been the blast-furnace-cum—basic oxygen steel making process, where residual levels are relatively lower compared to electric furnace steelmaking, where recycled scraps are used. Most structural steel standards specify such requirement of residual control, in addition to steel chemistry and mechanical properties.

Thus, selection of structural steels—a common class of steels—may require consideration for steelmaking and rolling route along with the choice of grades for properties. As regards mechanical properties, various features such as: strength, ductility (or toughness), weldability, and fracture resistance properties of structural steels are obtained by appropriately balancing the composition and properties.

Table 7A-1 in Appendix to Chapter 7 provides few important structural grade steel compositions, their properties and applications, which indicate that:

- Structural steels are graded as per applications;
- Common structural steels contain some Cu (0.20% min) for resistance against atmospheric corrosion;
- Strength is the prime requirement for structural steel, leaving choice of heat-treatment open;
- For critical applications requiring high toughness value of the steel, alloying by Ni, Ni-Cr or Ni-Mo is used; and
- %Carbon in all grades is generally kept on the lower side—for maintaining better ductility and weldability.

Since structural steels are used in bulk, care is necessary to select steel grade which is not expensive and easily available. Therefore, carbon or carbon-manganese based structural steels are generally preferred, excluding those cases where service conditions demand special consideration. An example of the latter is the under-carriage frame parts in automobile and commercial vehicles where, these parts not only experience high load but also dynamic fluctuating load due to vehicle movement and road conditions. In such cases, steel of higher yield strength with very high ductility—such as the HSLA steel—is used. Moreover, use of HSLA steel for automobile frame parts may also allow uses of steel with lower thickness (taking advantage of higher strength) with consequent reduction of vehicle weight, aiding to energy efficiency.

7.5. Engineering Steels: Properties and Applications

The bulk of steels consumed is either of structural grade or of engineering grade steels. Engineering grade steels—as the name implies—is applied in the manufacturing of engineering parts and components, which are characterised by the application of mostly fluctuating or dynamic loads in uses or service. For example, applications such as bolts, nuts, shafts, gears, turbine blades, etc. which experience dynamic load in service. Engineering applications can range from a simple bolt to huge moving parts like turbine blades or ship hulls. The duty of the steel varies with its application, necessitating right choice of steel composition (including permissible residuals), internal quality, and rightness for response to heat treatment. Heat treatment is central for developing the required properties in engineering grade steels—hence heat treatability of the chosen composition and quality has to be considered for selecting engineering grade steel. As such, choice of right composition and right steelmaking process—for cleanliness and internal quality—is important for engineering grade of steels.

The focus of engineering steels is heat treatability for developing right combination of mechanical properties. Unlike structural steels, all engineering steels are used after some kind of heat treatment— e.g. annealing, normalising or hardening and tempering. Hence, a special requirement of engineering steel is its ability to respond to required heat treatment. Table 7-7 shows few examples of popular grades of engineering steels of British Standards specifications along with their standard properties and areas of applications, and Table 7-8 shows an illustrative map of different groups of engineering steels for different applications.

Depending on the application, composition can range from simple plain carbon to steel with complex alloying element combination in order to ensure heat treatability. The gradation starts with lower carbon varieties—e.g. EN 3, 5, 8 etc.—where normalising is the common heat-treatment given for developing a combination of strength and ductility through refinement of grain sizes. Low strength nuts, bolts, screws etc. are the examples of uses of these grades of steel. If high surface hardness is required for any application, the low-carbon steel (e.g. EN 32C, 36A, etc.) can be selected and subjected to case carburising for enriching the surface area with higher carbon and then quenched and tempered for getting the surface hardness. If the application requires higher strength and toughness, choice of engineering steel then rests on the medium carbon grades with or without alloying.

Effects of alloying elements on the mechanical properties of steels have been depicted in Table 7-1, which guides about the choice of alloying and corresponding grades of steel. Alloy system in engineering steel grades is chosen for imparting application specific properties to the steel. Nonetheless, cost of the steel is equally important in industries. Hence, there is a need for choosing alloy system in the steel based on exactly—what properties are required for the uses and application and how that can be met at minimum cost. However, cost must not over-ride the importance of functional properties, sacrificing the reliability of performance. Table 7-8 attempts to map out different types of engineering steel grades that are used for different applications/purposes.

As regards cost of the steel, manganese is the cheapest of all alloy (after carbon) with high potential to contribute to strength. Hence, a combination of carbon-manganese alloying is used for common

Table 7-7. Examples of few British Standards (BS) grade engineering steels of different descriptions and their properties.

Grade	Description/Areas of Application	Heat-treatment (For End-use)	Limiting Hardness (HB)	Standard Properties*			
				TS (N/mm²)	YS (N/mm²)	%RA (Min)	Charpy Impact (J)
EN8 (080M40)	Medium carbon engineering steel	N (Can be A or H&T)	150–210	510 550	245 280	17 16	16
EN9 (070M550	Medium carbon engineering steel (high carbon than EN8)	N (Can be A or H&T)	170–250	600 700	310 355	13 12	—
EN18 (530M40)	Medium carbon alloy steel (Cr = 0.90/1.20)	H&T	230 or 250 max	*850 1000	680	13	40
EN19 (708M40)	Chrome-Moly high tensile engineering steel.	H&T	220 or 250 max	*700/850 850/1000	525 680	17 12	50 50
EN24 (817M40)	Nickel-Chrome-molybdenum high tensile steel.	H&T	300 or 350 max	*850/1000 850	650	13 11	50925/1225 40
EN30B (835M30)	Medium carbon 4¼% Nickel steel.	H&T	440 max	*1550	1160	7	16
EN31 (535A99)	1.0% Cr—Ball bearing steel	H&T	229 max (Softened)	— (Hardened and tempered to 60 ± 2 HRc.	—	—	—
EN40B (722M24)	3% Chromium molybdenum nitriding steel	H&T	300 max	*850/1000	650	13	42
EN41B (905M39)	Chromium aluminium nitriding steel	H&T	200/277	*700/850 850/1000	525 585	17 15	50 42
EN36A (655M13)	Low-C Nickel chromium carburising engineering steel	Carburised + H&T	60 HRc min	Core strength: 1000 max	—	—	36

* Varies with ruling section and the tempering temperature.

grades of engineering steels for increased strength (see Table 7-8). C-Mn grade steel can also be made *free-cutting* type (i.e. allowing easy and smooth machining) by addition of extra sulphur to the steel during steelmaking called re-sulphurisation of steel (e.g. En 1A). Extra sulphur in the steel gets combined with Mn to form MnS, and thus eliminate the ill-effects of sulphur on through-thickness toughness of steels, simultaneously promoting free machining due to presence of abundance of sulphide inclusions (refer Chapter 3). Carbon-manganese or plain carbon steel have the flexibility to develop a range of properties by simple heat-treatment, like annealing for softening and machining, normalising

Table 7-8. Different types of engineering steel grades that are used for different applications/purposes.

Applications ➡					
Carburising Steels		**Machined parts**	**H&T steels**	**Special steels**	**Stainless steels**
				➡	Stainless steels, Wear resistance steel,
Low C/alloy carburising steels.	Plain C steels: %C: 0.12 to 0.50.	C-Mn and C-Mn-S Free cutting steels. (%Mn: 1.0 to 1.5)	Low-alloy steels: %C: 0.35 to 0.50	Spring steels: %C 0.45 to 0.75 plus alloy of Si, Mn, Cr, V etc.	Heat resistance steel: Generally of high alloy content.
E.g. EN 351, 354 etc.	E.g. EN 3, 8, 9, 32, etc.	E.g. En 14, 15	Alloy: Cr, Ni, Mo, V etc.—as per grade	Special Steels: Low to medium C + special alloy like B, Mo etc.	Stainless steel: Cr-Ni-Mo
			E.g. EN 18, 19, 24, 26, 100, etc.		Wear resistance: Cr, V, Mo
					Heat resistance: Cr, Mo, W, Co etc.

for increase of strength by grain refinement and hardening (quenching and tempering) for attaining even higher strength and toughness through martensitic transformation route.

The bulk of engineering steel is used after hardening and tempering (H&T), including the plain carbon grades, like En8/En9. If the steel is to be used after H&T, it is necessary to consider the following factors for the choice of carbon and alloy types in the steel:

1. Carbon increases the hardness (i.e. strength) of steel, but decreases the toughness as well.
2. Excess carbon makes the steel prone to crack or distortion on quenching. Hence, higher carbon steel should be oil quenched due to less quench severity of oil compared to water, in order to reduce the distortion in steel parts.
3. Depth of hardening achieved by increasing carbon is rather shallow, because of cooling rate sensitivity of carbon steel in the formation of martensite, requiring faster quenching. Hence, it has limited use in higher 'ruling section' jobs where a steep temperature gradient sets up between surface and centre during quenching, producing non-martensitic product inside. Hence, for optimum depth of martensitic structure in the component, a combination of carbon and some alloying (e.g. Cr and Mn) should be considered.
4. For higher ruling section, alloy steel is always desirable due to its ability to increase the depth of hardening and reduce quenching distortion by permitting oil quenching, which has much lower quench severity than water. However, for heavier section steel parts, appropriate combination of alloying elements has to be chosen for ensuring the right depth of hardening and toughening of the central portion.

In general, the thumb-rule is to go for higher carbon in the steel, if higher surface hardness is to be obtained, but if the same hardness is to be obtained at deeper depth of the steel body (i.e. for depth of hardness) consider for sufficient alloying in the steel. *Carbon up to 0.80% (or the eutectoid composition) in steel helps in achieving higher hardness on the surface and all alloying elements, excepting cobalt, help in achieving deeper depth of hardening.* This concept and approach plays a very important role in terms of economy and cost effectiveness in the selection and heat treatment of engineering steel. It implies that expensive alloying for engineering grade steels should be used only if, the hardenability or toughness required for the application is not adequate in carbon steels. No doubt, carbon steel has limitation for uses where higher strength and toughness is required, but uses of expensive alloy steels–containing Cr, Ni or Mo–should be critically examined from the point of structure, strength and toughness, and should be used where it is necessary.

It is in this respect that hardenability of steel assumes critical significance in steel; hardenability, in effect, determines the obtainable final structure, strength and toughness of the steel. Effect of carbon and different alloying elements on the depth of hardening of steel (i.e. hardenability) is indicated in the following table:

Elements	Qualitative Effect on Hardenability	Remarks
Carbon	Increases—moderate	Further influenced by grain size; coarser grains increase.†
Manganese	Increases—strong	
Silicon	Increases—moderate	
Nickel	Increases—moderate	
Chromium	Increases—strong	
Molybdenum	Increases—strong	
Vanadium	Increases—moderate	
Cobalt	Decreases	An exception. Used for increasing hot-hardness
Boron*	Increases—very strong, but simultaneously produces fine grain size which partly nullifies the effect	Used for special purposes—like high strength fasteners: nuts and bolts

* Present in traces: 0.0005 to 0.003%
† Effect of G.S. is clubbed with carbon for all hardenability calculations.

Grain size is an important factor in hardenability; fine grains effectively lower the hardenability of steel. This effect is more pronounced as the grain sizes become finer below ASTM size 8. Hence, fine grained steels produced by micro-alloying are not popular for heat treatment.

Thus, engineering steel calls for balancing of carbon content and alloying elements for getting the desired hardening effect in a given 'ruling section' (i.e. the section thickness that rules how deep the hardening must reach; more about hardenability is discussed in Chapter 9). Increasing ruling section by increasing carbon is not always desirable, because (a) higher carbon might introduce cracks/distortion during quenching, and (b) microstructure achieved by carbon upon quenching might not be tough enough even after tempering. Hence, for heavy ruling section, a combination of medium carbon (0.35–0.50) with appropriate alloying (e.g. Cr, Mn, V etc.) is recommended.

The other important grades of engineering steel includes: (a) spring steel and (b) high-alloy wear, corrosion and heat resistance steel. Springiness of steel require high yield to tensile ratio (>80%). This necessitates the use of higher carbon steel (ranging between 0.45%C to 0.70%C) with appropriate alloying—such as Si-Mn (e.g. En45 and 45A) or Cr-V (e.g. En47) types. Steel for springs must have at least 0.45%C to attain required hardness. In general, plain carbon steel, such as AISI 1045, 1060, 1074,

Table 7-9. Some popular spring steels of different standard.

Grade	Typical Chemistry					Remarks/Uses
	%C	Si	Mn,	S & P	Others	
EN 43	0.50	0.30	0.70	0.015	nil	Carbon spring steel rounds and bars for general duty
En 45 (SAE 9260)	0.60	2.0	0.90	0.015	nil	Carbon spring steel flats for automobiles and railways. Oil hardening
EN 47	0.50	0.40	0.70	Cr =1.0%	V = 0.10%	Special automobile springs-coils and flats, Oil hardening
DIN 67K	0.70	0.35	0.70	S & P: 0.015		Plain carbon spring flats; oil hardening type
IS C-80	0.80	0.35	0.70	S & P: 0.015		Used after cold rolling for strips
C-95	0.95	0.25	0.70	S & P: 0.015		Used after cold rolling to thin strips

1080 or even 1095, can be used for springs. Plain carbon steels are generally used for flat springs. But, for critical applications involving fluctuating and sudden loading, alloy spring steels, such as AISI 6150, 9260 and 8650 are recommended, because of their higher toughness. Generally, Si-Mn spring steels are preferred for automobile spring flats (i.e. leaf springs) and Cr-V spring steels are preferred for coils and rounds (i.e. coil springs). While these alloy spring steels are used after heat-treatment, plain carbon steel (ranging between 0.45%C to 0.95%C) is used for springs after cold rolling or cold drawing. The purpose of cold rolling/drawing is to further increase the yield strength and yield to tensile ratio necessary for springiness and resilience of springs in action. Some popular spring steels of different standard are shown in Table 7-9.

Engineering steels may also embrace wear, corrosion and heat resistance steels, shown in the last column in Table 7-8. Wear resistance steels are characterised by the presence of higher chromium and some molybdenum content (e.g. En 56, 58 etc.), which are known to produce complex carbide precipitates in the steel after heat-treatment and, thus, impart high wear resistance. Corrosion and heat resistance steels are high alloy steels, which generally fall under stainless steel groups. This will be discussed next.

7.6. Stainless Steels: Properties and Applications

Unique property of not corroding, rusting or staining at room and moderate temperatures makes stainless steel a very attractive material for many structural and engineering applications. Stainless steel differs from other groups of steel, namely carbon and alloy steel, by the amount of chromium present. Stainless steels contain high percentage of chromium (>12%), making the steel stainless. Sufficient chromium in the stainless steel helps to form a passive film of chromium oxide, which prevents further surface corrosion and blocks corrosion from spreading into the metal's internal structure by stopping the ingress of oxygen to the metal-oxide interface. Furthermore, when the stainless steel is austenitic in structure, it has high degree of structural stability over a wide range of temperatures, making the steel heat resistant even at higher temperature. *Thus, stainless steels have the unique properties of corrosion resistance as well as good heat resistance due to their unique composition and structure.*

There are three major classes of stainless steel, depending upon the composition. They are classified as per structure: austenitic, martensitic and ferritic stainless steel. Each of the class of stainless steels again has a number of varieties. Applications of stainless steel are as wide as its varieties. There are over

60 varieties of stainless steels which include: sheets, plates, bars, rods, wires, coils, cookware, cutlery, household hardware, surgical and precision engineering appliances, industrial equipment, vessels, structures, building materials etc. Table 7A-5 in the appendix of this chapter provides the list of popular stainless steel grades and their applications.

Structurally, stainless steel are grouped as follows:

Austenitic stainless steel: Austenitic stainless steel contain lower carbon and higher chromium-nickel alloys; 16–26% chromium (Cr) and 6–22% nickel (Ni). Presence of Ni along with Cr stabilises austenite at room temperature and also contributes to corrosion resistance, especially against attack by acids. This steel is non-magnetic and not heat-treatable by quenching and tempering; it can be appreciably hardened only by cold-working. Popular example of austenitic stainless steel is AISI Type 304 (S30400) or "18/8" (18% chromium 8% nickel) stainless steel grade, which is widely used for many structural, utility and engineering applications.

Martensitic stainless steel contain variable carbon with 10–18% Cr and some Ni and Mo, if required. Carbon in martensitic stainless steel is higher than austenitic stainless steel because of the necessity of hardening for producing martensite. The steel can be hardened by quenching and tempering, producing the typical martensitic structure which is very strong and tough. This steel is magnetic the context of its properties. Martensitic grades are strong and hard; hence difficult to form and weld. Popular example of martensitic stainless steel is the AISI Type 410 (S41000), which is commonly used for making knives and many surgical instruments as well as in the manufacturing of pumps and valves.

Ferritic stainless steels contain 11–27% chromium and very low carbon; it is free from expensive Ni as an alloy. The steel is magnetic in contrast to austenitic steel, and has inferior corrosion resistance property due to the absence of Ni. AISI Type 420 (S42000), containing 12–14%Cr, is the example of ferritic stainless grade, which is extensively used for utensils and different domestic appliances. Ferritic stainless steel parts can be hardened by quenching and tempering, if necessary, by appropriately adjusting carbon in the steel.

There are two other types of stainless steel which are used for special purposes; namely duplex structured steel and precipitation hardened stainless steel (PH-stainless steel). Typical chemical compositions of some stainless steel grades are shown in Table 7-10. Duplex stainless steel has a mixed structure of (austenite + ferrite) due to lower nickel (4%–5%) and higher chromium (18%–28%) content. Additional alloying of molybdenum (0.4%–0.6%) is also used in duplex steel for making the steel stronger and heat resistant for applications like heat-exchanger equipment. Properties of duplex steel are somewhere between austenitic

Table 7-10. Typical chemical compositions of some popular stainless steel grades of different structures.

No.	Grade	C max, %	Mn max,%	Cr,%	Ni,%	Mo,%	N, %	Cu, %	Cb +Ta,%
AISI 201	Austenite	0.15	6.0	17.0	4.5	-	0.25 max	-	-
AISI 304	Austenite	0.08	2.0	19.0	9.5	-	-	-	-
AISI 316	Austenite	0.08	2.0	17.0	12.0	2.5	-	-	-
AISI 430	Ferritic	0.12	1.0	17.0	-	-	-	-	-
AISI 410	Martensitic	0.15	1.0	12.5	-	-	-	-	-
AISI 2205	Duplex	0.30	2.0	22.0	5.0	3	0.14	-	-
17-4PH	Precipitation hardening	0.07	1.0	16.5	4.0	-	-	4.0	0.30

and ferritic stainless steel, offering higher strength than fully austenitic steel and better weldability, but inferior corrosion resistance. Common applications of duplex steels include—desalination equipment, marine equipment, petro-chemical plant parts and heat-exchanger parts.

Precipitation hardening stainless steels have very high strength, good weldability and fair corrosion resistance, making the steel suitable for applications in areas of pumps and shafts working in marine or hazardous environment, turbine blades and many kinds of aero-space engineering equipment. The steel gets its strength after the special heat treatment of 'solution treatment' and 'ageing' at low to moderate temperature.

Most popular of stainless steels are the austenitic stainless steels (200 and 300 series) containing Cr and Ni (8% or more) as major alloying elements. This group of steel has highest corrosion resistance, ductility and weldability. Moreover, austenite being a stable phase over a wide range of temperature, it retains its strength and other properties at elevated temperatures as well. Since there is a tendency of chromium carbide to form along the austenite grain boundaries when reheated—which deplete the matrix with chromium needed for corrosion resistance—the steel needs to be stabilised at times for some application by the addition of other carbide forming elements like titanium, niobium etc. Austenitic grades are not heat-treatable; they can be strengthened only by cold working. This condition limits the engineering applications of austenitic stainless steel, which is mostly limited in areas requiring moderate strength at higher temperatures and subjected to corrosion e.g. furnace parts, heat exchangers, etc.

Martensitic grade of stainless steel is stronger and heat-treatable. As such, it finds lot of engineering applications—like blades, pumps, turbines etc. However, martensitic steel are lower in ductility compared to other stainless steel and poorer in welding. Typical mechanical properties of some popular stainless steel grades, under different conditions of treatment, are indicated in Table 7-11.

Much of mechanical and physical properties of stainless steels are nearly stable over a range of temperatures—for instance up to $350°C$—which makes the steels unique for higher temperature applications. Hence, these steels—particularly the martensitic grades—are popular for moderate to high temperature engineering applications, like nuts, bolts, pumps, valves, shafts, etc. Uses of stainless steels range from rail coaches and elevators to cutlery in structural sectors and from surgical implants to

Table 7-11. Mechanical properties of some popular grade of stainless steels.

AISI Grade/Type	Modulus of Elasticity (GPa)	Hardness (RB)			Tensile (Mpa)			Yield Strength (Mpa)			Elongation (%)		
		A	½ H	CD	A	½ H	CD	A	½ H	CD	A	½ H	CD
AISI 201: Austenitic	197	90	—		790	—	1030	380	—	760	55	—	10
AISI 304: Austenitic	193	80	—	35 R_C	586	—	1100	241	—	760	55	—	10
AISI 316: Austenitic	193	80	—	91 RB	586	—	620	241	—	415	55	—	45
AISI 430: Ferritic	200	82		90	517	—	586	310	—	483	32	—	20
AISI 410: Martensitic A	200	82			520			275			30		
H&T	200	24 R_C			834			721			21		
AISI 2250: Duplex	200	31			750			510			25		
AISI 17-4 PH A	196	33			1030			760			10		
H&T (Aged)		44 R_C			1380			1230			12		

Symbol: A = annealed; CD = cold drawn; and H&T = hardened and tempered.

Table 7-12. Summary of the main advantages of the various types of stainless steel.

Type	Examples	Advantages	Disadvantages
Ferritic	410S, 430, 446	Low cost, moderate corrosion resistance and good formability	Limited corrosion resistance, formabilty and elevated temperature strength compared to austenitics
Austenitic	304, 316	Widely available, good general corrosion resistance, good cryogenic toughness. Excellent formability and weldability	Work hardening can limit formability and machinability. Limited resistance to stress corrosion cracking
Duplex	2250	Good stress corrosion cracking resistance, good mechanical strength in annealed condition	Application temperature range more restricted than austenitic steel
Martensitic	420, 431	Hardenable by heat treatment	Limited corrosion resistance and formability compared to ferritic and austenitic steels. Weldability is also limited.
Precipitation Hardening	17/4PH	Hardenable by heat treatment, but with better corrosion resistance than martensitics	Limited availability, corrosion resistance, formability and weldability restricted compared to austenitics

turbine blades in engineering sectors, proving it to be the most versatile material of all. Stainless steel is not a one-type steel—as the name may suggest to laymen—but a multi-faceted material having solution for many tricky structural or engineering applications. The success in the use of stainless steel lies in choosing the right grade from the numerous grades available today, subject to justification of extra cost for the application. Table 7-12 provides some comparative advantages and disadvantages of uses of different types of stainless steels.

Thus, stainless steels represent a wide group of materials parallel to the conventional structural and engineering steels for special applications, subject to the justification of cost. Its uniqueness of its properties lie in its excellent all-weather corrosion resistance compared to the conventional steels and its ability to withstand moderate to high temperature in applications without any appreciable loss of strength. Aesthetically, stainless steel components look smoother, brighter and stain free, giving them a mirror-like finish.

However, different grades of steels are not exactly competing material to each other; they are complimentary to each other in their properties and uses. This demands discretion and judgment in the selection of steel for intended applications with regard to:

- Appropriateness of properties;
- Cost and availability;
- Flexibility in the process of attaining the component/composite properties
- Customer preferences and choices for 'value' to the product; and
- Conservation and contribution to the environment and ecology management—e.g. recyclability, energy efficiency in production and uses, resource control and conservation etc.

A comprehensive approach to steel selection with regard to properties, prices and processes—i.e. process of transforming steel to components and equipment—*vis-a-vis* other available alternative materials is, therefore, required for judicious selection and application of steels.

SUMMARY

1. The chapter discusses various aspects of steel composition and character influencing the properties of steels for manufacturing as well as for end-use applications. Based on these influencing factors, properties and characteristics of structural, engineering and stainless steels have been discussed *vis-à-vis* their applications.

2. Influence of carbon and different alloying elements on various properties of steels have been illustrated and discussed at length, because they form the foundation of all steel properties.

3. It has been further pointed out that other than composition, steel quality and properties are also dependent on (a) internal quality of the steel (e.g. homogeneity, cleanliness etc.), and different treatments it gets during steelmaking, rolling and heat-treatment.

4. Microstructure, which is an integral part of steel, is the product of steel composition and the treatment it gets. Therefore, for a given application, properties of steels in terms of internal quality of steel and the microstructure, including the effect of grain sizes and micro-inclusions, should be evaluated.

5. Discussions in the chapter focused on highlighting various special properties and attributes required for successful application of structural, engineering and stainless steels in their respective fields. In addition to the strength and strength related mechanical properties, importance of weldability, atmospheric corrosion resistance and low-temperature properties of structural grade steels have been highlighted, and their coverage in different standards has been illustrated.

6. Requirements of different properties of engineering steels have been discussed with reference to their processing needs (heat-treatment) and applications. Based on different uses and applications, suitability of different grades of engineering steels have been mapped out and illustrated. It has been shown that requirement of adequate hardenability is of paramount importance in engineering steels, because most steels of this class are used after hardening.

7. Properties and applications of different grades of stainless steels have been discussed, including their structure, and their relative merits have pointed out. Stainless steel grades, composition, structure and their applications have been illustrated with reference to popular standards. Further information and data on popular grades of these steels have been provided in the Appendix to this chapter in Table 7A-4.

FURTHER READINGS

1. ASTM Standards for Steel. *American Society of Metals*. 2001
2. British Steel Corporation. *Iron and Steel Specifications*. London: 1978.
3. Reed-Hill, Robert E. *Physical Metallurgy Principles*. Van Nostrand Reinhold Co.: New York, 1973
4. UK Steel. *Steel Specification Handbook*, 13th edition. London: 2010.
5. ASM International, USA. ASM Handbook, Vol. 01: *Properties and Selection: Irons, Steels and High-Performance Alloys*.1990.

APPENDIX TO CHAPTER 7

Table 7A-1. List of some common ASTM grade structural steels and their properties.

ASTM Structural Grades	Chemical Composition (divide by 100)						Mechanical Properties					Remarks
	%C	Mn	Si	S	P	Other	T (Ksi)	Y	El* (2" GL)	Hardness HB	Impact	
A-36	25	80/120	15/40	05	04	Cu-20	58/80	36	23	—	—	Plates & bars
A-53	25	90/120	—	06	05		45/60	25/35	21(min)	—	—	Pipes & Tubes
A-105	35	60 /105	35	05	04		70	36	22	187 max	—	Forged & HT parts
A106 Gr. A	25	30/90	10 min	06	05		48	30	35(L)	Hot F/A		Seamless C-steel tube for HT
B	30	30 /105	10	06	05		60	35	30(L)	Hot F/A		
	35	30/105	10	06	05		70	40	30(L)	Hot F/A		
A131 Gr. A Structural steel for ships	23	80/110	35	05	05		58/71	34	24	Welding grade		HR/N
	Cu = 35 (See Standards for other G) (In some Gr)											
A181 Gr. 60 70	35	110		35	05	05	60	30	22	35%RA		Forgings-C-steel for Gen purpose
	35	110		35	05	06	70	36	18	24%RA		
A182 Gr.F1	28	60/90	15/35	045	045	Mo44/	70	40	20	143/192	30%RA	Forged/Rolled alloy steels for valves, flanges, pipe fittings
F2	21	30/80	10/60	04	04	"	70	40	20	"	"	
...	65											
(See Standards for more details)												
A192	06/18	27/63	25	06	05		47	26	35	137 max	HF/ CD/A	Seamless boiler tube for high pressure services
A203 Gr. A	17/23	70	15/40	04	035	Ni = 210/	65/85	37	23	N or HT		Pressure Vessel plates: Alloy/Ni steel
B	21/25	80	15/40	04	035	250	65/85	37	23	"		
C	(See Standards for others)											
....												
A204 Gr. A	18/21	90	15/40	04	04	Mo = 45/60	65/85	37	23	HR or N		Pressure Vessel plates, Alloy Mo-bearing
B	20/25	90	15/40	04	035	"	65/85	37	23	"		
C	23/27	90	15/40	04	035	"	70/90	40	21	"		
A213	10/20	30/60	10/30	045	045	Cr-Mo alloy, depending on the structure: ferritic or austenetic steels	60	30	30	163 max (Gen range)		Seamless austenitic/ ferritic alloy steel for boiler super-heater
A240 Gr. 202, 203, 304, 304L, etc.	Depends on grade: Low C, S and P plus necessary alloy (See Standards for details)						95	38	40	202 max, typical		Heat resisting Cr and Cr-Ni Steel plate, sheet and strip for pressure vessel

ASTM Structural Grades	Chemical Composition (divide by 100)						Mechanical Properties					Remarks
	%C	Mn	Si	S	P	Other	T (Ksi)	Y	El* (2" GL)	Hardness HB	Impact	
A283 Gr. A	Open S = 050 max, P = 040 max, Cu = 20 min						45/55	24	30	(Treatment as required)		Low and inter TS carbon Steel plates, Shapes & bars
B												
C							50/60	27	28			
D							55/65	30	25			
							60/72	33	23			
A366	15	60	04	035		Cu = 20 min	Carbon steel sheet: Cold rolled com quality					CR sheet

Note: 1 ksi = 0.70 Kgf/mm^2; see conversion table in Appendices at the end.

Table 7A-2. List of some common grades of AISI engineering steels and their composition.

Steel Grade	%C	%Cr	%Mn	%Mo	%Ni	%P	%S	%Si
AISI 1018	0.14–0.20		0.30–0.90					
AISI 1040	0.36–0.44		0.60–0.90					
AISI 1095	0.90–1.04		0.30–0.50					
AISI 4023	0.20–0.25		0.70–0.90	0.20–0.30		0.035	0.040	0.15–0.30
AISI 4037	0.35–0.40		0.70–0.90	0.20–0.30		0.035	0.040	0.15–0.30
AISI 4118	0.18–0.23	0.40–0.60	0.70–0.90	0.08–0.15		0.035	0.040	0.15–0.30
AISI 4140	0.38–0.43	0.80–1.10	0.75–1.00	0.15–0.025		0.035	0.040	0.15–0.30
AISI 4161	0.56–0.64	0.80–1.10	0.75–1.10	0.15–0.25		0.035	0.040	0.15–0.30
AISI 4340	0.38–0.43	0.70–0.90	0.60–0.80	0.20–0.30	1.65–2.00	0.035	0.040	0.15–0.30
AISI 5120	0.17–0.22	0.70–0.90	0.70–0.90			0.035	0.040	0.15–0.30
AISI 5140	0.38–0.43	0.70–0.90	0.71–0.90			0.035	0.040	0.15–0.30
AISI 8620	0.18–0.23	0.40–0.60	0.75–0.90	0.15–0.25	0.40–0.70	0.035	0.040	0.15–0.30
AISI 8640	0.38–0.43	0.40–0.60	0.75–1.00	0.15–0.25	0.40–0.70	0.035	0.040	0.15–0.30
AISI 8660	0.56–0.64	0.40–0.60	0.75–1.00	0.15–0.25	0.40–0.70	0.035	0.040	0.15–0.30

Table 7A-3. Structure and broad composition of steels and their properties.

Structure	Yield Strength	Tensile Strength	Impact Strength	Fatigue Strength	Ductility	Toughness	Cold Forming	Machining	Chemistry*
Austenitic	Low	Low	—	—	High	—	Good	Poor	Low C + High alloy of Cr, Ni, Mn
Ferritic	Low	Low	Good	Low	Good	—	Excellent	Poor	Very low C
Ferritic + Pearlitic	Moderate	Moderate	Moderate	Moderate	Moderate, decreases with increasing Pearelite	Ok, if HT	Moderate, decreases with increasing P	Good to moderate	Low to medium carbon + with or without low-alloying (Mn, Si, Cr, Ni etc.)
Pearlitic	Good	Good	Poor	Poor in un-HT condition.	Poor	Good when HT	Poor	Good	High C with or without Low-alloying
Bainitic (HT steel)	High	High	Moderate	Moderate	—	Good	Very Poor	Poor, improves with tempering	Medium C + Low-alloying (Cr, Ni, Mn) Mo, V etc.)
Bainitic + Martensitic (H&Tempered)	High	High	Good	High	Low	High	Very poor	Good	Medium, C + low-alloy (Cr, Ni, Mo)
Martensitic (H&Tempered)	High	High	High	High	—	High	—	Good	Medium C + Low-alloy (Cr, Ni, Mo,)
HSLA: Fine grained with very fine carbide precipitates	High to Medium	Medium	Good	—	High	—	Excellent	Poor	Low C + micro-alloy of Nb, V or Ti; thermo-mechanically rolled.

* Carbon Equivalent (CE) value of the chemistry will determine 'weldability', as and when required. But, many poor to medium weldability steels can be welded by taking proper precautions and additional measures—like pre-heat and slow cooling.

Table 7A-4. Guide for selection of steel based on 'yield strength'.

Yield strength distribution in different classes and grades of steels. (9.85 MPa = 1Kg/mm^2)

Table 7A-5. SAE designation of stainless steels & their applications.

100 Series—austenitic chromium-nickel-manganese alloys

- Type 101—austenitic that is hardenable through cold working for furniture
- Type 102—austenitic general purpose stainless steel working for furniture

200 Series—austenitic chromium-nickel-manganese alloys

- Type 201—austenitic that is hardenable through cold working
- Type 202—austenitic general purpose stainless steel

300 Series—austenitic chromium-nickel alloys

- Type 301—highly ductile, for formed products. Also hardens rapidly during mechanical working. Good weldability. Better wear resistance and fatigue strength than 304.
- Type 302—same corrosion resistance as 304, with slightly higher strength due to additional carbon.
- Type 303—free machining version of 304 via addition of sulphur and phosphorus. Also referred to as "A1" in accordance with ISO 3506.
- Type 304—the most common grade; the classic 18/8 stainless steel. Also referred to as "A2" in accordance with ISO 3506.

- Type 304L—same as the 304 grade but contains less carbon to increase weldability. Is slightly weaker than 304.
- Type 304LN—same as 304L, but also nitrogen is added to obtain a much higher yield and tensile strength than 304L.
- Type 308—used as the filler metal when welding 304
- Type 309—better temperature resistance than 304, also sometimes used as filler metal when welding dissimilar steels, along with inconel
- Type 316—the second most common grade (after 304); for food and surgical stainless steel uses; alloy addition of molybdenum prevents specific forms of corrosion. It is also known as marine grade stainless steel due to its increased resistance to chloride corrosion compared to type 304. 316 is often used for building nuclear reprocessing plants. 316L is an extra low carbon grade of 316, generally used in stainless steel watches and marine applications, as well exclusively in the fabrication of reactor pressure vessels, due to its high resistance to corrosion. 316Ti includes titanium for heat resistance; therefore, it is used in flexible chimney liners.
- Type 321—similar to 304 but lower risk of weld decay due to addition of titanium. See also 347 with addition of niobium for desensitization during welding.

400 Series—ferritic and martensitic chromium alloys

- Type 405—ferritic for welding applications
- Type 408—heat-resistant; poor corrosion resistance; 11% chromium, 8% nickel.
- Type 409—cheapest type; used for automobile exhausts; ferritic (iron/chromium only).
- Type 410—martensitic (high-strength iron/chromium). Wear-resistant, but less corrosion-resistant.
- Type 416—easy to machine due to additional sulphur.
- Type 420—Cutlery Grade martensitic; similar to the Brearley's original rustless steel. Excellent polishability.
- Type 430—decorative, e.g., for automotive trim; ferritic. Good formability, but with reduced temperature and corrosion resistance.
- Type 439—ferritic grade, a higher grade version of 409 used for catalytic converter exhaust sections. Increased chromium for improved high temperature corrosion/oxidation resistance.
- Type 440—a higher grade of cutlery steel, with more carbon, allowing for much better edge retention when properly heat-treated. It can be hardened to approximately HRc 58 hardness, making it one of the hardest stainless steels. Due to its toughness and relatively low cost, mostly display for only. Replicas of swords or knives are made of 440 stainless. Available in four grades: 440A, 440B, 440C and the uncommon 440F (free machinable). 440A, having the least amount of carbon in it, is the most stain-resistant; 440C, having the most, is the strongest and is usually considered more desirable in knifemaking than 440A, except for diving or other salt-water applications.
- Type 446—f0or elevated temperature service.

500 Series—heat-resisting chromium alloys

600 Series—martensitic precipitation hardening alloys

- 601 through 604: Martensitic low-alloy steels.
- 610 through 613: Martensitic secondary hardening steels.
- 614 through 619: Martensitic chromium steels.
- 630 through 635: Semi-austenitic and martensitic precipitation-hardening stainless steels.
- Type 630 is most common PH stainless, better known as 17–4; 17% chromium, 4% nickel.
- 650 through 653: Austenitic steels strengthened by hot/cold work.
- 660 through 665: Austenitic superalloys; all grades except alloy 661 are strengthened by second-phase precipitation.

Type 2205—the most widely used duplex (ferritic/austenitic) stainless steel grade. It has both excellent corrosion resistance and high strength.

CHAPTER 8

Heat Treatment and Welding of Steels

The purpose of the chapter is to discuss the metallurgical practices of different heat treatment processes of steel, and also to outline the basics of welding metallurgy as applicable to steel. These two subjects have been discussed together, because of their common roots to the austenite decomposition and transformation involving heating and cooling. Metallurgically, processes of heat treatment of steels are based on the principles of austenite decomposition and transformation, and so is the process of welding, which additionally involves solidification. Various aspects of steel solidification and phase transformation have been already discussed in Chapter 2. Therefore, without discussing the same again, discussions of this chapter is more focused to describe the practical aspects of heat treatment and welding processes, which are, in principle, designed to give effect to produce desirable microstructures by controlling the phase transformations in steels.

8.1. Introduction

Uniqueness of steel lies in the fact that it can be heat treated to a wide range of strength, toughness and ductility. Steel properties can be easily tailor made for an application by appropriate choice of composition and heat treatment, chosen from many available ones. Each of these heat treatment types— e.g. stress-relieving, annealing, normalising, hardening, etc.—alter either the state of stress (internal) or the nature of microstructures; thereby producing the characteristic properties. The metallurgical principles of heat treatment emerge from the mechanisms of austenite decomposition, which have been critically discussed in Chapter 2.

Welding, which involves hot jointing of two steel pieces, either of similar or different steels, also follows many of the rules of heat treatment, involving localised heating and cooling of the weld zone. Localised heating and cooling of weld zone produces local stress and microstructural changes in the same way as in heat treatment processes. Thus, these two subjects of steel technology are linked to a common thread and so need to be explained together.

Steel is an alloy of carbon and few other elements, present in the range of few decimal percentages to few hundreds of decimal percentages, which are intentionally added to the steel composition, in order to make it responsive to required heat treatments or to impart certain special properties (e.g. corrosion and oxidation). Amongst the compositional constituents of steel, carbon is the most important element in terms of developing mechanical properties by change in carbon distribution. Therefore, most high temperature heat treatment processes are based primarily on controlling the carbon distribution—necessary for producing different microstructures e.g. ferritic, pearlitic, martensitic etc.—and carbide formation, resulting in microstructural changes and consequent mechanical properties. Other alloying elements—like Cr, Mo, Ni, V, etc.—influence the heat treatment operation by altering the time and course of microstructural transformation. For example, presence of Cr, Mo etc. as alloying elements, delay the transformation of austenite—the high temperature form of steel—to ferrite and pearlite, on the one hand, and, thereby, facilitate formation of bainite and martensite, on the other hand. Presence of such alloying elements also changes the nature of carbides, making them more complex in composition and character (refer Chapter 2).

Therefore, control over the time and course of transformation is the main aspect of heat treatment processes. As a result, heat treatment cycles, in general, are concerned with how to control such courses of transformation and obtain the desired microstructures. Control over the time and course of transformation in steels is generally carried out by controlling the cooling rate of the job from higher heat treatment temperature, called austenitising temperature (A_3 temperature). For example, in the uses of DIN C-45 (or IS-45C8 or AISI 1045) steel, if softer microstructure is desired, *annealing* of the steel with slow cooling inside the furnace from austenitising temperature is resorted to, which produces coarser grains and softer microstructures with coarser pearlitic lamellar spacing. However, when higher than that strength is required, the steel is *normalised* by air cooling—faster than furnace cooling—to produce microstructures with finer grain sizes and finer pearlitic spacing. If the same steel is required with even higher strength than the previous one (normalising), the steel is *hardened* by quenching (a fast cooling process) to produce martensitic structure, which is subsequently tempered down to obtain required strength and toughness. Thus, in the same steel of appropriate composition, mechanical properties can be altered through the control of cooling rate and the resultant microstructural changes—the products of transformation process. This is precisely the task of heat treatment operations.

Welding is also a process of heating and cooling, like the heat treatment, and much of its success depends on how well the heating and cooling rate is controlled to produce desired microstructures in the weld zone area. Heating is required for ensuring proper fusion of metals for welding, and cooling is the natural process that follows the post-weld operation. If the heat input is not controlled, there could be coarser grains and microstructures, lowering the toughness of the steel joint or even burning of steel, and if cooling rate is not controlled, there could be cracking of the weld metal and distortion of parent metal. Like heat treatment, welding is also susceptible to development of harmful residual stresses, when the weld joint is heated very fast or cooled very fast.

Therefore, planning and precautions are necessary to control the heating and cooling in both heat treatment and welding processes in order to: (a) obtain certain planned microstructiural combination for desired mechanical properties, and (b) to avoid ill-effects of very fast heating and cooling for avoiding harmful residual stresses, distortion and cracking. This chapter aims to discuss these aspects of heat treatment and welding processes from the point of view of their metallurgical practices and applications. It does not however, attempt to deal with the details of all heat treatment processes or welding processes. Rather, it attempts to analyse and discuss the processes from the angle of metallurgical practices in order to bring out the essentials of metallurgical principles.

8.2. Heat Treatability of Steels

All steels are not equally heat treatable. Heat treatability of steels differs with differing composition and type of heat treatment. For example, low-carbon unalloyed steels—like AISI C-1010, C1015 etc.—can be annealed, normalised or stress-relieved, but not hardened by quenching and tempering. For steels to be hardened by quenching, it requires a minimum level of 'hardenability', which is a function of carbon plus alloying elements present in the steel. All types of steel can be subjected to stress-relieving, annealing or normalising heat-treatment, but for hardening, it must have sufficient 'hardenability' for adequate response. Hardenability is, popularly, determined by 'Jominy Hardenability' Test, conducted as per national or international standards (e.g. BS 4437).

8.2.1. Jominy Hardenability Test

Hardenability refers to the ability of steel to be hardened to a specified depth below the surface to a defined hardness level, which could be a pre-defined hardness value (e.g. 50 Rockwell in C scale) or criteria set by %martensite in the structure (e.g. 50% martensite). The term is related to depth of hardness below the surface and must not be confused with maximum hardness, which is measured on the surface. Hardenability is determined by a standard test called 'Jominy Hardenability' test (ref: BS 4437) where a 25 mm cylindrical test bar of length 100 mm is heated to normal hardening temperature (i.e. above its A_3 temperature) and then quenched at an end in a standard fixture by controlled water velocity. The free height between the end to be quenched and water jet is 62.5 mm. Thereafter, the specimen is cooled, a portion of the surface is flattened by slow grinding, and hardness in Rockwell-C scale is measured at regular interval from the quenched end, and plotted. The distance up to the point where the steel shows 50% martensite, (or the corresponding hardness value), is taken as the measure of hardenability of the steel. Figure 8-1 illustrates the Jominy hardenability test set-up as per BS 4437.

Following the Jominy hardenability test, Jominy hardenability curves are generated by plotting the hardness values measured against distance from the quenched face. Figure 8-2 shows a series of hardenability curves for different carbon and alloy steels, established by following the Jominy hardenability test.

If 500 HV (approx. $50HR_C$) is taken as the required standard hardness to be achieved and then the corresponding depth of hardness is obtained by measuring the distance of the $50HR_C$ point from quenched end. Since the cooling rate will progressively change from the quenched face backward

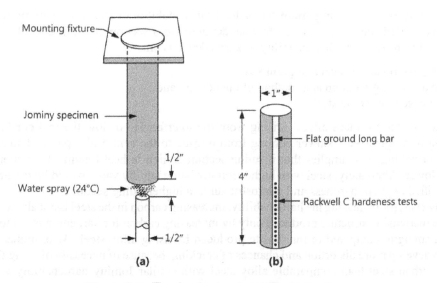

Figure 8-1. The Jominy hardenability test set up.

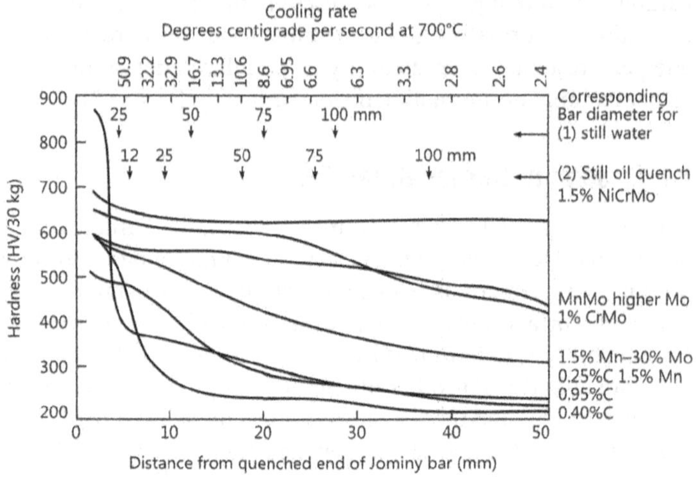

Figure 8-2. An illustration of the relative depth of hardening of different steels in Jominy test. Cooling rate shown above corresponds to cooling rate at the centre of the Jominy bar. *Source: Rollason, E.C. Metallurgy for Engineers. 1973.*

(where the rate of cooling is fastest), Jominy test can provide not only surface hardness data with varying cooling rates but also the resultant microstructure of different locations, if thin slices are cut at corresponding locations and examined under microscope. Thus, Jominy test provides a versatile test method for knowing in advance what would be the hardness and depth of hardness upon quenching the steel. Jominy hardenability measurements, represented in Fig. 8-2, show that plain carbon steels have lower hardenability than the alloy steels—sharply increasing with increase of alloy content like nickel, chromium and molybdenum.

8.2.2. Interpretation and Uses of Jominy Hardenability Data

Hardenability is the most important factor for heat treatability of steels; it determines if the steel can respond to a hardening operation and to what depth or cross-section, the steel could be hardened. In the practice of heat treatment, hardenability of steels depends on:

- Composition of the steel and grain size,
- The quenching medium and method of quenching, and
- Ruling section of the steel.

The latter is due to 'mass effect' arising from the interference of low thermal conductivity of steel, resulting in progressively slower cooling from outside to the centre of a piece of steel bar even with severe quenching. This implies that if ruling section is high in heat treatment, apparent hardenability will be lower. Alternately, steel with higher Jominy hardenability value would be required for attaining the specified depth of hardness and microstructure, in higher ruling section.

However, for increasing the hardenability, increasing carbon in the steel is not always recommended, because martensitic structure produced only by increasing carbon for hardening is not tough enough for a given strength, compared to martensite produced by using alloy steel. Also, higher carbon in steel tends to develop more distortion and chances of cracking, because of necessity of using faster quenching rate in carbon steel than comparable alloy steel with similar Jominy hardenability value. Therefore,

Table 8-1. The effect of main alloying elements in steel, including carbon, and grain size effect.

Alloying Element	Qualitative Effect on Hardenability	Multiplication Factor per %Concentration[†]	C + Grain Size Effect[‡]
Carbon	Increases (mild)	Carbon effect is combined with grain size (see next column)	At **0.40%C**: 0.230, 0.213, and 0.198 At **0.90%C**: 0.346, 0.321, 0.296, for respective grain size
Manganese	Increases (strong)	**0.40**: × 2.333, and **0.90**: × 4.000	
Silicon	Increases (moderate)	**0.40**: × 1.280, and **0.90**: × 1.63	
Nickel	Increases (moderate)	**0.40**: × 1.280, and **0.90**: × 1.321	
Chromium	Increases (strong)	**0.40**: ×1.864, and **0.90**: ×2.944	
Molybdenum[*]	Increases (strong)	**0.05**: ×1.15, and **0.40**: × 2.20	

* This is as per Grossman formula, and the whole range of factors for all other elements has been avoided.
† Grossman formula.
‡ Grain size ASTM #6, #7 and #8 respectively.

beyond a certain carbon level, it is best to obtain high hardenability by addition of suitable alloying elements, in cases where toughness requirement is high and tolerance to distortion is low. In principle, carbon in the steel should be adequate for attaining required surface hardness, whereas alloy contents should be adjusted for attaining required depth of hardening in the steel. This, however, does not mean carbon has no role in increasing depth of hardness; it only points out to the limitation. Hence, a judicious choice of carbon plus alloying elements is necessary for attaining the desired hardness as well as depth of hardening.

Metallurgical factors that contribute to hardenability are chemical composition and grain size, and they behave in the same way as they influence martensite formation as proposed by Grossman. Table 8-1 illustrates the effect of carbon and various common alloying elements on the hardenability of steels.

As far as hardenability of steels is concerned; carbon has relatively milder effect on hardenability but strong effect on the development of hardness. Hence, %carbon in steel should be designed to control more of surface hardness level in the steel than hardenability (refer Chapters-1 and 2), and alloy content should be designed to control the hardenability in the steel, but should be balanced in order to keep the alloying cost low and make the steel cost-effective.

Exact influence of carbon and alloying elements on hardenability of steels can be seen and compared from Fig. 8-3, which shows the hardenability curves of plain carbon and alloy steels of comparable carbon content (0.40%C) but with varying alloy contents. If 50% martensite is taken as the acceptable value of Jominy hardenability, then the horizontal line drawn through the figure at 50% martensite in Fig. 8-3 shows the difference in value of hardenability of these steels, as illustrated by the difference in hardenability (50% martensite criteria) of SAE 4140, SAE 5140 and SAE 1040 steels. There is marked difference in the hardenability value of these steels although carbon level is similar. Such experimental study clearly establishes that, for all practical purposes, carbon content determines the achievable hardness level in the steel, whilst alloying elements such as nickel, chromium, manganese and molybdenum determine the depth of hardening—corresponding to that hardness—for a given ruling section, which has been pointed out earlier. The Jominy hardenability value is typically expressed by the depth of hardness to which the steel can be effectively hardened to 50% martensite or any chosen criterion of hardness.

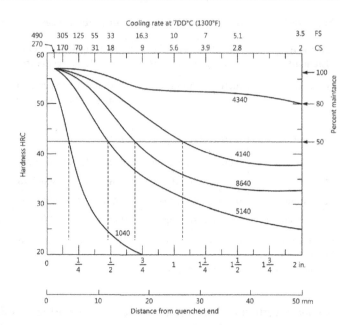

Figure 8-3. Jominy hardenability curves of 0.40% plain carbon and alloy steels. Note: Corresponding 50, 80 and 100% martensite hardness line has been shown on the right hand vertical axis.

Study of hardenability provides information about the ability of the steel selected to produce sufficient martensite in the structure so that the steel part undergoing heat treatment, develops required level of strength and toughness after tempering for a specific application. This information is necessary for right selection of steel and right choice of heat treatment for a given application. For example, referring to Fig. 8-3, it would be obvious that choice of AISI 1040—a plain carbon steel of 0.40% average carbon—will be sufficient for a thin section hardening, whereas an alloy steel of chromium or Cr-Mo containing such as SAE 5140 or 4140 respectively, will be required for hardening of heavier sections. Again, if the section to be hardened is within the reach of SAE 4140 or 5140 steels, choosing expensive alloy steel like SAE 4340, containing Cr-Ni-Mo and capable of producing much higher volume of martensite, might be unnecessary. Thus, hardenability value not only provides information about heat treatability, but also guides to the right selection of steel.

8.2.3. Cooling Rate Dependence of Hardenability of Steel

Laboratory hardenability test is a simulation test, where cooling condition and rate is simulated, as per given standard, by injecting water onto a face. This cooling rate may not be the same that steels encounter during practical heat treating. Hence, the hardenability value is comparative and not an evaluation of what it would be in actual practice, where quenching condition may vary. The actual performance of steels will depend on the quenching efficiency and effectiveness as practised in the industry.

Table 8-2 gives the quench severity of different quenching media and conditions, which is expressed by *H-value*. It can be observed that oil quenching with no external agitation has the H-value of 0.20 and water quenching with strong agitation has H-value of 1.5, in the relative scale. The table also demonstrates that agitation is an influencing factor for quenching effectiveness. Hence, there is a need for correlating the Jominy hardenability value with that of hardenability value obtainable by industrial cooling rate, having different H-value, for all practical purposes. This correlation can be established by referring to what is known as 'Ideal critical diameter' (D_I) *vis-à-vis* 'critical diameter' (D) corresponding to a particular quench severity value H. Ideal critical diameter is defined as: the hardened diameter that has 50% martensite at the centre under an ideal quenching condition i.e. when the surface is cooled at an infinite rate* (see Table 8-2)—this is what Jominy test attempts to simulate by cooling the test face by high pressure water jet.

Fortunately, there is good correlation between critical diameter of hardening by Jominy test—referred to as 'ideal critical diameter' (D_I)—and the 'critical diameter'

Table 8-2. An illustration of the quench severity of different quenching media and bath conditions.

Quenching Conditions	H-value
Poor oil quench—no agitation	0.20
Good oil quench—moderate agitation	0.35
Very good oil quench—good agitation	0.50
Strong oil quench—strong agitation	0.70
Poor water quench—no agitation	1.0
Very good water quench—strong agitation	1.5
Brine quench—no agitation	2.0
Brine quench—violent agitation	5.0
*Ideal quench (referring to DI value)	∞ (infinity)

(D) obtained in actual heat-treatment practice by following different industrial cooling practices which give rise to different 'quench severity', marked with H-value. Figure 8-4 demonstrates the correlation between the D_I and D obtainable by following a quenching system of definite H-value. In this respect, D_I value acts as reference point for comparison of D value by following a particular quenching system. It can be noted that D and D_I value nearly merge for highest H value (H = 5.0) and gradually decrease with reducing quenching severity (H-value). Thus, for effective hardening, steels require adjustment of

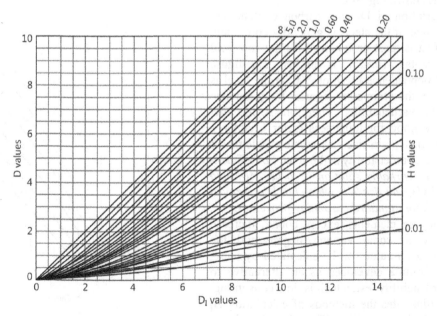

Figure 8-4. Relationship of the 'Critical diameter' (D) to the 'Ideal critical diameter' (D_I) for different rates of cooling (H-values).

composition along with selection of right quenching process for producing martensitic structure in the steel. This is the central concern of heat treatability of steel.

8.2.4. Practical Approach to Heat Treatability of Steel

For appropriate heat treatability, the steel composition, and also the steelmaking process for control of grain size, should be suitably modified, so that martensite can be formed at greater depth and at lower cooling rates—the latter is to avoid distortion and cracking in heat treatment. In this respect, carbon is required for the formation of martensite with a minimum level of hardness (see Fig. 2-12 illustrating hardness of martensite versus carbon), and alloying elements are required for increasing the depth upto which that hardness can be obtained in the steel after heat treatment. According to Table 8-1, chromium and molybdenum have strong effect on hardenability, whereas nickel has somewhat moderate effect. However, nickel containing steel tends to improve toughness of the resultant microstructure. Furthermore, presence of nickel and molybdenum in steel tends to promote minor amount of *retained austenite* spots—i.e. untransformed islands of austenite—which are good for arresting propagation of crack within the matrix, leading to higher toughness.

Thus, heat treatment and heat treatability of engineering steels is a tricky question of selecting steel with suitable carbon and alloying elements for maximising martensite formation upto a depth of hardening under a chosen cooling condition, but without causing any distortion—due to high residual stress—or cracking of the hardened parts. Grain size of the steel and steelmaking practice i.e. killing practice, also plays a part by influencing the martensite formation process (as discussed in Chapters 3 and 4). The effect of carbon and alloying elements on hardenability has been shown in Table 8-2, the effect of grain size and carbon content on the ideal critical diameter of steel—which is also a reflection of martensite formation in the steel—is shown Fig. 8-5.

The variation of D_I with carbon content and prior grain size of the steels is generally combined, and the effect is most pronounced till the eutectoid carbon percentage (0.85%) is reached. The figure shows that dependence of D_I on %C in the steel is influenced by the grain size—coarser the grain size better is the hardenability. Therefore, while fine grain size is good for toughness (refer Chapter 1 and 3), too fine a grain size may create problems for attaining good hardenability in the steel. *For efficient hardening, a workable solution is to use Al-killed steel of ASTM grain size 5–8, which has been proven good enough for standard hardening processes.*

All alloying elements, except cobalt, increase hardenability to some varying degree and the corresponding D_I value can be calculated by using Grossman formula developed as an alternative to Jominy hardenability test, which is time consuming. Table 8-3 illustrates the methods of calculating D_I value of a steel composition. The table also shows the strong effect of chromium and molybdenum on

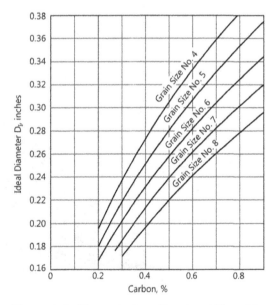

Figure 8-5. The variation of hardenability with carbon and prior grain size in steels.

Table 8-3. Grossman method of calculating hardenability of steel.

Method of calculation of D_I values for Steels
D_I value = f (C) × f (Si) × f (Mn) × f (Cr) × f (Mo) × f (Ni), where:
$f(C) = 2 \times \sqrt{(0.024 \times \%C)}$
$f(Si) = 0.7 \times \%Si + 1$
$f(Mn) = 3.35 \times \%Mn + 1$ (when %Mn <1.2%)
$f(Mn) = 5.00 \times \%Mn + 1$ (when %Mn >1.2%)
$f(Cr) = 2.16 \times \%Cr + 1$
$f(Ni) = 0.36 \times \%Ni + 1$ (when Ni <1.8%)
$f(Ni) = \dfrac{(Ni-1)^2}{3.35} + 1.5$ (when Ni <1.8%)
$f(Mo) = 3 \times \%Mo + 1$

Note: This is only an assessment and not a guarantee, like Jominy value. The assessment is based on ASTM grain sizes 5–8. Calculation of hardenability by using Grossman formula can be very useful in absence of otherwise ready metallurgical data.

hardenability. In this table, effect of molybdenum (Mo) has been shown at lower level than others, because Mo addition to steels is limited within the range of 0.05%–0.40% in most engineering steels, except stainless steels.

Boron is another element that has a very strong effect on hardenability (though not shown in the Table 8-3). The effect of Boron is appreciable even in the range of 0.0005 to 0.003%. Because, of this splendid effect of boron (B) on hardenability, boron containing low-carbon steels are extensively used for high-tensile 'fasteners' after heat-treatment, e.g. 8 K and 10 K grade 'nuts and bolts' for engineering applications.

8.3. Heat Treatment Types and Processes

Heat treatment is a process of controlled heating and cooling of materials (i.e. steel, in this case) for altering their physical and mechanical properties without any change of shape, size and dimensions of the product. Heat treatment processes can be grouped under the following three heads as per the process objectives:

- Softening processes—example: Annealing, Sub-critical Annealing and Normalising;
- Hardening processes—example: Quenching and Tempering, Induction Hardening; and
- Thermo-chemical processes—example: Carburising, Nitriding, etc.

These apart, there are other auxiliary processes such as, stress-relieving, ageing etc. which are lower temperature operations for changing the state of internal stresses and strength. Heat treatment can also be used to alter certain manufacturability, such as improving machining, formability by bending and pressing, restoring ductility after cold working, etc. Thus, it is a very versatile process that can—(a) improve manufacturing process by softening, stress relieving, ductility enhancement etc. and (b) improve product performance by increasing strength, toughness and other desirable characteristics of the steel.

In fact, heat treatment can very closely tailor the properties in steels necessary for forming as well as end-use applications. However, it should be noted that not all steels will respond to all heat treatment processes. The following table summaries the responses, or otherwise, of different steels to the different heat treatment processes.

Steel Types	Annealing	Normalising	Hardening	Tempering
Low carbon <0.3%	yes	yes	no	no
Medium carbon 0.3%–0.5%	yes	yes	yes	yes
High carbon >0.5%	yes	yes	yes	yes
Low alloy	yes	yes	yes	yes
Medium alloy	yes	yes	yes	yes
High alloy	yes	may be	yes	yes
Tool steels	yes	no	yes	yes
Stainless steel (austenitic e.g. 304, 306)	yes	no	no	no
Stainless steels (ferritic and others e.g. 405, 430 442)	yes	no	yes	yes

Common heat-treatment processes are annealing, normalising, and hardening and tempering. And other processes that are used in conjunction with these or independently include: stress-relieving, induction hardening, carburising, nitriding, ageing etc. But, each of them imparts certain characteristic properties to the steel which are beneficial for either processing (manufacturing) or end application of the steel. The commonality between these processes is that they all have to be carried out with reference to standard Fe-C diagram or relevant T-T-T/C-C-T diagram (see Figs. 2-2, 2-5 and 2-8 in Chapter 2). Details of metallurgical aspects of hardening and tempering and the means and methods of controlling the processes for desired mechanical properties from the angle of austenite decomposition have been discussed in Chapter 2 (refer Sections 2.4 to 2.8). This chapter will, therefore, focus more on the practices and precautions of heat treatment from the metallurgical practice point of view. Some features and benefits of various heat treatment processes are described below.

8.3.1. Annealing

Annealing is a softening process carried out by heating above the critical temperature, holding at that temperature for soaking for a time (as per cross-section), and then slowly cooling (inside the furnace or a closed area) for causing changes in properties, such as hardness and ductility (elongation). Figure 8-6 shows the range of annealing temperatures vis-à-vis steel composition, which shows that for steel composition up to 0.85%carbon (eutectoid carbon), full annealing is carried out by heating above the corresponding A3 temperature of the steel and above this carbon level, annealing is to be done above the lower critical temperature (A1). The generally recommended full annealing temperature is about $(A_3 + 50°C$ min.). This being a high temperature operation, full annealing is always associated with total phase changes, resulting in the change of microstructure and resultant properties. For example, full annealing produces coarser lamellae of pearlite and coarser grain sizes, which results in lowering the strength. Annealing can be also carried out just below the A_1 temperature, but that is *sub-critical annealing*—designed to induce ductility to cold worked steel, soften material, relieve internal stresses

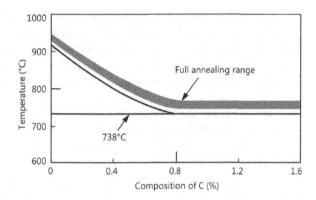

Figure 8-6. Fe-C diagram showing the range of annealing temperatutes *vis-a-vis* steel composition.

and improve cold working properties. Full annealing is used for softening the steel and for homogenising the structure.

Annealing of steel takes place in three steps: (1) recovery of any remaining stresses, (2) recrystallisation of the steel and (3) grain growth. Full softening occurs when temperature and time allowed are sufficient for completion of these steps. However, for sub-critical annealing, which is widely practiced in industries, stages up to recrystallisation are only allowed. *Sub-critical annealing* is applied to soften cold-worked steel parts, castings and wire rods by heating between $550°C$ to $700°C$, i.e. below the lower critical temperature, for several hours in order to produce some degree of recrystallisation and spheriodisation of carbides in pearlite. Since the process uses lower temperature, the process is cost effective and saves energy, provided it serves the purpose.

Full annealing process induces more softness in the steel by helping the iron carbides to ball-up (i.e. globularise) and by grain growth. The process is very effective for softening steel forgings, steel castings, cold-worked steel sheets and plates, bars and wire rods, and tools. However, care is necessary to control the full annealing temperature and time in order to avoid over-heating, burning or under-annealed structures, which are highly objectionable structures for all industrial applications.

8.3.2. Normalising

Normalising also consists of heating the job above upper critical temperature—for steels up to eutectoid composition—of the corresponding steel composition, but lower than the full annealing temperature of the steel. If annealing temperature is $A_3 + 50°C$, normalising temperature is $A_3 + 30°C$ of the steel. The normalising process involves holding for temperature equalisation, soaking for a specific time, and then cooled in the air—as against furnace cooling in annealing. This treatment is especially given for homogenising the composition and for producing finer grain size.

The main difference with full annealing is that normalising is carried out by faster cooling (air cooling) than annealing. Such cooling process gives rise to finer grain size and finer pearlite structure, resulting in stronger properties. Since the normalising process involves faster air cooling, there could be variation of structure between surface and the centre, because of slower cooling of centre due to thermal lag between surface and centre, when the steel bar is of larger diameter. Therefore, properties of normalised steel might vary with the section size—higher the section lower are the strength properties of the given section. In contrast, full annealed structure is more uniform throughout a cross-section. Normalising produces, in general, higher tensile strength, yield strength, reduction in area, and impact strength than the annealing process. Higher strength of normalised structure comes from finer ferrite grain size and finer pearlitic spacing compared to annealing.

The primary purpose of normalising is to obtain uniform composition and finer structure, resulting in higher strength than annealing. Other objective of normalising is to counter the effects of prior working of the steel and inhomogeneity in structure by processes like casting, forging or rolling—it is especially useful after forging of steels, for promoting good machining and uniform response in hardening. Normalising can refine the existing non-uniform composition and structures into a more homogeneous composition and uniform structure, which enhances machining, forming or can even produce higher and uniform properties after hardening. However, there is a risk of surface degradation by oxidation and scaling in both annealing and normalising processes. Hence, the heat treated parts require shot-blasting or chemical pickling for cleaning of surface before being put to use for further operations e.g. machining, forming, welding etc.

8.3.3. Hardening

Hardening of steel occupies the centre stage of steel heat treatment, being the most important and critical process of all heat treatment processes. The hardening process requires:

- A minimum level of hardenability of the steel as per required structure and properties;
- Good heating and quenching facilities to match the required control over the heating and cooling cycle;
- Facility for tempering the jobs within a close range of temperature;
- Facility for surface cleaning by non-chemical processes, like shot-blasting; and
- Testing/Inspection for cracks and distortion in the parts after finish heat-treatment.

Hardening operation is carried out with reference to corresponding CCT diagram of the steel, which has been discussed in Chapter 2. An illustrative CCT diagram is presented here in Fig. 8-7 for present discussions. With reference to CCT diagrams see Figs. 2-7, 2-8 and 8-7), hardening process steps can be listed as follows:

- Heating the job above the upper critical temperature of the steel i.e. above A_3 temperature—referred as austenitisation temperature;
- Holding the job at that temperature for full austenitisation—referred to as soaking time;
- Quenching appropriately commensurate with steel composition and its hardenability in a medium; Cooling rate of quenching should be sufficient to directly cool to martensite formation temperature range without cutting into any other territory of the CCT diagram;
- Holding in the quenching bath for 100% transformation to martensite; and
- Tempering the martensite—with minimum time gap between quenching and tempering—for adjustment of strength and toughness.

Figure 8-7 brings out the reality of cooling in materials, indicating that core of a body always cools at slower rate than the surface due to lower thermal conductivity of the material, causing thermal time lag between the surface and core. This necessitates that cooling rate for quenching has to be designed with respect to the centre of the job in order to get 100% martensite in the bulk. Any slackness in cooling

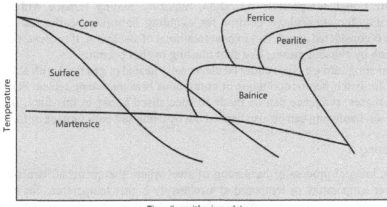

Figure 8-7. The CCT-diagram of steel indicating different fields of phase transformation and a set of practical cooling curves in which cooling rate between surface and the core differs.

rate might give rise to non-martensitic structure inside—as would be evident from the core cooling in Fig. 8-7 which enters the bainite filed. However, faster cooling rate than what is required for 100% martensite is also undesirable, because of chances of cracks and distortion under faster cooling.

Hardening is a precision heat treatment operation, requiring close control of heating, holding and cooling. Heating for austenitisation is done at temperature range of $(A_3 + 25°C–30°C)$—higher temperature mighy cause higher distortion. Soaking time for 100% austenite conversion will depend on the composition of the steel—plain carbon takes lower time than steel with complex alloy carbide forming alloying elements. Generally, the range is 30–45 min/1 inch (30 mm) section, which is taken as the thumb rule in general. However, care has to be taken so that the heating quality and time in the furnace does not induce excessive decarburisation or oxidation of the steel surface by the reaction of carbon in the steel and oxygen in the air inside the furnace.

Hardening of steel is synonymous with the process of 'quenching and tempering'. Common quenching media are water, oil and salt-bath, with or without agitation, and each having its characteristis cooling rate. Table 8-2 indicates the H-value (quench severity) of some cooling media that are commonly used for steel hardening—higher H-value of the media might produce more martensite in the structure but might also cause higher distortion of the job, which is objectionable.

Quenching involves quick transfer of the heated job to the cooling tank (quenching bath) containing appropriate medium in sufficient volume. Sufficient volume of the quenching medium is required for ensuring efficient transfer of heat from the quenched jobs to the fluid in the tank without abnormal rise in bath temperature. Any appreciable rise in quench bath temperature will slow down the cooling rate and, thereby, may induce formation of non-martensitic product in the structure. After allowing short time for temperature equalisation of quenched jobs and for completion of transformation, the jobs are taken out and tempered. The time of holding in the quenching bath should be so controlled as to cool the job below the martensite transformation finish temperature (M_F). Thereafter, the job can be cooled in the air or normally.

Hardening is a high temperature operation, requiring heating furnace with adequate capacity and having facility for atmosphere control for avoiding harmful decarburisation or deoxidation. Decarburisation is considered as a defect in heat treatment of steels and, if present, needs to be removed from the finish job by suitable means, like shot blasting or shot peening.

Modern hardening furnace types could be electrically heated or gas fired with atmosphere control or vacuum system for batch, semi-continuous or continuous heat treatment cycles. For more information on hardening furnaces, reference can be made to specialised books in this filed. In addition to these common processes, hardening can be also carried out by using salt bath furnace or fluidised bed furnace.

8.3.3.1. Tempering

Tempering is an integral process of hardening of steel where the quenched structure is either 'stress relieved' at lower temperature or tempered at a relatively higher temperature for a specific time. The purpose of stress relieving process is to release the locked-in quenched stresses without much change in the structure. As such the process is carried out at a relatively lower temperature e.g. between $150°C$ to $220°C$. But, the purpose of tempering is to induce toughness in the steel by change in the structure, such as by precipitation of carbide out of the strained as-quenched martensitic structure. Hence, temperature used for tempering is in the range of $300°C$ upward, depending on the degree of hardness to be retained and toughness to be improved. Details of tempering stages in steel have been discussed later.

Tempering is a relatively lower temperature operation and requires a furnace with high convective heat for temperature pick up and equalisation, such as air circulating furnace or salt bath furnace with fluid circulating mechanism. An important feature of tempering operation is that tempering should be carried out as early as possible after hardening in order to avoid the cracking and distortion of jobs due to internal stress produced during quenching. Hence, very often, hardening and tempering facilities are built alongside each other.

Tempering can be carried out over a range of temperature, but below the Ac_1 (the lower critical temperature) to avoid phase transformation. Tempering operation must avoid allotropic phase transformation. Objective of tempering is to (a) relieve locked-in internal stresses from the structure due to fast quenching and (b) controlled softening for inducing toughness in the hardened martensitic steel structure. Controlled softening is accomplished by heating at relatively higher temperature—e.g. $400°C$ and above as per degree of softening required—where diffusion rate of carbon atoms are sufficient in the matrix, allowing the carbon atoms to come out of the strained martensitic structure and precipitate nearby as carbide—e.g. Fe_3C or alloy carbide as per steel composition—by combining with the iron or alloying elements in the matrix. If it is plain carbon steel, the carbides are only Fe_3C (plain iron carbide) and tempering temperatures could be lower because diffusion of carbon is easier. But, if the steel is alloy steel, tempering temperature will be higher, because of relatively higher temperature and/or time required for diffusion of carbon and alloying elements to form alloy carbides.

Martensite produced after quenching is very strained and very hard. By tempering, carbon atoms from the distorted and strained martensite lattice (in fact, it is distorted ferrite type lattice with high tetragonality) starts migrating out and form iron carbide (if plain carbon steel) or alloy carbide (in alloy steels) by combining with iron or alloys in the matrix. Tempering takes place in stages, depending on the temperature. The stages of tempering are in general:

1. Below $100°C$, carbon slowly diffuses to areas of lattice defects (e.g. dislocation sites).
2. Between $100°C$ to $200°C$, carbon atoms start migrating from their unfavourable martensitic lattice and dislocation sites to the matrix and precipitation of fine epsilon-carbide ($Fe_{2.4}C$) starts, which is a transitional phase. At this stage most of internal stresses get released.

3. When temperature is increased further, this process is accelerated. At about 300°C–400°C, epsilon carbides coalesces and remaining carbon atoms stars leaving the martensitic lattice and forms normal carbides (Fe_3C). This causes lowering of strength and increase of ductility (toughness).

4. At temperature over 400°C, martensite structure begins to appreciably soften by coalesecence of precipitated fine carbides.

The process is faster for plain carbon steel, and for alloy steels the process is slower, requiring longer time or higher temperature. If alloying elements are chromium, molybdenum, vanadium or tungsten, the process is even slower, requiring higher tempering temperature for attaining similar strength and toughness. Molybdenum as an alloying element gives rise to further hardening of the steel (called secondary hardening) when heated above 550°C due to formation of special complex alloy carbides. As a rule of thumb, higher the tempering temperature lower is the final hardness and strength, but greater is the toughness. However, some alloy steels of type Cr-Mo, Cr-V exhibis increase of strength in the steel when tempered above 550°C. This phenomenon is known as 'secondary hardening'.

Generally, tempering is conducted in the temperature range of 150°C–650°C, depending on the type of steel, and the process is time dependent as the microstructural changes occur relatively slowly. However, care is necessary in tempering for avoiding, what is known as 'temper brittlenes'. There are two forms of this kind of brittleness; one is known as temper embrittlement that affects both carbon and low alloy steels when either they are cooled too slowly from above >575°C or are held for excessive times in the range of 375°C–575°C. This embrittlement effect can, however, be reversed by heating to above 575°C and rapidly cooling. The other embrittlement is known as 'blue brittleness' that affects carbon and some alloy steels after tempering in the range 230°C–370°C. This effect is not reversible and longer time tempering in this temperature range should be avoided. Steels suspected to have induced blue brittleness should not be used in applications in which they face shock loads.

8.4. Surface Hardening

The hardening and tempering process discussed in the preceding sections is referred to as 'bulk hardening process' or 'through hardening' process, where the aim is to produce as much martensitic structure as possible (or required) in the cross section of the job for bulk strength and toughness. There are other hardening processes which are not through hardening but selective hardening of the surface area. This can be accomplished either by using a hardenable steel as for induction or flame hardening or by selective alloying of the surface area by thermo-chemical processes such as, carburising, nitriding, etc. which will be discussed in the next section, and induction or flame hardening will be briefly discussed here.

In these processes, a portion or the surface of the job—which has to be hardened for carrying the load or withstand the wear—is heated by induction heating or flame heating to temperature above AC_3 and then spot cooled by using a suitable medium like forced air, water or oil. Principle of flame and induction hardening is same, but induction hardening is more precise and controllable than flame hardening. Flame hardening is carried by heating the selected area by appropriate nature of flame (e.g. reducing flame) and then cooling fast to produce martensitic structure in the heated zone. Flame hardening lacks precision and sophistication compared to induction hardening. Induction hardening is carried out by heating the job under a suitable inductor where heat in-put can be controlled by controlling the power frequency. Figure 8-8 illustrates a typical set up for induction hardening of a steel shaft.

Inductor

Figure 8-8. Practical induction hardening set-up for a shaft hardening where inductor encases the job and the bottom wider cooling ring has fine pores for forced air cooling. The inductor travels upward or the job travels downward for progressive hardening.

Power frequency required for induction heating is higher for lower depth of hardening, decreasing with increasing depth. Typical frequency range used for different case depth and bar diameter is illustrated below.

Case Depth (mm)	Bar Diameter (mm)	Frequency (kHz)
0.8–1.5	5–25	200–400
1.5–3.0	10–50	10–100
	>50	3–10
3.0–10.0	20–50	3–10
	50–100	1–3
	>100	

Note: This table is indicative only; one needs to balance between power densities and frequency for optimum result.

Induction or flame hardened jobs also require stress relieving or tempering as for bulk hardening process, but, stress relieving is the general norm. Purpose and principle of stress relieving is also same as for bulk hardening i.e. to relieve the locked-in (internal) stresses that arise from the fast cooling (quenching) process used in hardening. Flame or induction hardened steels are generally stress-relieved by heating at temperature between 150°C to 180°C and holding for a short time (e.g. 30–45 min as per section thickness).

The other important surface hardening processes are carburising, nitriding carbo-nitriding, boronising etc., which fall under 'thermo-chemical heat treatment processes'.

8.5. Thermo-Chemical Processes of Hardening

Thermo-chemical process of heat treatment are involved with thermally assisted diffusion of alloying elements—e.g. carbon, nitrogen or a combination of them—to a predetermined depth below the surface, which upon diffusing into the steel body changes the steel composition of the surface area and helps to form hard martensite or a hard chemical compound—e.g. nitride or carbide—when cooled appropriately from the upper temperature. The diffusing elements—generally C and N, but it can be Boron also for special application—are supplied by external environment around the job being hardened. The environment is made rich with the diffusing elements either in gaseous, liquid or powder form. The process results in a surface which is a radically different composition from the bulk. Depending on the nature and character of major diffusing elements, these processes are called 'carburising', 'nitriding', carbo-nitriding', etc. Similar processes can be also used for 'boronising'—a process that causes to diffuse boron to the surface and produces hard boron carbide on the surface. Boronising can be used for highly wearing parts in engineering assemblies. These processes follow the thermo-dynamical rules of Fe-C diagram for diffusion, where the process is carried out either above the upper critical temperature, A_3, of the steel—e.g. for carburising and boronising—or below the lower critical temperature, A_1, of the steel—e.g. for nitriding.

8.5.1. Carburising

Carburising requires steels of initially leaner carbon content in order to ensure that (a) there is sufficient chemical potential for carbon diffusion from the carbon rich environment to carbon lean steel surface and (b) the core, where no changes in the composition takes place, remain tough while hardening the carbon rich surface layers. Figure 8-9 shows a schematic diagram of carbon diffusion into iron lattice for enriching the surface area with carbon in carburising process.

Carburising process can be carried out using solid, gas or liquid (salt bath) medium, and they are accordingly called *pack carburising*—packed in solid carbon rich medium; *gas carburising*—exposed to gaseous carbon rich medium; and *salt bath carburising*—immersed in liquid salt of appropriate composition. The process of carburising is carried out above A_3 temperature in the range of about $900°C$–$1000°C$, in general, for a length of period commensurate with the temperature, required depth of carburising and method of carburising. In general, 'pack carburising' process using solid carbon rich powders takes longer time (about 1–2 full days), gas carburising process using carbon rich gaseous atmosphere for carbon diffusion takes about half or less of pack carburising time (about 8–12 hr), and shortest time is taken is by liquid salt bath carburising (about 2–5 hr) but for lower case depth. Figure 8-10 shows the trend of case depth obtainable by different carburising processes.

Among these carburising processes, gas carburising process, which is widely used in industries, can be easily made fully automatic and controllable for precision working. The gas carburising processes are now carried out almost universally in 'seal quench furnace' and subsequent hardening heat treatment is carried out immediately on-line without taking the work out of the furnace. This arrangement provides handling convenience and reduces the cost of handling and improves quality of the out-put products as well.

However, in a general manual set-up, after the process is over, the jobs or charges are taken

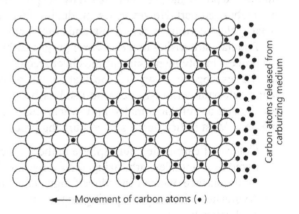

◄—— Movement of carbon atoms (•)

Figure 8.9. A schematic illustratios of the process of carbon diffusion into the iron lattice.

out of the furnace and rapidly cooled to stop further diffusion of carbon inside the job while hot so that *case depth*—depth up to which carbon diffused and changed the structure—remains within the stipulated and controlled limits. The jobs or charges are then re-heated in another furnace—with control of atmosphere for avoiding any decarburisation—and heated to above A₃ temperature corresponding to surface carbon—which is generally about 840°C–860°C. Thereafter, the charge is sharply quenched in oil or salt bath for hardening; case carburised jobs are never quenched in water due to high risk of cracking and distortion. The quenching process produces hard martensite at the carbon rich area of the surface and a relatively softer and tougher microstructure at the core. Since as-quenched martensite is hard but brittle, the jobs are then lightly tempered (at about 180°C–220°C) to relieve stress without losing much of its hardness. Generally, surface

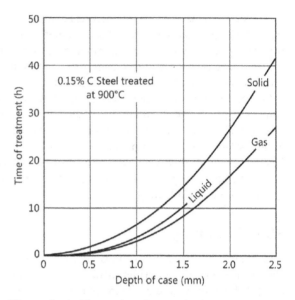

Figure 8-10. Time taken for producing a finite case depth by different carburising process.

hardness of 58–62 HR$_C$ is aimed at on the carburised surface and core strength of 35–45 HR$_C$, depending upon the application. Common starting point for carburising process is to start with normalised and finished machined parts. Carburising and hardening is the finish heat treatment operation; any correction of finish dimensions can be done only by light grinding, whenever necessary.

Carburising is a popular process for many engineering applications, involving wear and fatigue resistance. The process not only imparts high surface hardness beneficial for wear resistance, but can also produce favourable *surface compressive residual stress,* which is very beneficial for countering fatigue load in service. Effect of tensile stress part of fatigue cycle which causes the fatigue crack to initiate and propagate gets diminished by the presence of such residual compressive stress on the surface of a carburised part. However, for producing favourable compressive residual stress on the carburised surface, close control of case-depth and core strength is necessary in the corresponding steel.

Popular carburising processes are: (a) gas carburising or (b) salt-bath carburising. *Pack carburisin* —a process involving packing of jobs in charcoal or similar C-bearing material—is now a near obsolete process, excepg for a very big job that cannot be economically handled in a gas carburising furnace. While salt-bath carburising can be used as batch type furnace for mass production, gas carburising process can be fully automated and synchronised for continuous production of jobs required in large numbers. Most critical factors in carburising process (including hardening after carburising) include:

- Control of carburising atmosphere in the furnace to a constant level of carbon potential in order to avoid over or under carburising and formation of undesirable microstructure—e.g. angular carbides on the surface, which can induce fatigue crack on the hard surface in service;
- Avoidance of scaling and pitting on the surface of heated jobs—adverse for wear as well for fatigue application);

- Avoidance of decarburisation of surface area—adverse for wear and fatigue due to softer structure at those spot);
- Avoidance of distortion of the heated or quenched jobs, which is detrimental for most applications due to causing mismatch of load-bearing contact points amongst the matching parts in service; and
- Careful handling of heat treated jobs to avoid any dents and cuts on the surface, which can act as point for fatigue crack initiation from those spots in service.

Typical areas of application of carburising are automotive gears and shaft pinions where high wear resistance along with high fatigue strength are required.

8.5.2. Nitriding

Nitriding is also a diffusion controlled process like carburising, but involves lower temperaturew <A_1 temperature) than carburising. In this process, nitrogen in atomic state is made to diffuse onto the steel surface at temperature of about $550 \pm 50°C$. Ammonia serves as a source for atomic nitrogen, which diffuses into the steel and occupies interstitial positions in the iron lattice in the same way as carbon atoms, see Fig. 8-9. The job being nitrided does not require quenching and hardening like carburising, because the effect is not due to formation of martensite but due to nitride precipitation, which takes place while the process is on and during normal cooling from the nitriding temperature. Since the temperature is low, rate of diffusion is also slow and nitriding process takes much longer time for completion than carburising. Figure 8-11 shows the trend of nitride case depth (by gas nitriding) with time at the nitriding temperature of 500°C. However, requirement of depth of nitrided layer for most applications is quite thin compared to carburising; it ranges between few microns to 200 microns, depending on the end-use requiremets and adopted process. Nitriding process could be gas nitriding or salt-bath nitriding—the latter process generally gives more diffused layer but lower hardness than gas nitriding process.

In general, diffusion rate of nitrogen is much slower at the temperature where the process is carried out; hence it produces rather thin layer of nitride compared to carburising, which is carried out at higher temperatures. Because of lower temperature of operations, nitriding process does not cause any distortion of the jobs. However, for successful nitriding, steel should have nitride forming alloying element, like the aluminium (Al) and Titanium (Ti). Hence, engineering steels for nitriding are specially designed with Al content (e.g. EN 40 steel). Typical obtainable hardness by nitriding of different steels is indicated below:

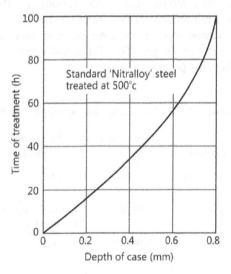

Figure 8-11. The progress of nitride case depth in gas nitriding process with time at 500°C. The figure refers to Nitralloy steel, an aluminium containing alloy steel noted for developing very hard surface.

Steel Composition					Surface Hardness (Typical Value)	Characteristics and Uses
%C	Cr	Mo	V	Al		
0.5	1.5	0.2	—	1.1	1100 VPN	High surface hardness with high core strength
0.4	3.0	1.0	0.2	—	900 VPN	Crankshaft, Ball-races, aero-screw etc. where high
0.4	3.0	0.4	—	—		surface hardness with bulk strength is required

8.5.3. Carbonitriding

Because of the obvious advantages of carburising and nitriding, there is an intermediate process known as' carbonitriding' which is carried out at an intermediate range of temperature e.g. 650°C–780°C. For carbonitriding not only carbon is diffused into the steel but also some nitrogen is diffused for nitride precipitation in the surface zone. In contrast to nitriding, carbonitriding requires quenching for hardening and some tempering for getting the effect of martensite hardening from extra carbon on the surface. Effect of martensitic hardening is more if the temperature of operation is higher, but the effect of nitrogen will be less. However, in general, the carbonitrided layer is thinner than the carburised layer obtained by standard process. The combined effect of carbon and nitrogen diffusion produces a thin but hard layer, which is low in coefficient of friction and useful for some special applications, where distortion control of the job is very critical. A typical example of carbonitriding application is automotive crankshaft where the hardened job is carbonitrided in a gaseous atmosphere for a thin layer of very hard surface for wear resistance.

Similarly there is another modified process called 'nitrocarburising' where not only nitrogen is diffused into the steel, some amount of carbon—using carbon monoxide or hydrocarbon as source—is also introduced. Temperature used is about the same as nitriding i.e. 550°C–580°C and in this range of temperature, carbon diffusion is very slow, concentrating only on surface region while nitrogen penetrates deeper. The Table 8-4 summarises the processes and process characteristics of thermo-chemical processes in vogue.

Table 8-4. Methods and characteristics of 'thermo-chemical' processes

Process	Temperature (°C)	Diffusing Elements	Methods	Processing Characteristics
Carburising	900–1000	Carbon	Gas, Pack, Salt Bath, Fluidised Bed.	Care needed as high temperature may cause distortion
Carbo-nitriding	650–780	Carbon/Nitrogen (mainly C)	Gas, Fluidised Bed, Salt Bath.	Lower temperature means less distortion than carburising.
Nitriding	500–650	Nitrogen	Gas, Plasma, Fluidised Bed.	Very low distortion. Long process times, but can be reduced by plasma technique.
Nitro-carburising	560–570	Nitrogen/Carbon (mainly N)	Gas,Fluidised Bed, Salt Bath.	Very low distortion. Cannot be machined after processing.
Boronising	800–1050	Boron	Pack.	Coat under protective shield. All post coating heat treatment must be in oxygen free atmosphere. No post coating machining possible.

Table 8-5. Appropriate steel grades and potential applications of different thermo-chemical processes.

Process	Case Characteristics	Suitable Steels	Applications
Carburising	Medium to deep case: Oil quenched to harden case. Surface hardness 700–820 HV (58–62 HR$_C$) after tempering.	Mild, low carbon and low alloy steels.	High surface stress conditions Mild steels for small sections <12 mm. Alloy steels large sections
Carbo-nitriding	Shallow to medium to deep case.	Low carbon steels. Oil quenched to harden case. Surface hardness 700–820 HV (58–62 HR$_C$) after tempering.	High surface stress conditions.
Nitriding	Shallow to medium to deep case. No quenching. Surface hardness 700–1150 HV (58–70 HR$_C$).	Alloy and tool steels which contain sufficient nitride forming elements e.g. chromium, aluminium and vanadium. Molybdenum is usually present to aid core properties.	Severe surface stress conditions. May infer corrosion resistance. Maximum hardness and temperature stability up to 200°C.
Nitro-carburising	10-20 micron compound layer at the surface. Below, nitrogen diffusion zone.Hardness depends on steel type: carbon and low alloy 350–540 HV (36–50 HR$_C$), high alloy and tool steel up to 1000 HV (66 HR$_C$).	Steels from low carbon to tool steels.	Low to medium surface stress conditions. Good wear resistance. Post coating oxidation and impregnation provides better corrosion resistance.
Boronising	Thickness inversely proportional to alloy content >300 μ on mild steel and 20 μ on high alloy. (Not to exceed 30 μ if part is to be heat treated). Hardness >1500 HV typical.	Most steel from mild to tool steel except austenitic stainless grades.	Low to high surface stress conditions depending on substrate steel. Excellent wear resistance.

8.6. Development of Special Heat Treatments: Martempering and Austempering

Quenching from higher temperature—used in through hardening—always introduces internal stresses into the steel—which can be sufficiently large at times—to cause heavy distortion or cracking of the jobs. This is particularly applicable for alloy steel with deeper hardenability. To avoid such difficulty in hardening of deep hardenability steel, two other types of quenching methods can be employed—(a) martempering, and (b) austempering. Figure 8-12 depicts these two special heat treatment cycles.

Martempering involves initial fast cooling from the higher temperature to an intermediate temperature and then holding the steel isothermally in a quench bath maintained at or around the M_S temperature of the steel. The purpose of holding at this temperature is to allow temperature equalisation across the section and then cooling further for martensite transformation. Cooling from the intermediate temperature just above M_S produces more uniform and less stressed cooling, eliminating chances of high residual stress or distortion or cracking. A characteristic martempering process cycle is illustrated in Fig. 8-12(i).

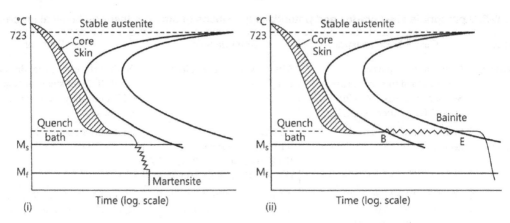

Figure 8-12. Schematic isothermal conditions of (i) martempering and (ii) austempering of steels.

Incidentally, M_S temperatures of high alloy steels are also very low; around $180°C$ or below, which facilitate closer control of quench bath temperature and release of thermal stress due to high temperature quenching. Further cooling from such low temperature, leaves very little scope for developing any further residual stress in the job. Therefore, martempering of steel is applied to high hardenability steel for avoiding distortion and cracking by following the process steps as listed here in the following:

- Quenching from the austenitising temperature into a hot fluid medium—for instance, hot oil, molten salt, fluidised sands, etc.—at temperature slightly above the M_S temperature of the steel (M_S temperatures of high hardenability steels are generally low);
- Holding unto the fluid media until the temperature throughout the steel body is substantially uniform;
- Cooling from this holding temperature at a moderate rate (usually air or warm water); and
- Tempering, if necessary, at an appropriate temperature for improving toughness in the steel.

The process aims to minimise thermal stresses and stress gradient within the body of steel for avoiding development of high internal residual stresses or cracking of higher carbon or higher hardenable steels. In general, martempering is applied to alloy steels that are prone to distortion by standard water or oil quenching method. The grades of steel that are commonly martempered to full hardness include: SAE 1090, 4130, 4140, 4150, 4340, 300M (4340M), 4640, 5140, 6150, 8630, 8640, 8740, 8745, SAE 1141, and SAE 52100. Carburizing grades such as SAE 3312, 4620, 5120, 8620, and 9310 are also commonly martempered after carburising. Occasionally, higher-alloy stainless steels, such as AISI type 410, are also martempered, but this is not a common practice.

Austempering also involves similar cooling from higher temperature but holding the job isothermally just above the M_S temperature of the steel for a longer time than martempering, in order to allow total temperature equalisation and transformation to lower bainite (see Fig. 8-12 (ii)). The job is then cooled to room temperature at an intermediate rate, usually in air or warm water. These steps help to produce a relatively stress free lower bainite microstructure and do not require further tempering. This process is widely used in high alloy steels and in alloy cast irons for producing a tougher martensitic or bainitic structure. Austempering is also used for high carbon steel strips and wires for springs or similar parts.

In austempering, quenching is interrupted at relatively higher temperature than the martempering and the process does not necessarily combine with any requirement of another tempering. Because of holding pattern, the product is mostly 'lower bainite'—unless the process is specially designed to produce martensite—which has high strength combined with proper toughness. Such a structure is very often used for many industrial applications and a popular one for combining strength and toughness in SG iron casting, where the product is called 'austempered ductile iron' (ADI). ADI castings are now economically replacing many steel forging applications (e.g. lower duty crankshaft, camshaft etc.) due to ease of manufacturing and cost advantage.

8.7. Welding

Welding is a jointing process between two or more parts where the parts are joined together by fusion or bonding of respective materials under applied or generated heat, which could be with or without pressure and with or without additional filler material. The purpose of heat or pressure is to render the interfaces plastic or liquid, whereby bonds can be established between the parts. ISO defines welding as: "An operation by which two or more parts are united, by means of heat or pressure or both, in such a way that there is continuity of the nature of the material between these two parts". This condition of *continuity of nature of material* often necessitates use of a filler material whose nature and melting point is of similar order as the parent materials.

Process-wise, welding can be of various types, but the following types nearly describe all the important metallurgical features associated with welding of steels. These welding types include: (a) Fusion welding, (b) Electric resistance welding and (c) Pressure welding. The idea of a perfect welding is to establish a perfect continuity of matrix between the parts joined, such that every part of the joint is indistinguishable from the metal to which the joint is made. However, ideal weld seldom occurs in practice due to occurrence of phase changes with heating and presence of microstructural inhomogeneity, which might cause variation of mechanical properties over the weld area. This problem is more relevant to fusion welding, which involves higher heat input for fusion of the base metals, using external heat. Nonetheless, imposition of the condition of continuity of matrix on the welding, as far as possible, makes it necessary to design the process with sufficient precautions and care to ensure:

- Weld metal is as close in composition as possible to parent metal;
- Weld metal is cooled in controlled manner to avoid any shock, crack or microstructural inhomogeneity and brittleness; and
- There is sufficient heat in-put (or heating) during welding for either bonding under pressure or fusion of associated metal parts for continuity of structure across the joint, but not over-heating.

The source of heating can be external heat from a suitable source of flame—e.g. oxy-acetylene flame—or electrical resistance heat or by arcing over the point of welding. Electric arc is produced under high current and low voltage conditions between the points in contact. Thus, the heat can be supplied by electricity or by gas flame.

By far the most popular methods of welding include:

- *Fusion welding*—e.g. arc welding, gas welding, laser welding etc.;
- *Electric resistance welding*—e.g. resistance welding, flash butt-welding, spot welding etc.; and
- *Solid phase welding*—e.g. pressure welding—under high frequency, forging under heavy force, explosive welding, friction welding, ultrasonic welding etc.

Other than welding, there are other jointing methods like 'brazing' and 'soldering', which are carried out by using an intermediate metal (or alloy) of lower melting point, which binds the metals together but do not produce similar strength or metal matrix as welding.

All steels are not equally weldable, especially where fusion of steel is involved. Fusion creates a molten metal pool in the jobs, having high heat content—temperature reaching the level where steel is molten. When this molten pool of a given chemistry cools after welding, the rate of cooling must be controlled to avoid cracking and brittleness due to coarse grains, absorption of gases (hydrogen) from the exposed environment, and formation of highly strained martensite that gets produced on fast cooling. Thus, there are chances of cracking and brittleness in fusion welded joints unless proper care is taken in advance. The controlling factors are:

1. The carbon equivalent (CE) of the steel—a higher CE raises the risk of more cracking.
2. How the molten pool of metal is protected from the open atmosphere; to avoid gas absorption, especially the hydrogen, which causes the phenomenon known as 'hydrogen embrittlement'.
3. The cooling rate of the molten pool of steel and subsequent cooling of surrounding area that gets heated during welding; to avoid formation of crack and brittleness in the weld and heat-affected zone (HAZ) due to phase transformation during cooling.

Carbon equivalent provides the weldability index of the steel in the same way as the hardenability of steel provides idea about heat treatability of steels. Any part of the steel which has undergone heating and cooling during welding will be affected as regards hardness and hardenability in the same way as hardening of the steel. Higher carbon content or carbon in combination with alloying elements—such as manganese, chromium, silicon, molybdenum, nickel, copper etc. increases the CE, decreases the weldability and produce higher hardness than the matrix structure, leading to incongruity in the matrix. However, their influence is not uniform; they differ in terms of magnitude as indicated by the formula below. Widely accepted carbon equivalent formula for good welding is as follows:

$$CE = \%C + \left(\frac{\%Mn + \%Si}{6} \right) + \left(\frac{\%Cr + \%Mo + \%V}{5} \right) + \left(\frac{\%Cu + \%Ni}{15} \right)$$

And, the level of carbon equivalent for good weldability as per this formula is shown below:

Carbon Equivalent Level (CE)	Weldability Index
Up to 0.35 CE	Excellent
0.36–0.40 CE	Very good
0.41–0.45 CE	Good
0.46–0.50 CE	Poor

Steels with CE value above 0.45 need to be welded with extreme precaution to avoid cracking, such as by pre-heating and post-heating of the weld zones or by using filler rod or electrodes having lower melting point and requiring lower heat input. However, this table of CE and weldability index does not mean that steel with CE above 0.45 cannot be welded—it implies that precautions are necessary for such steels for avoiding cracks and distortion after welding. In fact, challenges of good welding technique arise from this standpoint i.e. how to design and control the welding processes for higher

CE steels such that the job does not crack or distort in welding; and not avoiding welding of high CE steels altogether.

Among the different types of welding processes, there are many sub-divisions of welding types as per their process characteristics. The following is the map listing of such main processes and sub-divisions within types of welding:

Fusion Welding	Electric Resistance Welding	Solid Phase Welding
Arc welding (using coated electrodes)	Resistance butt-welding	High frequency pressure welding
Gas welding (using filler rods)	Flash butt-welding	Explosive welding
Electro-slag welding (special sub-merged welding process)	Spot welding	Friction welding
Laser welding	High frequency resistance welding	Ultrasonic welding
Electron beam welding		

Listing and explaining all these processes is a subject matter of a 'Welding Handbook' and cannot be covered here. For more details of such processes, books from the list of Further Readings and Bibliography may be referred. However, arc welding, a part of fusion welding, using coated electrodes, is a widely used process for steel and very aptly demonstrates all the metallurgical tricks and tactics of welding process. Hence, some aspects of arc welding techniques will be further discussed here.

8.7.1. Metal Arc Welding

This is a fusion welding process where heat required for fusion is generated by the electric arc formed between a metallic electrode and the steel base plate. Figure 8-13 illustrates a setup of metal arc welding process where the top part shows the setup with sources of power and bottom part shows the fusion welding process in progress on a base steel plate. The electric arc—struck between the metal electrode and the work piece—is an ideal source of heat as the arc can produce very high intense heat of temperature >5000°C. The key to good welding is to focus this arc on a small area over the welding line for fraction of a time, so as to melt the electrode and deposit the droplets on the heated steel plate being welded at correct and uniform rate. Any overheating of the plate being welded will cause burning and cracking. However, the process is flexible enough to allow control over the welding

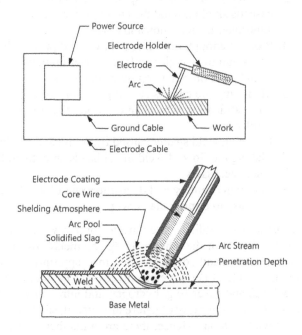

Figure 8-13. A setup of a metal arc welding process and the working details.

speed—determined by the rate of travel of electrode over the work piece—and restrict heating to a narrow zone; thereby restricting metallurgical disturbances of the whole plate structure.

The electrode used in arc welding has a base core wire, generally composed of low carbon steel, which is coated with suitable fluxes that facilitate arc stability under the applied current. The flux coating also provides gas shielding by arc cover over the working area and acts as carrier of alloy addition, if necessary. Depending on the steel core wire size, electrodes are sized as per 'screw gauge' (SWG) and appropriately chosen for an application as per steel to be welded and precautions to be taken. Generally, lower gauges—i.e. higher diameter of core wire—electrodes are used for thicker plates and *vice-versa*.

Quality of flux coating of welding electrodes is very critical for successful applications. Fluxes are chosen from hosts of inorganic chemical compound—e.g. silica, limestone, fluorspar, rutile, titania, alumina etc.—as per the tasks of welding and electrodes are accordingly graded and classified—e.g. cellulose, rutile, acid, basic etc. The flux coating thickness of manual arc welding electrodes varies with the core wire diameter and is usually expressed as 'coating factor', which is the ratio of coating diameter to core wire diameter. Electrodes are classified depending upon the type of flux coating, performance characteristics and potential to give all-weld mechanical properties. Each class of electrodes is then coded as per recognised standards by using group of letters and numbers—in order to help the user to choose the right type for right application. For example, E6010, E6011, E6012, etc., meaning high cellulose sodium; high cellulose potassium; high titania sodium, etc., respectively.

Flux coating of arc welding electrodes perform very critical role in the welding process. Their major functions can be summed up as:

1. Help to strike arc and maintaining it uninterrupted during welding.
2. Generat gases which displace the oxygen and nitrogen of the surrounding air and provide a gaseous shield around the arc to protect the molten liquid beads as well as molten weld pool from atmospheric aggression and reaction.
3. Produce enough fluid slag to cover and protect the molten pool of weld from further atmospheric attack and to provide insulating cover on them for controlled cooling.
4. Help to deoxidise and refine weld metals.
5. Modifying the chemistry and providing alloying elements, wherever necessary, to the weld metal for modification of mechanical properties; thereby matching the continuation of matrix structure or similar one.
6. Reducing weld spatter caused by arcing and gas bursting.
7. Making the flow of weld metal beads smooth and uniform, and facilitating all positional welding where necessary.
8. Improving fluidity and deposition rate of weld metal, yet maintaining uniform cover on the weld and easy to remove slag cover.

In fusion arc welding, core wire for ordinary welding is simply a low carbon rimmed or semi-killed steel and most of the other welding and metallurgical functions are performed by engineering the flux characteristics and covering. Therefore, appropriate flux covered electrode is necessary for countering the ill-effects of certain elements present in steel (e.g. S and P in steels which tend to promote weld cracking) and gases (e.g. hydrogen and nitrogen gas absorption can cause HAZ crack). However, for ensuring the continuity of nature of material, steel of widely differing chemistry would require different core wire material. Hence, there are a number of groups of electrodes for welding different grade of steels, namely: mild steel electrodes, low-alloy steel electrodes, stainless steel electrodes, electrodes for cast irons, etc. These are all covered by respective national standards of a country or the ISO.

There are many other welding processes including gas welding, electro-slag welding, electron beam welding, submerged arc welding, tungsten inert gas (TIG) welding, metal inert gas (MIG) welding etc., which find place in the practice of steel welding. All of them have their respective attraction, merits and uses according to the situational necessity or diverse metallurgical requirements. Since welding is a metallurgical process, involving melting, solidification and transformation of the metal, metallurgical requirements of the process and the product play the most critical role in choosing the welding method, though the actual welding process is executed by engineering excellence and innovativeness.

8.8. Metallurgical Aspects of Steel Welding and Precautions

Welding of steel involves joining of two or more steel pieces together by heat, where the heat can be supplied or introduced into the system by different means and methods such as—electric arc, gas flame, electric resistance, pressure, friction, etc. Some of these heating methods—like the arc welding, gas welding etc.—involve melting of filler materials/electrodes for providing extra material to fill the gap and fusing the joint for sound and strong jointing. Thus, basically steel welding involves: (a) solidification of molten filler material, (b) cooling of the weldment—i.e. result of solidification of molten steel pool—and surrounding 'heat affected zone' area, and (c) associated transformation due to heating and cooling.

Solidification of molten pool of steel resembles to, in principle, all the features and characteristics of steel casting and solidification, as discussed in Chapter 4, including the necessity of controlling grain size and removal of slag as impurity (inclusions) from the molten metal. Cooling of weldment and HAZ area calls for all the precautions and controls necessary for controlling the ensuing phase transformation and associated issues of crack and distortion (see discussion on these issues in heat treatment of steels discussed earlier, and phase transformation discussed in Chapter 2). In fact, the filler material (electrode compositions) and the cooling condition of the weldment is so designed as to produce the weld metal structure that matches the mechanical properties and chemical composition of the steel being welded, as closely as possible in order to maintain continuity of the nature of material. Any drastic and sharp change in composition and structure of the weld metal and the matrix is not at all desirable as it might act as 'similar to notch' and can cause sudden fracture originating from such sharp interfaces, especially when the interface has chance of having long columnar structure due to directional cooling during welding.

Molten filler material depositing onto the welding area of steel plate gives rise to solid–liquid pool and the solid–liquid interface tends to cool in highly directional manner towards the surrounding cold metal. Hence, the weld metal acquires a distinctly columnar structure, in which the grains are very long and columnar in nature. Solid-liquid interface also adds to the woes by providing sites for heterogeneous nucleation and very little super-cooling effect. The combined effect leads to producing large coarse and columnar grains, which are brittle. Precautions are necessary in fusion welding to correct this situation as far as possible, which is done by introducing some oxidising elements such as Si and Al in the flux coating for grain size modification (refer Chapter 4) and by pre-heating to reduce the directional cooling. Another critical feature of fusion welding is the 'gas-metal reaction'—in the same way as in steel making—which can make the weld metal porous with gas pin-holes and bubbles; thereby affecting the mechanical properties of the weld joint. Hence, the weld pool must be protected from the ingress of gases like oxygen, nitrogen and hydrogen during welding. Hydrogen has the additional effect of making the solidified steel brittle (hydrogen embrittlement). Precautions to protect the arc and the liquid metal from the atmosphere is, therefore, very important in welding. This is accomplished by flux covering and flux design so that the positive pressure of gasses formed from their burning produces a gas envelope that resists aggression from outside atmosphere.

The other important purpose of flux is, to produce a slag cover over the newly deposited weld metal and slow down the cooling rate, in order to avoid welding crack and distortion due to rapid cooling from liquid state of the steel. However, this arrangement might be good enough for welding of steel with lower carbon equivalent, but for higher CE value of 0.45 CE and above, the steel plate being welded needs appropriate pre-heating and/or post-heating for controlling the cooling rate so that no cracks or distortion take place due to rapid cooling and associated phase transformation of the weld and HAZ area. Heat-affected zone *is demarked as the area of the base metal lying next to the fusion line of the weld which had not melted but whose microstructure and mechanical properties got altered due to heating during welding.* Figure 8-14 illustrates the heating pattern (i.e. temperature distribution) of the liquid pool and HAZ in fusion welding of lower carbon steel (0.30% C steel).

Figure 8-14 shows that the centre of the weld, which is liquid, can reach as high as $1700°C$ (see dotted black line) and the solid-liquid interface (see point 'd') can have temperature above $1400°C$. This heat gets dissipated through the surrounding solid, heating the HAZ area (represented by bold arrow). This shows that many points of HAZ will have temperature above A_3/A_1 temperature. This means that any point having higher than A_1 temperature will have austenite spots, which, if cooled faster, can form martensite and other non-equilibrium products, leading to brittleness and cracking. It is this non-equilibrium cooling and the associated phase transformation that poses the problem of HAZ crack, necessitating control of the cooling rate of this region by taking some extra precautions to reduce the cooling rate.

The cracks that arise in 'heat affected zone' of the solid steel (called HAZ crack) may not be instantaneous with the phase transformation; they might occur after the weld had been cooled, due to extra residual stresses and strain in the matrix. HAZ crack can also occur due to hydrogen embrittlement, especially when there is chance of hydrogen absorption from the environment. The common methods of dealing with weld cracks are:

- Using low CE steel;
- Using pre-heat of the base plate in order to reduce cooling rate after welding, if the steel CE is a suspect;

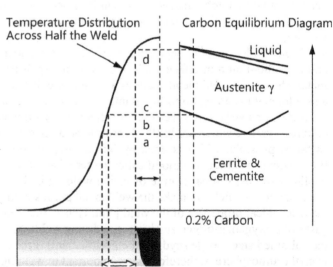

Figure 8-14. Temperature distribution of fusion arc welding using coated electrode *vis-à-vis* the relative Fe-C diagram portion of the base steel, indicating the potential phase changes.

- Ensuring that welding electrodes are freshly pre-baked in a furnace to expel all absorbed moisture from the flux coating;
- Using low-hydrogen special electrode; and
- Post heating of the weld and HAZ, where necessary.

Microstructure of the HAZ is determined by the (a) composition of the steel i.e. CE, and (b) the thermal cycle of the welding process. Higher the CE of the base steel, which can be due to higher carbon and/or alloy content, higher the chance of having harder and non-equilibrium microstructures (e.g. martensite or bainite) in the HAZ area, causing distortion and crack. Cooling of HAZ, influencing the microstructural changes and cracking tendency, is dependent on:

1. The thickness of the plate being welded; cooling rate increasing with increasing thickness;
2. Preheat temperature; higher preheat will decrease the HAZ cooling rate, and
3. Size of the weld bead, which, in turn depends on the electrode's size and type; larger bead will cause decrease in cooling rate.

Weld metal so produced cannot be conventionally heat treated to alter the microstructure and mechanical properties. Hence, the entire process—including the method of welding and nature of consumables—needs to be correctly designed and executed. Only post-weld treatment that can be applied after welding is 'post-heating' of the weld area, which relieves the residual stresses and can soften the microstructure to a certain extent. Post-heating is widely adopted for welding of high hardenable steels, like the high-carbon or alloy steels, for avoiding cracks by relieving residual stress and controlling the mechanical properties by introducing some tempering effect. However, the actual mechanical properties of the weld metal will depend on (a) the base steel composition, (b) the microstructure of the weldment and (c) alloying element, including grain refiners, introduced into the weld through the electrode and fluxes. Influence of carbon, alloy content and heating on microstructural development will follow the same rules as for phase transformation discussed for bulk steels in Chapters 1 and 2. Testing and evaluation of weldments also follow the same methods and procedures as for bulk steels discussed in Chapter 5.

Welding is a vast subject, comprising of metallurgical parts and complex technology that are to be adopted for executing the welding processes. It also requires high degree of skill for the personnel engaged in welding jobs, requiring extensive theoretical and practical training. However, now-a-days, many automatic welding processes are being adopted for coping with the complexity of weld joints, location of welding (including under-sea welding of marine structures), and the processes. For further details of any particular welding technique, reference could be made to different specialised welding books and handbooks; some have been listed at the end of this chapter.

SUMMARY

1. The chapter discusses the metallurgical aspects of heat-treatment and welding processes and practices from the angle of steel uses and applications. It has been pointed out that both heat-treatment and welding of steel is linked to the principles governing the phase transformation in steel. Therefore, success of both processes is dependent on how well the heating and cooling rate is controlled to—(a) produce desired microstructures for required mechanical properties, and (b) to avoid ill-effects of very fast cooling that is known to produce harmful residual stress, distortion and cracking.

2. The chapter outlines the necessity of adequate hardenability for right response of steel for hardening and highlights the various factors influencing the heat-treatability of steels, including the effects of different alloying elements and grain sizes. Methods of theoretically calculating the hardenability value from composition, using Grossman formula has been mentioned.

3. For systematic discussions and analysis, heat-treatment processes have been grouped under three heads as per their process objectives, namely: (1) Softening processes—examples: annealing, sub-critical annealing and normalising, (2) Hardening processes—examples: quenching and tempering, induction hardening, and flame hardening, and (3) Thermo-chemical processes—examples: Carburising, Nitriding and Carbo-nitriding.

4. Each of these heat-treatment processes has been discussed and analysed from the angle of metallurgical principles and practices and their relative merits and limitations. Development of specialised heat-treatment processes like martempering and austempering for deep hardening steels have been discussed highlighting their merits.

5. The process of welding has been discussed with reference to its process objectives, based on its accepted definition and condition of 'continuity of the nature of materials between two parts'. In this regard, three types of welding process have been dealt with, namely: (a) Fusion welding, (b) Electric resistance welding, and (c) Pressure welding.

6. Applicability and working principles of each of these welding processes have been highlighted. The metallurgical aspects of fusion welding have been demonstrated with reference to the 'Metal arc welding' using coated electrodes, which virtually encompass all the care and concern necessary for good welding, following the principles of metallurgical phase transformation and related issues.

7. Metallurgical functions of fluxes and the purpose of using flux-covered electrodes have been mentioned along with the classification system of electrodes.

8. Finally, the cause and effect of 'heat affected zone' (HAZ) cracking and hydrogen cracking in weldments have been discussed and precautions necessary for avoiding such defects have been mentioned.

FURTHER READINGS

1. ASM International, USA. ASM Handbook, Vol. 4: *Heat Treating.* 1991.
2. ASM International, USA. ASM Handbook, Vol. 6: *Welding, Brazing and Soldering,* 1994
3. Totten George E., ed. *Steel Heat-Treatment Handbook,* 2nd edition. CRC Press, 2006.
4. Brooks, Charlie R. *Principles of Heat-Treatment of Plain carbon and Low-Alloy Steels.* ASM International, USA, 1996.
5. *Stahlschlussel (Key to Steel).* VERLAG, Germany, 2006.
6. Nadkarni, S.V. *Modern Arc Welding Technology.* Oxford & IBH Publishing Co. Pvt. Ltd., 1988.
7. Sharma, Ramesh C. *Principles of Heat-Treatment of Steels.,* Delhi: New Age International, 2007.
8. Unterweiser , P.M. *et al.,* ed. *Heat Treaters Guide: Standard Prectice and Procedure for Steel.,* Ohio: ASM, 1982.

APPENDIX TO CHAPTER 8

Table 8A-1. Guide to some standard heat treatment processes.

Heat Treatment Process	Austenitisation/ Soaking/Treating Temperature	Holding Time	Cooling	Precautions (Finishing)
Annealing	A$_3$ temperature of the steel + at least 50°C	As-Cast: 4–5 hr or longer Rolled/Forged: 1–1 ½ hr/25 mm section thickness	Closed furnace	Avoid over-heating and excessive scaling (Shot-blast cleaning).
Normalising	A$_3$ temperature of the steel + at least 30°C	Rolled/Forged: 1 hr/25 mm section	Air cooling	Avoid excessive scaling (Shot-blast cleaning).
Hardening	A$_3$ temperature of the steel + 20°C–30°C	30–45 min/25 mm section. Longer soaking for alloy steels	Quenching by: water, oil, brine, salt-bath etc. depending on the steel composition	Strict tempreture control, avoid scaling and decarburisation, Immediate tempering(Shot-blast cleaning, grinding etc.)
Induction hardening	Heating the surface above A$_3$	Generally, no holding time	Fast quenching; either water or forced air with low time gap between heating and cooling	Avoid over heating or fast heating. (Grinding/buffing)
Nitriding	Heating to between 550°C to 670°C, depending on gas or liquid salt-bath nitriding	20 hr and above, depending on case depth, process and alloy content in steel	Normal	Avoid loose white layer formation (Buffing)
Carburising	Heating to between 900°C to 1000°C, depending on the process. (For gas, it is around 930°C)	4 hr and above, depending on process. (Liquid process takes shorter time but less case depth). Gas takes about 8–12 hr depending on case depth.	Furnace cooling/air cooling after carburising, but quenching in oil or liquid bath after hardening from about 830°C–860°C with required soaking.	Avoid decraburisation, scaling, pitting etc. (Shot-blasting and grinding)

Table 8A-2. Guide to calculating different 'critical temperatures' of alloy steels for facilitating correct heat treatment design.

USEFUL EQUATIONS FOR HARDENABLE ALLOY STEELS

Ae_1 (°F) ~ $1333 - 25 \times Mn + 40 \times Si + 42 \times Cr - 26 \times Ni$ .... (1)

Ae_3 (°F) ~ $1570 - 323 \times C - 25 \times Mn + 80 \times Si - 3 \times Cr - 32 \times Ni$ .. (2)

Ac_1 (°C) ~ $723 - 10.7 \times Mn + 29.1 \times Si + 16.9 \times Cr - 16.9 \times Ni +$
$290 \times As + 6.38 \times W$ .... (3)

Ac_3 (°C) ~ $910 - 203 \times \sqrt{C} + 44.7 \times Si - 15.2 \times Ni + 31.5 \times Mo + 104$
$\times V + 13.1 \times W$ .... (4)

•→ Ms (°F) ~ $930 - 600 \times C - 60 \times Mn - 20 \times Si - 50 \times Cr - 30 \times$
$Ni - 20 \times Mo - 20 \times W$...................................... (5)

M_{10} (°F) ~ $Ms - 18$.. (6)

M_{50} (°F) ~ $Ms - 85$.. (7)

M_{90} (°F) ~ $Ms - 185$.. (8)

Mf (°F) ~ $Ms - 387$... (9)

Bs (°F) ~ $1526 - 486 \times C - 162 \times Mn - 126 \times Cr - 67 \times Ni -$
$149 \times Mo$ .... (10)

B_{50} (°F) ~ $Bs - 108$ .... (11)

Bf (°F) ~ $Bs - 216$ .... (12)

Carburized Case Depth (in.) ~ $.025\sqrt{t}$, for 1700°F .... (13)

Carburized Case Depth (in.) ~ $.021\sqrt{t}$, for 1650°F .... (14)

Carburized Case Depth (in.) ~ $.018\sqrt{t}$, for 1600°F .... (15)

•→ (t = time in hours)

Note: Each equation above is subject to the chemistry limitations under which it was developed.

1 & 2: R. A. Grange, Metal Progress, 79, April 1961, p 73.
3 & 4: K. W. Andrews, JISI, 203, 1965, p 721.
5: E. S. Rowland and S. R. Lyle, Trans. ASM, 37, 1946, p 27.
6-12: W. Steven and A. G. Haynes, JISI, 183, 1956, p 349.
13-15: F. E. Harris, Metal Progress, 44, August 1943, p 265.

CHAPTER 9

Selection and Application of Steel

Role of Quality, Cost and Failure Analysis

The purpose of this chapter is to distinguish between the selection and application of steel, and to highlight the processes and measures for improving the 'quality value' of steel and reliability of performance in application. Selection of steel is based on identifying the right relationship between 'composition, structure and properties', but right application of steel is additionally concerned with identifying how 'quality value' of steel can be improved and reliability of performance can be assured. In this regard, this chapter briefly discusses the processes and measures for cost, quality and reliability optimisation, on the one hand, and analyses the service failure modes and mechanisms for prevention of failures in application, on the other.

9.1. Introduction

Application of steel is the ultimate goal for all steel selection exercises. Selection of steel involves understanding of manufacturing processes and required properties, on the one hand, and application specific end-use properties, on the other. But, setting goals on properties do not necessarily assure reliable and cost effective application of steel. Property is the means and not the end for ensuring reliability of performance. For improved and reliable performance, it may not be enough to focus only on the development of properties (as discussed so far in this book), but also on factors that can impair or improve the behaviour of steels in service.

Factors that are required to be considered for improved and cost-effective application of steels include:

- Choice of right grade and properties of steel;
- Identifying ways and means to improve 'quality value' i.e. to reduce quality loss function (QLF) as pointed out by Taguchi;
- Understanding of how components fail in service, and designing components for fail-safe performance as per set target;

- Understanding of what can impair or improve the component integrity—a measure of risk factors that influence the performance in service; and
- The preventive measures and steps that can be taken to prevent pre-mature failure in service.

In practice, the above mentioned factors should be integrated in the design and manufacturing programme of components for reliable performance at optimum cost. Cost is an important parameter for selection and application of steel not only for economic reasons but also for the necessity of reducing social and national waste of 'value'—value of money and resources spent versus the value of technical benefits obtained or obtainable from the chosen steel. Focus of steel application should be to maximise the gain in value by taking necessary steps for quality-value optimisation in design and by prevention of pre-mature failure by employing superior manufacturing processes with high process capability. Manufacturing quality is equally important like design quality; both are to be at an optimum level for best possible results. For example, despite a good design, leaving behind sources of stress raisers or manufacturing defects in the components will lead to early failure of the component, no matter how good the design or selection of steel was. Thus, the application of steel calls for a holistic approach of 'total quality' that includes cost, quality and reliability in performance, where reliability can be built through component integrity—the process of analysing risk factors that can impair the performance of a component under actual service conditions. In the context of component integrity, failure modes, mechanisms, nature and factors leading to pre-mature failure have been discussed in this chapter.

9.2. Introduction to Reliability in Performance

Components made of steel must not fail pre-maturely in service. Pre-mature failure is highly expensive not only in terms of replacement of parts but also down-time of machine tools and loss of customer orders. However, no system, or a part in the system, can be made fail-safe to run infinitely, no matter how costly the steel is. Components will fail sooner or later due to wear and tear. What is needed for effective application of steels is a knowledge-based decision for ensuring fail-safe life i.e. it should not fail unexpectedly and pre-maturely. To make the system fail-safe (i.e. reliable), designers approach the problem from 'reliability engineering' point of view, where the components are designed for reliable performance for a certain period of finite life. Reliability is built into the system by considering the following fact:

$$\text{Reliability, } R = \text{Function [Design quality}$$
$$+ \text{ Component integrity}$$
$$+ \text{ Manufacturing accuracy}$$
$$+ \text{ Maintenance quality]}$$
$$\times \text{ [Environment + Service conditions].}$$

This implies that assurance of reliable performance in engineering services is much more than selecting steel with right properties; it is about total quality of all inputs that go into the production and application of a component, including the cost for the quality. Amongst these inputs, *selection and quality of steel are no doubt constitute the 'hub in the wheel' for reliable performance,* but that is only one important part of the steel application. Application has wider dimensions than steel selection—it includes the role and contribution of design and manufacturing quality, steps for prevention of faults and failures, and the cost at which these are being accomplished. Such an integrated approach for steel application requires consideration of setting right quality at right cost, understanding the nature and cause of faults and failures of steel parts, and actions to prevent those faults, based on field failure data and analysis.

This chapter, therefore, aims to discuss the ways and means to optimise quality, cost and performance for a given application by considering:

- Optimisation of quality goals and cost for the quality,
- Means and methods of controlling quality and quality variance for cost effectiveness,
- Areas of concern for design and manufacturing faults, giving rise to sources of faults and failures,
- Analysis of failure features and failure modes in service, and
- Identification and implementation of preventive measures against cases of failure.

Single focus on steel grade and steel quality is not the answer to the challenge in steel applications. Such an approach is prone to over reliance on a single factor of steel quality, which might not work or may lead to selection of very expensive steel without commensurate economic benefits because of other deficiencies. Selection and application of steel must justify economic return *vis-à-vis* value of the service obtained from the application. Such economic justification demands that quality specification should be examined in the light of Taguchi's 'Quality Loss Function' (QLF). For the performance to be cost-effective, over specifying quality by the designer or selecting higher grade steel is not the answer. Instead, as per Taguchi, quality should be built into the component by controlling the 'controllable factors' that put limitations on performance, e.g. control of variations of dimensions and mechanical properties, design short-comings, manufacturing faults, etc. *By exercising control over variance from the steel making stage to the subsequent manufacturing and heat treatment processes, 'quality value' of steel component can be substantially improved.* Increasing 'quality value' of steel should be the crux of all steel developments or approach for steel application.

In fact, in many cases, quality value of so-called cheaper grade steels can be made good enough by controlling the variances in terms of quality, properties and dimensions, enabling substitution against higher grade steel used at higher cost. Therefore, the purpose of this chapter is to bring out these economic considerations for steel selection and applications, along with other measures for prevention of faults and pre-mature failure of steel parts.

Prevention of failure by fault analysis is also the yard stick for measuring or improving *component integrity*—which attempts to examine the capability of the component to meet the demands of reality, where reality involves examination of the impact of: applied load, its types and fluctuation, environmental effects and required safety margin for safe operation. In fact, *study of component integrity is a kind of 'risk analyses' for finding out the critical quality and manufacturing parameters required for fail-safe operation.* Such analyses help to improve the quality value of steels, and should form the part of steel metallurgy dealing with selection and application of steels for reliable performance.

9.3. Quality Goals and Quality Cost

Quality is not absolute; it is need based. It should be right for the application but cost effective too; neither less nor more. According to Taguchi, swings in quality either way add to loss. Following Taguchi's QLF theory, quality within the required specification is fine, but anything beyond the specification is a 'total loss'. This can be visualised from the following quality loss function diagram in Fig. 9-1.

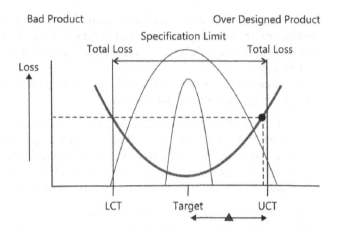

Figure 9-1. Quality target *vis-à-vis*
Taguchi's 'quality loss function' concept.

In Fig. 9-1, Taguchi's QLF—i.e. probable loss for not maintaining the quality at correct levels—is shown by bold red curve and the original specification has been superimposed on it. The original specification is bound by two limits, namely (a) lower control specification (LCT) and (b) upper control specification (UCT), as shown in Fig. 9-1.

Conventional concept is to set quality as specified by customer or user with acceptable range (i.e. specification) and reject anything that falls below or outside the specification limit i.e. LCT for fear of failure. In conventional practice, tolerances beyond the level of UCT are however, accepted without much questioning, considering that it does no harm. But, a critical analysis of process and events will demonstrate that maintaining quality entails cost—(a) cost of raw-material, (b) cost of processing and manufacturing, and (c) cost of failure comprising to conform to specified properties after manufacturing and in services. For the steel selection and application to be cost effective, these sources of cost involvement should have to be balanced. In other words, there is a loss and gain situation in maintaining quality, which needs to be optimised for any application.

According to Taguchi, quality outside the specification range either way entails 'loss to the society'—e.g. customer, users, producers, etc.—due to either poor quality or over-specified quality, which is not necessary for the function. Such losses will continue to occur until the spread of property can be centred on a target value where loss is zero—i.e. loss due to maintaining quality equals gain from the quality. Any spread of property beyond the target will cause some economic loss with respect to function it can perform. Hence, *there has to be a realistic quality target for an application, beyond which either way cost will not be justified.* Practically, quality target should have a range, because maintaining quality exactly on target without any tolerance is not realistic—every process has statistical spread and cannot be exactly controlled to a single value. Therefore, accepted approach by following Taguchi's observation is to control the process for lower variance of property by improving the process capability to make the quality loss optimal—equal to cost of producing quality and gain in product performance and, thereby, make the overall production and application economical and effective.

Referring to Fig. 9-1, components with quality below LCT will fail prematurely and beyond UCT the cost will not be justified, leading to economic loss without commensurate gain in the required performance. As shown in the loss function diagram (Fig. 9-1), better quality (with lesser loss) means decreasing variance from the aim or target, where target is the mean of specification. This situation has been superimposed in the loss function diagram of Fig. 9-1; the narrow thin curve at the centre with lower spread of property is of better quality than the other wider curve with higher spread, though both are within the specification. Thus, *quality is not exactly specification, but relates to the variance within the specification.* The degree of tolerable variance for an application should be set by considering the criticality and necessity of application, and should not be either less or more stringent than what exactly is required, in order to avoid loss due to quality. This translates to the following tasks for the selector of steels:

1. Can the variance in chemical composition of the steel be reduced with agreement of steel producers?
2. Can the steel grade selected be procured with closer range of properties, justifying cost?
3. Can the manufacturing quality and range of dimensional tolerance level be reduced with improved process capability in the manufacturing shop?

The purpose of Taguchi's QLF analysis is to make the selector and designer aware of the cost involved in setting quality specification, on the one hand, and emphasising the need for building 'process capability' in a production shop, on the other—their purpose is to avoid unnecessary loss arising due to wider quality specification or variance in the processes. In the absence of such approach, an attempt to select and apply right steel for right purpose can be frustrated.

Variance in quality and performance behaviour is inherent in all materials and processes. It could be due to the following facts:

- Material defects (macro-defects),
- Structural defects (micro-defects),
- Inhomogeneity and excessive inclusion in materials,
- Lack of process capability in manufacturing,
- Manufacturing defects—like steps, rough surface, tool marks, micro-cracks, etc., and
- Variation or fluctuation in processes due to external interference or interfering events such as lubrication failure, power failure, operator negligence/absence, etc.

If any one of the above factors is bad and others are good, the result is bad. It is the task of application engineers to harmonise the total 'in-put–out-put system' for getting the right result at right cost. Selection of steel has to be approached from the angle of 'quality goal' and 'quality cost' optimisation in a given set-up of manufacturing.

Drawing from Taguchi's approach, there is scope for cost and quality optimisation with respect to an application, and the right solution for such cases could be:

1. Identify key functional parameters in the performance under a given situation.
2. Examine the controllable factors which contribute to failure and necessitate better control of properties or use of expensive material For example, control of case depth, core strength, microstructure of the case, avoidance of surface decraburisation, etc. for a gear application.
3. Analyse and identify key controllable factors—by Pareto analysis of failure data, if available—that have most pronounced effect on component performance.
4. Act upon and improve those key controllable factors by reducing process variance.
5. Assess the improvement and reconfirm—if necessary by laboratory trial or limited field trial.
6. Incorporate process chart steps into the process for precise control on the basis of trial results.

Such an approach to solution for an application can substantially reduce the 'quality-function-loss' and also inculcate a culture of working through process capability improvement programme in a manufacturing shop. No matter how good the steel grade is, it cannot work unless manufacturing system is 'quality capable'.

Another aspect of application is that no component or system needs to be designed for infinite period of service i.e. be made fail-proof. Components will fail, as a matter of natural course of events, after a certain period by following the laws of probability. The probability of failure can be, however, decreased by increasing the specified 'statistical confidence limit' through design improvement, quality improvement, process improvement, etc.; but parts cannot be made absolutely fail-proof. They will fail after a certain period of time despite the quality being on target (see Fig. 9-3). This is because of the involvement of many other parameters in determining reliability of performance of a system or component.

Therefore, there is a diminishing return for the cost beyond a certain level of quality improvement. The optimum level of quality is thus dependent on the cost or the economic benefit that the improvement efforts can derive, where cost includes direct cost and indirect cost due to consequences of failure. This situation has been depicted in Fig. 9-2.

The cost of improving quality or reliability by choice of exotic steel quality beyond a level—except for highly safety-critical parts (e.g. nuclear plant parts, aero-engine parts etc.) may not be justified. High cost to attain high degree of quality and reliability for all applications is not the best economic solution or approach. Hence, *a trade-off between unit cost for building quality or reliability of performance and the unit 'cost of failure' often guides the selection and application process of materials in industries.* The trends in variation of cost *vis-à-vis* cost of improving quality or reliability and the 'cost of failure' is illustrated in Fig. 9-2. Figure 9-2 indicates necessity of trade-off between cost and quality for economic and realistic solution, as against the tendency to build lots of safety margin in the material specifications for furthering the life of components. However, quality must be functionally appropriate at any cost.

Figure 9-2 simply reminds that over acting on quality enhancement may not have commensurate return for the cost—i.e. the same conclusion that QLF diagram points out (see Fig. 9-1). On the contrary, it suggests that the better way to control cost of reliability in performance is to reduce the 'cost of failure' to a lower level—i.e. bring down the cost of failure curve—so that trade-off point for 'cost versus value' can be shifted to the left (see Fig. 9-2). To secure an established design and product, the approach to reduce the cost of failure is not by increasing the product quality further, but by ensuring fault-free manufacturing and preventive maintenance for a given design. This is the key to successful application of steel.

In sum, there is an optimum level of quality up to which the product is cost effective; beyond that level, loss (cost minus return) due to increased quality increases. Also, components in engineering applications have finite life due to wear and tear in the system. Hence, quality level should be right and

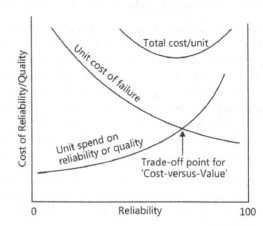

Figure 9-2. Schematic presentation of degree of reliability/quality and cost.

need not be superfluously high, unnecessarily increasing cost. A practical solution is to improve reliability in performance by improving manufacturing, processing and assembly quality, and thereafter, focusing on better maintenance of the system or components. Care for quality in the entire chain of design, production and maintenance of the parts is essential for ensuring reliable performance—deficiency in any one area could negate the merits of other factors and might cause loss due to underperformance or pre-mature failure. However, the level of adequacy of these parameters has to be worked out by 'cost-versus-value' optimisation and by 'component integrity' evaluation and, thereafter, effectively closing or ending the risk factors. Only when the part is meant for extremely safety-critical operations, such as the nuclear plants, aero engines etc. that fail-safe reliability factors have to be built into the total system by resorting to 'quality and reliability engineering' approach.

9.4. Characteristics of Failure in Service Life

Every component has its own life cycle, depending on the quality of design, manufacturing, service conditions and environment, and care in maintenance. Life cycle of an engineering part can be prolonged but cannot be made infinite. Components will fail, sooner or later. Generally, failure pattern of engineering parts typically follows a 'bath-tub' curve as shown in Fig. 9-3. The figure shows that failure of steel parts can occur very early in the service life or after prolonged service life, but the fact remains that steel parts do fail sooner or later.

Bath-tub curve plots failure rate over time, and shows three stages of failure of components in service:

- *Stage (1), early failure*—this occurs very early in the component life and the rate of failure could be initially high, but comes down sharply with time. Cause-and-effect analysis of such incidents of failure have indicated gross lapse of manufacturing quality or deviation from designed parameters, including material quality, leading to early life failure.

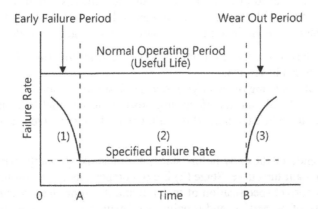

Figure 9-3. Illustration of the failure rate curve (Bath-tub curve) showing three stages of failure rate. Stage 1: early failure rate which rapidly decreases with time followed by stage 2: steady or constant failure rate followed by stage 3: accelerated failure rate at the end due to wear and tear.

- *Stage (2), the steady state failure rate*—once component has lived the early life cycle (OA in the time scale of Fig. 9-3), the component shows normal operating behaviour as per design (stage 2). Failure rate at this stage is low and steady, occurring due to such factors as accidental overloading, maintenance failure, change of service environment, etc., and
- *Stage (3), wear-out phase*—after prolonged or steady state service under stage (2), parts start failing again due to normal wear and tear of service i.e. stage (3). With the on-set of this stage, failure rate sharply increases with time.

Therefore, efficient application of steel requires some understanding about how and why steels fail in service. The purpose of such understanding is to prevent (or delay) the failure by taking certain precautionary and pre-emptive measures that can help to improve performance of the component, and not just opting blindly for enhanced design quality, making its production difficult and expensive. For example:

1. If steel part fails early in service life, look for manufacturing quality improvement.
2. If steel fails during normal operating period but below the expected life, look for design improvement, including uses of better material.
3. If there is inconsistency in the operating life, look for improved quality of processing the steel or better maintenance.
4. If life cycle has reached the 'wear-out' phase—as indicated by frequent machine break-down, plan for replacement or re-engineering of the parts.

Experience shows that bulk of early failure is due to manufacturing lapses for components of established design, and once such defects have been weeded out, steel parts can give a satisfactory service life if the design is correct vis-à-vis service conditions and environment. Failure rate in stage (2), if high, is a reflection of design shortcomings and poor quality of operating discipline—e.g. service abuse, change of environment etc. Any up-gradation of material, design etc. is justified if only stage two service life can be economically extended—i.e. benefits of up-gradation more than off-set the cost of up-gradation.

Despite such efforts, components would reach stage (3) due to wear and tear of service environment. Once component or the system has reached stage (3) of failure curve, rate of failure would be high in natural course, and attempt to prolong the life could be expensive, both in terms of direct cost of maintenance and down-time of the machine or equipment. Hence, the best solution to the problem is to plan for early replacement or re-engineering of the parts. These stages and states of failure indicate that:

- Selection of steel has to be based on understanding and information about why and when such components have failed in service,
- There is no point of designing or specifying steels to last forever; and
- The right kind of material and manufacturing process ensures that there is no premature failure and also the ability to endure the normal operating period under a given set of application load and environment.

Once steel parts have endured the early failure stage; it is most likely that the part will also successfully endure the stage (2) of bath-tub curve. Stage (2) life of components can be prolonged by correctness of design—that includes correct specification of material and its properties—accuracy of manufacturing, reduction in variability of properties, and timeliness of maintenance actions. In other words, reliable service life of engineering components depends on:

- Quality and appropriateness of design;
- Improving and ensuring quality of manufacturing, including conformance to metallurgical, dimensional and surface related properties;

- Improving process capability for reduction in variance of properties—including variation of composition in steel—and dimensions;
- Design and operational care for maintainability and flexibility in service;
- Preventive care and maintenance for prolonging steady state service life; and
- Sensing, monitoring and replacing the parts in stage (3) before other collateral damages occur.

Therefore, the task of ensuring reliable performance of a component or assembly is a composite one, where material selection is only one part of it. Nonetheless, if material selection is wrong, then failure would surface up early in service life, entailing loss of service, costly rework and replacement, and customer dissatisfaction. Similarly, if the processing of the steel is wrong—e.g. forming, heat treating, welding, etc.—the component will fail early in service life, no matter how good the steel is or the other factors.

In sum, to prevent premature failure of steel parts in service, the following steps of care are necessary:

- Correct assessment and estimation of load (magnitude) and load types—e.g. static, fluctuating (fatigue load), impact, abrasion etc. in service;
- Correct assessment of service environment—e.g. hot and humid, fluctuating temperature, low temperature, corrosive, abrasive etc.;
- Analyses of service failure and causes in similar applications;
- Right selection of steel—for the process of manufacturing as well as for the purpose of end-use;
- Precise manufacturing quality—including machining, heat treating, finishing and assembling of parts; and
- Effective and preventive maintenance practices.

An important step in preventing pre-mature or early failure is to understand the cause and effect relationship between these factors by assessing the practical situation and demands for a fail-safe operation. Figure 9-4 represents a tentative 'Cause-and-Effect' diagram of component failures, which must be examined before starting process of the selection of steel or examining the solution to a problem. Cause and Effect is a relationship, which is based on the concept that reliability of performance which is a combined function of—design quality + manufacturing accuracy + component integrity + maintenance quality—factored by the environment or service conditions. Such a situation of multi-factor intervention for an event is easy to visualise, but difficult to predict or prevent in a complex system where a number of factors work together and influence each other for the assigned performance or job. Hence, a common approach is to draw the relevant 'cause and effect' diagram and identify areas and focus necessary to prevent premature failure.

The 'Cause and Effect' diagram is useful to plan actions for preventing pre-mature failure due to design, manufacturing and maintenance. It is a spread-sheet of the issues and concerns directly related to the effect, which is 'failure' in this case. The other source of help in the process of selection and application of material is the analysis of past instances of failure, identifying the major actionable areas by 'Pareto' analysis. It is based on 80:20 rule and can indicate the areas of major actions needed for prevention of failure, and thereby helping to focus on where to act and what to act upon. For example, as per Pareto analyses—it is said that 80% of all cases of engineering component failures is due to fatigue. Hence, for preventing premature failure of steel parts, focus on factors that can control fatigue failure could be considered first for preventive action.

However, Pareto analysis can give a broad idea about the primary cause, which can then be sub-divided between specific causes. For example, fatigue failure could be due to material weakness or extra stresses working on the component in service (e.g. due to stress concentration factors), etc. Hence, further examination of causes and sources of failure is necessary for right design and to identify measures for prevention of failure which is discussed in the later part of this chapter.

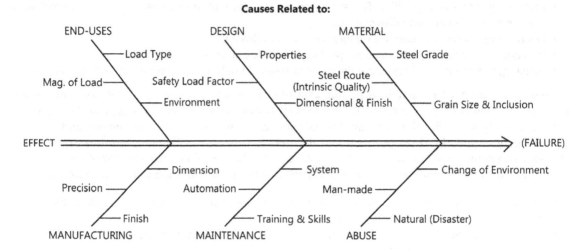

Figure 9-4. A 'Cause and Effect' diagram of component failures.

9.5. Sources of Faults and Failure: Role of Stress Raisers and Stress Concentration in Steel

Failure is a time-bound event, where the component stops performing its assigned functions after some time. Such failure can be after a short period of time due to some transient event or inherent defect, leading to 'early failure'. Failures may also occur at random after functioning for a certain period of time (see Fig. 9-3, stage (2) failure). All kinds of failure in engineering components are due to some kind of fracture or rupture, which leaves its trail and foot-prints from which the causes of failure can be established. Hence, it is useful to understand what could lead to fracture in materials in order to undertake the necessary preventive measures while planning an application.

The *fracture* of a material should theoretically depend upon the cohesive forces that exist between the atoms or plane of atoms in the crystal structure of the material. Because of the forces that exist between the atoms or plane of atoms, there exists a theoretical strength of material—called cohesive strength—which needs to be overcome by force to cause the fracture in the material. Nonetheless, experimentally measured fracture strength of a material is found to be much lower—10 or more times below the theoretical strength value. This loss of strength is due to the presence of many small atomic flaws or micro-cracks or micro-voids found either on the surface or inside the material body.

These flaws causes the stress surrounding the flaws to be amplified where the magnitude of amplification is dependent upon the orientation and geometry of the flaw—i.e. sharper the flaw, higher is the stress amplification, leading to stress concentration at the tip of the defect or flaw. Crystalline solids are full of such atomic level flaws/defects, causing sharp lowering of observed shear strength compared to theoretical cohesive strength. Figure 9-5 schematically illustrates such stress amplification around a flaw inside a material.

Due to presence of such atomic level defects, elastic–plastic stress level of material comes down and the material can start yielding and plastically flow at lower stress level than theoretical strength. However, the progress of plastic flow—leading to fracture—is the function of the ductility of the matrix which can cause stress absorption ahead of progressing shear and fracture. The more ductile the material is the more difficult it is for shear face to progress due to higher absorption of stress, leading to fracture. This is the reason for delayed fracture in ductile or tough steels.

However, the effect of atomic level defects that brings down the shear strength of materials from the theoretical value is accounted for in the design by considering the strength values based on practical tests in reality. Hence, this is not the concern of designer or users of steel in practice—unless it is introduced in the steel due to design or manufacturing or human faults, such as internal voids, ruptures, steps, cut-marks on the surface etc.

The effect of stress amplification (i.e. concentration) is more significant in practice when the defects are present at micro and macro levels. Examples of micro level defects in steel include: the presence of inclusions, micro-

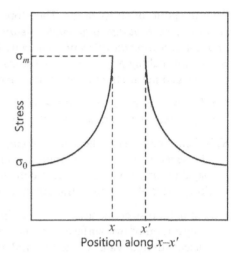

Figure 9-5. Schematic stress profile along the line X–X$'$ demonstrating stress amplification (from σ_0 to σ_m) at flaw tip position. Where X–X$'$ represents a flaw inside.

cracks, grain boundaries, voids, etc.; and the macro level defects are the cracks, laps, surface dents and cuts, steps, etc. Their deleterious effect on stress concentration depends on the size, shape and interfacial characters. State of such defective features in a material could be comprehensively studied by component integrity evaluation. Component integrity suffers from the presence of all types of micro and macro level defects and cracks present inside or outside the material, which are capable of causing appreciable stress concentration and can cause crack initiation. These defects are termed as 'stress raisers' for practical applications in industries.

The concerned stress raisers, giving rise to stress concentration, are then defined as the flaws having the ability to amplify an applied stress in the locale. Presence of stress raisers, which act as points of stress concentration—be that on the surface or inside the body—play a critical role in the initiation and propagation of fracture, thereby limiting the performance; no matter what the magnitude of applied load or amplitude of a fluctuating load is—e.g. in fatigue which may take place at stress levels lower than the tensile strength. The existence of stress raisers and stress concentration factor—which depends on the defect profile and length—can cause magnification of stress at the tip of a crack or defect and initiate fracture and failure from that point, i.e. if stress concentration value is high enough to overcome the shear stress. Hence, it is necessary to control the presence of such stress raising flaws in materials and components for satisfactory service; no matter what grade or quality of steel has been selected. Prominent and frequent among such stress raisers are fillet radius, section change point on the surface, key-holes, machine marks, cuts, dents, pits and cracks on the surface, and cracks and voids inside the body.

However, totally avoiding stress raisers in engineering parts is difficult as all engineering components will undergo some changes in section and shape, leading to stress concentration. Common examples are shoulders on shafts, key ways, screw threads, etc. Any such discontinuity will change the stress distribution in the locality, giving rise to stress concentration, which will help initiation of fracture from the point of stress raiser. Therefore, *designers have to consider redistribution of stresses in the vicinity of such designed geometric features and specify the strength properties of steels after taking into consideration the effect of such unavoidable stress raisers*. In this regard, the following relationship is used for making provision for the stress concentration factor:

The geometric stress concentration factor K_t or K_{ts} is calculated by using the following relationship:

- K_t = max direct stress/nominal direct stress, where K_t is the 'stress concentration factor', and
- K_{ts} = max shear stress/nominal shear stress, where K_s is the shear stress concentration factor.

K_t is then used to relate the actual maximum direct stress at the discontinuity to the nominal stress.

Presence of stress concentration over the applied nominal stress can influence the initiation and propagation of fracture from such defective location, depending on the location position and axis of working stress. The steps in fracture initiation from such defective sites of stress raisers are:

a. Initial stress concentration at the tip of a defect initiate opening up of a small crack—a micro-separation of two surfaces—at the tip of the defect, due to magnified applied stress exceeding the shear strength value of the material at that point.
b. As a result, further stress concentration occurs at the tip of this fine crack, where its value is dependent on the crack length—longer the crack higher is the stress concentration.
c. This concentrated stress at the crack tip then exerts further pressure in front of it for the crack to propagate and cause further fracture opening, subject to ductility of the steel/material.
d. If the crack is very sharp and long, the stress concentration value might be large enough—relative to applied shear stress value of the material—to cause further propagation of the crack, leading to final fracture, especially if the material is brittle.

Figure 9-6 illustrates some typical stress concentration curves of different geometric configurations, depicting variation of stress concentration in the shoulder of a shaft with increasing D/d, where D is the larger diameter and d is the smaller diameter.

Once the crack initiating the fracture has formed, crack propagation through brittle matrix is quite fast, because brittle material cannot absorb the fracture stress in the matrix. This means that more brittle a material, more pronounced is the effect of stress concentration factor (K_t); hence more severe is the effect of notches or stress raisers in such brittle material.

However, not all materials are brittle. If the material has sufficient ductility, a part of this concentrated stress at the tip of a crack gets absorbed in the matrix, making the crack wait for a while for further stress accumulation before it can propagate further. Thus, a ductile material (e.g. toughened steel) can delay the crack propagation and consequent fracture. This is the reason why

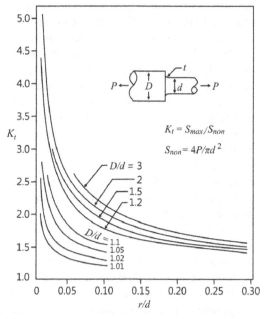

Figure 9-6. Stress concentration factor K_t for a stepped shaft in tension.

ductility or toughness of steel is considered important for performance. However, this is a simplistic view of fracture initiation, which is not exactly valid for a geometric shape such as thick plate where the corresponding *fracture toughness* property of the material plays the decisive role. Fracture toughness is a property which describes the ability of a material containing a crack to resist fracture along certain crystallographic planes. It is one of the most important properties of any material for all design applications. The linear-elastic fracture toughness of a material is determined from the stress intensity factor (K), at which a thin crack in the material begins to grow. It is denoted by K_{Ic}.

This characteristic of steel about how much stress it can absorb ahead of a crack or notch before allowing it to propagate further is dependent on *notch sensitivity* (q) of the steel. Empirically, notch sensitivity, q, is defined as:

$$q = (K_f - 1) / (K_t - 1)$$

where value of q varies between 0 and 1, K_t is normal stress concentration factor as per geometry and K_f is the reduced stress concentration factor due to the probability of experiencing such concentrated stress absorbed in the ductile matrix. A value of $q = 0$ (or $K_f = 0$) indicates no notch sensitivity, whereas $q = 1$ (or $K_f = K_t$) indicates full notch sensitivity. Therefore, steps for correct designing for performance in a notched component of ductile steel should have to consider the following:

- Finding out K_t from the geometry of the component,
- Specifying the material as per other design considerations,
- Looking up the notch sensitivity value, q, for the specific notch radius of the respective material from published data,
- Determination of K_f—the reduced stress concentration factor—using the equation:
  $K_f = 1 + q(K_t - 1)$, and
- Finally, ascertaining the right design specification and properties of the steel, based on stress concentration factor or reduced stress concentration factor.

Since brittle material cannot absorb any stress ahead of the crack, crack propagates very fast in such highly notch sensitive material leading to sudden fracture. This type of fracture is without any reduction in local area i.e. without any sign of deformation and is called *brittle fracture*. As against this, ductile materials can absorb some stress ahead of the crack, causing what is analogous to blunting of the crack tip. The blunt crack tip with reduced K_f value has to wait for further stress build-up to occur if the crack has to propagate further. Thus, in ductile material crack propagation and resultant fracture gets delayed and appearance of crack shows signs of elongation or deformation before fracture. This is called *ductile fracture*.

However, if the loading is alternating—as in fatigue—the stress build up at the tip of a notch can be due to alternating stress amplitude, where propagation of such cracks take place in alternate cycle—cycle in tension will cause crack to propagate and cycle in compression will cause the crack tip to close, thus leaving a trail of fine crack propagation marks. Such propagation of crack can take place at stress level lower than the yield strength of the material. Each cycle of deformation at the tip of a crack can cause 'work-hardening' of the locale, whereby residual ductility in the material gets exhausted. When work-hardening is enough, the tip of the crack can no longer absorb stress and give way to further crack propagation and consequent stress relaxation. This cycle of crack propagation in alternate loading continues till the residual section of the material is too weak to support the applied load, giving way to final fracture. This type of fracture is called *fatigue fracture*, which leaves trail of minute deformation marks behind (see Fig. 5-6 in Chapter 5).

Thus stress concentration factor and notch sensitivity, contribute to the process of fracture and failure, and points to the need for avoiding such stress raisers in the components; otherwise performance of the component in application will suffer.

In practice, stress raisers can be present in a component due to:

- Defects due to geometric shape and design, e.g. radius, steps, corners, holes, section change etc., and
- Defects introduced or left over from manufacturing processes, e.g. cuts, scratches, tool marks, heat-treatment cracks, grinding cracks, pits etc.

These are externally introduced sources of stress raisers, affecting the fracture behaviour and performance of engineering parts. Hence, their presence has to be eliminated or their effects have to be neutralised or minimised for getting desired component life—no matter how good the selection of steel is. High local stress can cause objects to fail more quickly—so one important task of a designer is to design the geometry of engineering parts with least possible points of stress raisers. *If the steel is free from internal defects and no external defect has been introduced in it by faulty designing or manufacturing, the steel component is supposed to run in service for time (t) without failure.* If failure occurs before time *t*, it could be early life failure (see Fig. 9-3) due to faulty manufacturing e.g. faulty heat treatment; or a random failure in the systematic stage (i.e. stage 2) due to design deficiency or inadequate material properties.

Thus, for successful application of steel, both right selection of steel and fault-free design and manufacturing are necessary for ensuring desired component life. Both of these have to be right to get the right result, no matter how good is the other one; it is not either or, it is both. For example, if radius (or shoulder) of a component is left sharper than the design specified, it will lead to higher stress concentration at the shoulder and may lead to early failure by fatigue or sudden fracture (as the case may be) originating from such weak spots. Similarly, if any tool mark is left over in the component in stress bearing areas, then the tool mark can act as starting point for cracks. Alternately, if the steel is not of right grade or it is not properly heat treated to produce the right microstructure and strength, the component will fail early despite the manufacturing process being correct. Therefore, the task of making a component work satisfactorily for the time *t*, is a combined one where:

1. A designer has to design accurately after taking into consideration the service load, environment and safety factor along with geometric design that does not allow any avoidable stress concentration factor or unnecessary stress raisers.
2. Selector of steels has to rightly choose the steel that is capable of developing the specified properties after processing and that which is free from harmful internal defects—like heavy inclusions and segregations or internal voids and bursts, which can impair processing or service behaviour.
3. Manufacturer of the components has to ensure fault-free manufacturing, leaving no spots for stress concentration beyond the permissible limits set by the design, including surface finish.
4. Operator of the system—where component is a part—has to monitor and maintain the components and system in orderly and regulated manner so that there is no change in the condition of operation, environment or lubrication—an integral part of engineering operation to control friction due to metal-to-metal contact under load.

It is to be appreciated that in a system, most parts function in sync with each other—i.e. condition of one part might affect the performance of the other parts in the system. Hence, maintenance of the system has to be appropriate.

The total system is often assessed and checked by a *component integrity* study which assesses the risk in the system for failure and identifies corrective measures for the component to perform. 'Component integrity' means, in simple terms, suitability and ability of a component made out of a chosen material to withstand the stresses and strain of a given application without premature failure i.e. examining the integrity of the component structure and features for withstanding the service conditions and associated fluctuations that the component might face in reality. Therefore, determination of component integrity involves checks and balances of material properties, as assumed in the design, with that of:

- Probable property loss due to stress raisers and other internal defects;
- Impact of load types: tensile, fatigue, creep, impact, etc and corresponding fracture behaviour;
- Impact of residual stresses arising from faulty heat treatment or working;
- Impact of service environment: corrosion, oxidation, creep, low-temperature impact, wear, etc.;
- Requirement of safety in services and adequacy of safety factor in the design;
- Capability of manufacturing system for fault free manufacturing and holding to the design parameters; and
- Maintainability and maintenance system for continuing performance.

The purpose of component integrity study might be to evolve a predictable model for an operation or case by case assessment of the aforementioned factors in order to either enhance the quality wherever necessary for improving the probability of success in application or to eliminate risks of failure in application.

However, component integrity study requires elaborate study centre with instrumented facilities for checking and balancing the inter-connected factors. Hence, instead of component integrity study, a practical way to address the concern is to—(a) examine the root-cause of failure or short comings by using 'cause-and-effect' diagram; (b) study the failure modes and mechanisms of steel under similar condition in service, using Pareto analysis for identifying and taking preventive actions in areas where they are needed; and (c) ensuring structural and surface integrity by right selection and processing of steels, including heat treatment and welding. Further discussions in this chapter and subsequent chapter will follow this line of approach for making the steel selection and application cost-effective and need-based.

9.6. Failure Modes and Mechanisms: Learning Points for Steel Applications

Engineering steels fail by some kind of fracture—termed as 'fracture mode'. There are three basic fracture modes in steel: *ductile, brittle,* and *fatigue fracture*. Fracture mode and type help to identify probable causes of failure and steps for prevention by taking appropriate action either at the stage of initial design or for improvement of an existing design and manufacturing. Knowledge of fracture mode and its cause is necessary for correct selection and application of steels. Hence, understanding of failure modes and mechanisms forms an integral part of decision making process for steel selection and application.

Ductile failure involves plastic deformation of the fracture face due to overload and leaves a permanent change in shape. Ductile fracture surfaces are dull in appearance, and when observed under a microscope, they appear rough and pitted with tiny pits (called micro-voids) on the fracture surface. In fact, coalescence of these voids leads to ductile fracture. By contrast, *brittle fractures* leave no permanent plastic deformation, and have a smooth or woody or silky appearance of the fracture face.

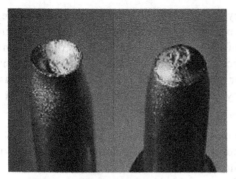

(A) Ductile Fracture   (B) Brittle Fracture

Figure 9-7. Depiction of the ductile and brittle fracture under tensile load.

If examined under microscope, brittle fractures are either smooth, when the fracture flows through the grains or faceted like rock candy, when the fracture follows the grain boundaries. Figure 9-7 depicts the ductile fracture (on the left) and brittle fracture under applied tensile load (on the right) showing grainy fracture.

Some characteristics of brittle fracture are:

1. There is no gross, permanent deformation of the material.
2. The surface of the brittle fracture tends to be perpendicular to the principal tensile stress—although other components of stress can be factors for any change of course.
3. Characteristic marking of crack advancing fronts frequently point to where the fracture originated.
4. The crack path of fracture depends on the material's structure, generally trans-granular in nature.

Some characteristics of ductile fracture include:

1. There is permanent deformation at the tip of the advancing crack that leaves distinct patterns.
2. As with brittle fractures, the surface of a ductile fracture tends to be perpendicular to the principal tensile stress, although other components of stress can be some factors.
3. In ductile fracture of crystalline metals (e.g. steel), it is the resolved shear stress that is operating to expand the tip of the crack.
4. The fracture surface is dull and fibrous in appearance.
5. There has to be a lot of energy available to extend the crack through the ductile matrix.

Ductile and brittle fractures are the two principal modes of fracture, where components may fail with signs of these fracture modes either singly or in combination. *Fatigue* mode of fracture—the other important fracture mode—takes place without any elongation or reduction of the cross section but with lots of fine serration marks on the plane of fracture, described as 'beach marks'. These are tiny signs of foot-prints left by progressing fracture. Hence, a fatigue fracture resembles brittle fracture in looks, despite the material having good ductility or toughness. A typical fatigue fracture is shown in Fig. 9-8. Ductility in material helps to arrest the fatigue crack propagation by absorbing the crack tip stresses and, thereby blunting it and delaying the failure.

Fatigue of metals take place under cyclical or alternating stress and the failure can occur at stress level below the tensile strength of the material i.e. in the elastic stress range. Fatigue failure (already discussed in Chapter 5) is the most dominant mode of failure of all engineering (mechanical) parts—over 80% of all mechanical part failures in industries are due to fatigue. Hence, an understanding of fatigue fracture, and the corresponding preventive measures, is an important step in the direction of improvement of component performance.

Fatigue is a phenomenon which results in the sudden fracture of a component after a period of cyclic loading in the elastic regime. Fatigue fracture is the end result of a process involving the initiation and growth of a crack, usually originating at the given site of stress concentration on the surface or at a fault on the surface or just below the surface—e.g. site of presence of a heavy inclusion patch below the surface. Eventually the cross sectional area under the propagating crack is so reduced that the component ruptures under a normal service load, which could have been satisfactorily withstood otherwise. Figure 9-9 illustrates typical fatigue fractures—illustrating where the crack initiates, how it progresses, and area under or affected by final fracture.

Fatigue fracture faces always exhibit from where the crack had actually started and the direction of

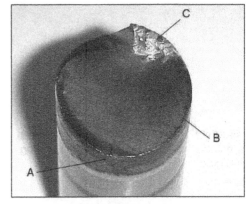

Figure 9-8. Illustration of a fatigue fracture surface with beach marks. Note the crack initiation point marked A and the shear region of fracture opposite to it where the fracture process ends, marked C. Beach marks are shown by B, which is typical of bending fatigue failure. Shear fracture in the region C indicates separation due to over load corresponding to remainder of the cross section.

propagation, and the end. Even, the nature of and inter-space between beach marks can indicate the operative stress level. Thus, it leaves behind ample information as to the cause of failure, which should be used for further improving the performance of a component in service.

Because of the prevalence of fatigue mode of fracture in engineering components, it is necessary for the designer and selector of steels to consider 'fatigue strength' (σ_F) value of the steel at the specified strength levels and consider the same for design stress calculation along with the effect of stress raisers present in the design. If there are points of stress raisers in the design—like the radius, cuts, grooves,

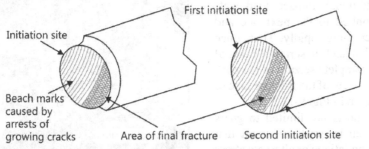

Figure 9-9. Sketches of fatigue fracture showing single point crack initiation (left side figure) and presence of second crack initiation point during reverse loading (right side figure). Area of final fracture shows a ductile fracture (shear fracture) band across the remaining section. When the applied stress is high, fatigue failure can show multiple sites of crack initiation.

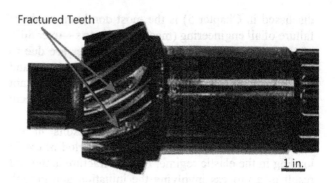

Figure 9-10. A failed shaft pinion wherein consecutive teeth have failed by bending fatigue.

notches etc.—fatigue fracture is most likely to start from those points due to notch sensitivity in fatigue. Therefore, such factors have to be taken into consideration while specifying steel types and strength. Steel with higher toughness at a given strength level will have higher fatigue life.

In principle, components designed with the applied stress levels that do not exceed the fatigue strength level—also called 'Endurance limit'—should not fail in service, unless there is localised stress concentration present or introduced during service which can lead to fatigue crack initiation. The shape of the component, surface condition, applied stress level and quality of assembly are few important parameters that determine whether fatigue failure is a possibility.

A frequently observed failure mode in engineering steel parts is the *bending fatigue* where the dynamic load acting on a body also produces 'bending moments' at some critical part of the body. A pinion shaft where dynamic load acts on the tooth tip and simultaneously produces a bending moment at the tooth root is one such example. Figure 9-10 shows a typical bending fatigue failure of gear teeth in a shaft pinion.

Bending fatigue fracture got initiated in this component from the tooth root where the acting stress is high due to bending moment produced at the root, while the actual stress had been applied towards the tooth tip. Presence of root radius further magnifies the bending moment at the root. Once a drive tooth fails in such gear arrangement, additional impact load gets generated and applied to the next one and that one fails even more rapidly. Thus, once failure has started, it leads to series of cases of tooth failure and complete seizure of the part. A typical enlarged view of the fracture face of gear failure is shown in Fig. 9-11.

Bending fatigue is not limited to gears only; it can happen in any component that experiences rotating fatigue or alternate stress reversal, producing bending moment at certain points by the application of load. Examples of such cases are rotating shaft, rods, bolts, leaf springs and coil springs in automobiles, etc.

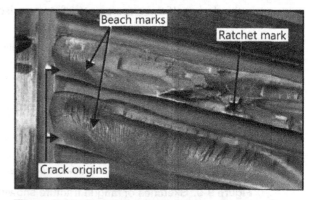

Figure 9-11. Bending fatigue fracture surface of a wind turbine gear. The crack originated at the bottom tooth first which shifted additional load on to the next tooth (top tooth) leading to failure.

Figure 9-12 is a view of a classical reverse bending fatigue fracture of a bolt. The arrows point to the initiation sites of the fatigue crack at the thread root. The small lines or striations on the metal surface show, how the crack advanced from the surface to the inside of the bolt. The rutted grey area in the middle of the bolt is the area of final fracture where the bolt cross-section was reduced and the bolt could not carry the load anymore. Such instances of failure can be a result of design deficiency as well as improper assembly of the parts, when due to looseness of the joint a jerky stress can act on the part, producing higher bending moment.

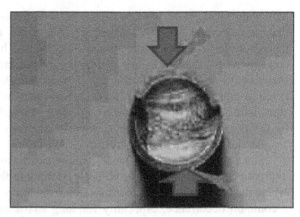

Figure 9-12. Reverse bending fatigue fracture in a bolt. Arrows show point of fatigue initiation on either side.

The other important type of fracture is the fracture under 'torsion'. Torsional fractures occur due to twisting effect, where one end is fixed or two opposing forces are applied from two opposite ends of a shaft. If the steel is soft, it fails in ductile manner with twist marks on the surface, as shown in Fig. 9-13, with an eye at the centre of the fractured face, indicating the last portion to shear. If the steel lacks ductility, it fails by brittle torsion (see Fig. 9-13). Cases of stress acting on a torsion bar are generally high, as the stress is applied from one end (drive end) and gets multiplied by lever action at the opposite end (the driven end). Hence, for torsional application, it is necessary to ensure: (a) high strength of the steel, (b) good toughness in the material, and (c) freedom from surface defects which can act as opening point for crack i.e. freedom from pits, dents, slag patch, inclusion stringers etc. on the surface or sub-surface area.

To meet such quality requirements for heavy duty shafts, high strength steels with good toughness— e.g. AISI 4150 grade—and low inclusion count are used after heat treatment and surface crack testing. Common steel grades for torsional applications are Cr-Ni-Mo or Cr-Mo or Cr-V grade alloy steels, with low inclusion content, which are used after suitable heat treatment. High strength and toughness level can be reached by using steel with higher carbon and sufficient alloying—e.g. DIN 50CrV4—and by heat treatment to produce tensile strength level of 120–140 Kgf/mm^2 and elongation of about 8% and more. Alternately, steel with adequate carbon content can be surface hardened by induction hardening or a low carbon Cr-Ni-Mo or Cr-Mo steel can be chosen for processing through case-carburising and hardening.

Ductile Torsion Failure

Brittle Torsion Failure

Figure 9-13. Depiction of torsional ductile and brittle fractures of a transmission shaft.

Reason for making such choices is the necessity to have high surface hardness (strength) to counter the torsional load, which is highest on the surface. The choice depends on the areas of working where the stresses could be high and properties are most critical. For example, if it is a straight shaft, through hardening steel might work well, but if it is threaded and that part requires selective hardening—in order to keep threads softer and area under thread less affected by stress concentration—then induction hardening of selected parts might be the answer. And, if the part is of complicated shape (such as pinion gears), use of lower carbon case carburising steel with control of case depth and core strength might be desirable for attaining high surface hardness as well as tougher core.

Since most engineering parts experience dynamic and alternating stresses in service, failure by torsional fatigue—where fluctuating applied stress is torsional in nature—is also a common occurrence, especially for long shafts and rods. A typical torsional fatigue failure of a shaft is shown in Fig. 9-14. A typical foot-print of torsional fatigue is a sign of fatigue crack propagation from multiple-points, leading to tear off of the last part, which generally coincides with the

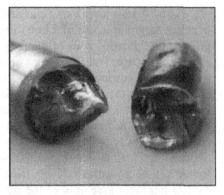

Figure 9-14. Picture of 'torsional fatigue' fracture. Note the multi-point fatigue cracks merging towards the centre.

middle or near middle portion towards which such fatigue cracks tend to grow and merge.

Fatigue and torsional fatigue stresses are also influenced by the presence of stress raisers i.e. from the stress concentration effect. Such effect in fatigue and torsional fatigue is described as 'notch fatigue' effect. The fatigue notch factor is described in terms of the material notch sensitivity as:

$$Kf = 1 + q\,(Kt - 1),$$

where q is the fatigue notch factor. However, notch effect increases with increasing strength of steels—it is generally higher in higher strength steels than softer strength steels. Also, fatigue notch factor is usually lower than theoretical stress concentration factor, because of stress relief at the crack tip due to plastic deformation involved in fatigue crack propagation.

The aforementioned discussions and illustrations of failure modes and causes point to the necessity of taking decisions for steel selection and application after due consideration of how steel components can fail in service. The final choice of steel should be based on ability of the steel to withstand the specific service conditions and delay fracture under the acting stress type. For instance, if components experience torsion in application then steels of higher strength and toughness should be used than what should have been selected for bending fatigue resistance. In this regard, it would not be enough to consider the steel quality and its properties alone—design and manufacturing quality must also be considered along with steel properties. They are integral, influencing the effects of each other. Preventive action based on such integrated approach should form a part of the exercise for application of steels.

9.7. Prevention of Failures: Ultimate of Steel Application

Summing up the different types of failures, it can be observed that cases or instances of failure may take the form of physical or chemical failure or a combination of both. For instance, failure could be due to fracture—a physical form (e.g. fatigue or tensile fracture), corrosion—a chemical form (e.g. thinning of metals or pitting), or a combination of load and environment, like the creep or corrosion-fatigue—the former being the effect of load and temperature and the latter being the effect of fluctuating load (fatigue) and environment (corrosion). While corrosion is a chemical phenomenon which continually weakens the steel part with its progress inside the steel body, presence of load (either static or alternating like the fatigue) accelerates the failure due to fracture initiation and propagation from damaged sites (e.g. corrosion spots) under corrosive environment. The process is termed as *stress corrosion cracking* when the load is static and *corrosion fatigue* when the load is cyclical or dynamic. Analyses of these cases of failure allow further grouping of cases of failure caused by: fracture (e.g. ductile fracture, brittle fracture, fatigue fracture etc.), by deformation (e.g. bend and buckle in structures) or by damage (e.g. by corrosion, oxidation etc.). Examples of failure by deformation or by damage are plenty in structural applications. Thus, for prevention of failure, it is necessary to identify the nature of failure and then seek for its cause and cure by taking preventive action. To illustrate further:

1. If failure is caused by ductile tensile fracture, one should opt for higher strength steel with adequate ductility. For instance, if a bolt of strength 80 kgf/mm^2 tensile strength with corresponding yield strength of 64 kgf/mm^2 and elongation of 14% (8 K bolt quality) fail by ductile extension and fracture, higher strength bolt of say 100 kgf/mm^2 with 80 kgf/mm^2 yield strength and minimum 8% elongation (10 K bolt quality) should be used for that joint.

2. If the failure is by brittle fracture under tension, then one should go for material of sufficient strength but with sufficient elongation (e.g. 8% elongation), and simultaneously check if there is any cause for embrittlement of the steel during processing (e.g. hydrogen embrittlement). For example, a common experience is the brittle failure of ship hauls due to faulty welding practices, resulting in hydrogen embrittlement of the structure. In brittle fracture, sudden drop in operating temperature below room temperature can also play an important role.

3. If the failure type is by fatigue, then identify the nature and source of fracture origin. For instancee, is it bending, torsional or corrosion fatigue, and then examine the failed face for signs of crack's origin.

   - If fatigue failure shows signs of early life failure, check for correctness of material strength and quality—e.g. high inclusions, internal cracks or voids etc.
   - If a fatigue failure surface shows up signs of crack starting from a sharp radius or groove, corrective step should be taken to liberalise the radius for lesser stress concentration.
   - If the failure has originated form multi-points, check whether torsional load component of the application is too high. If so, identify sources of torsional effect and eliminate the same or increase the strength level for withstanding the torsional load.

4. If failure is caused by deformation (bend, twists, etc.) of a structure, increase yield strength of the material for resistance to yielding and bending.

5. If failure is caused by corrosion fatigue, ensure that steel has proper alloying or coating for adequate pitting resistance or corrosion resistance—along with sufficient fatigue strength and/or lowering of magnitude of cyclical load.

Fatigue is undoubtedly the single most important cause for steel failure in service. Though there could be several types of fatigue load in service, there is a common thread of causes that contribute to fatigue failure and needs to be taken care of in design and manufacturing. Table 9-1 outlines such factors, which demand attention for prevention of fatigue.

Barring corrosion-related failure and failure due to faulty manufacturing, other kinds of steel failure are essentially related to some metallurgical factors such as: (a) strength, (b) elongation/toughness, (c) inclusions, (d) segregations, (e) voids, surface defects, etc. As far as mechanical failure of steels is

Table 9-1. List of factors for controlling/preventing fatigue failure in steels.

Fatigue Failure Factors	Attention for Prevention of Fatigue Failures
Cyclic stress state	Depending on the complexity of the geometry and the loading, one or more properties of the state of stress need to be considered, such as: stress amplitude, mean stress, bi-axiality, in-phase or out-of-phase shear stress, and load sequence.
Geometry	Notches and variation in cross section throughout a part can lead to stress concentration where fatigue cracks initiate.
Surface quality	Surface roughness causes microscopic stress concentration that lowers the fatigue strength. Hence, fatigue application calls for better surface finish after hardening. Compressive residual stress can also be introduced on the surface by shot peening or controlled rolling to increase fatigue life (see Fig. 9-16 for effect of surface finish in fatigue).
Material type	Fatigue life, as well as behaviour during cyclic loading, varies widely for different steels, e.g. Cr-Ni-Mo steel with tougher matrix has higher fatigue resistance property than plain carbon or Cr-Mn steel.
Residual stress	Welding, cutting, casting and other manufacturing processes involving heat or deformation can produce high levels of tensile residual stress, which decrease the fatigue strength.
Size and distribution of internal defects	Casting defects such as internal voids, shrinkage, macro-inclusion, etc. can significantly reduce fatigue strength.
Grain size	For most metals, smaller grains yield longer fatigue life.
Inclusions	Inclusions of all types are harmful for fatigue. Hence, inclusion content in steel must be controlled—finer and even distribution of inclusions is less harmful than coarser and clustered inclusions.
Environment	Environmental conditions can cause erosion, corrosion or hydrogen embrittlement, which all affect fatigue life. Corrosion fatigue is a problem encountered in many instances of aggressive environment.
Temperature	Extreme high or low temperatures can decrease fatigue strength.

concerned, the following parametric features should be borne in mind while choosing a particular grade of steel and steel properties for prevention of failure:

a. *In static load applications* (as in structural applications), higher the strength of steel, better is the deformation resistance and fracture (including *shear fracture*) resistance; but some minimum amount of ductility (measured by % elongation) in the steel should be ensured in order to avoid *brittle fracture.* This is achieved by choosing an appropriate grade of C-Mn steel or HSLA steel (refer Chapter 7 for structural steel specifications) with good fracture toughness property.

b. *In dynamic load,* as in *fatigue failure,* a combination of high strength and toughness—measured by the ability to absorb deformation energy at the tip of propagating cracks—is necessary in order to delay the failure. These types of properties in steel can be introduced by choosing an appropriate alloy steel grade and subjecting it to hardening and tempering process.

c. *For wear resistance,* steel surface has to be hard, which can be developed by hardening steel containing suitable amount of carbon and hard carbide forming element.

d. *For a combination of fatigue and wear*—as in the case of gears etc. where two load-bearing surfaces slide over each other—both *high hardness of the surface* and *strength with toughness at the core* are required. This can be achieved by special heat treatment of suitable steel, such as by case-hardening, surface induction hardening, nitriding, etc.

However, steels can still fail despite having these properties in adequate measure, because of inclusions and internal defects. Influence of inclusions in steels on their mechanical and forming behaviour has been discussed earlier in Chapter 3. Nonetheless, as regards fatigue, properties of steel are not the last word—fatigue behaviour of steels can be effectively and economically improved by improvement in design in order to:

- Eliminate or reduce stress raisers by smooth designing of the part,
- Avoid sharp surface tears resulting from punching, stamping, shearing, etc.,
- Prevent the development of surface discontinuities or decarburising during processing or heat treatment,
- Ensure smoothness of surface by proper grinding or polishing,
- Reduce or eliminate tensile residual stresses caused by manufacturing, heat treating, and welding, and
- Improve the quality of fabrication and fastening procedures.

Whatever could be the mode of failure, notches are the major sources of failure initiation, and they ought to be taken care of either in design or in processing to avoid premature failure. Figure 9-15 illustrates the change in notch sensitivity (as indicated by K_f factor) with steel hardness in bending fatigue and torsion. The diagram indicates that no matter how good the steel is, the component will suffer from high stress concentration and would have an adverse effect in service; if the designer is not careful either to specify liberal radius or leave some stress sensitive area selectively softer.

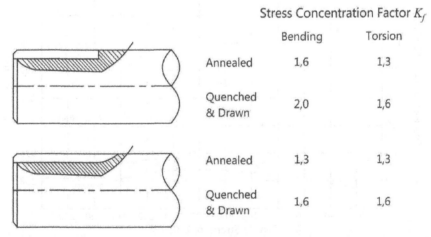

	Stress Concentration Factor K_f	
	Bending	Torsion
Annealed	1,6	1,3
Quenched & Drawn	2,0	1,6
Annealed	1,3	1,3
Quenched & Drawn	1,6	1,6

Figure 9-15. Illustration showing the effect of key-hole notch on stress concentration factor in rotating fatigue in annealed and hardened steels. Key holes are cut into a shaft reduce its normal torque carrying capacity. It is a generally accepted fact that for a standard keyway (width approx. 25% diam. and depth approx. 12.5% diam.) the design load carrying capacity is reduced to 75% of the normal working strength.

Notches or stress raisers can arise in a component from various sources, other than having radius or key-hole in the design. They can arise from for example:

- Rough machined surface, having deep machine marks or groove,
- Dent and tool marks on the surface,
- Abrupt end of a hardened zone, creating sharp transition of microstructures,
- Presence of sharp edged oil-hole,
- Presence of sharp cut edge and burrs, and
- Presence of threads on the surface, etc.

Stress risers and residual stresses are often overlooked in the manufacturing process of components, but they are frequent sources of failure in service. Instances of residual stress can occur in components due to heavy machining, grinding, working, etc. as well as due to faulty heat treatment processes. Stress raisers are eliminated by taking care of design and surface finish, e.g. grinding, polishing etc. And cases of surface residual stress are taken care of by either stress relieving the component after machining, heat treating or by selectively introducing compressive residual stress by shot-peening, cold rolling etc. Machining, grinding etc., which are used for finishing the engineering components, can produce surface residual tensile stress lowering the fatigue strength. Figure 9-16 illustrates the way fatigue endurance limit of steel can get influenced by the quality of surface finish in steel parts. The effect is more pronounced in higher strength steels.

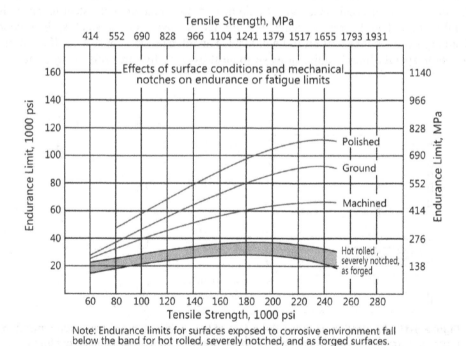

Note: Endurance limits for surfaces exposed to corrosive environment fall below the band for hot rolled, severely notched, and as forged surfaces.

Figure 9-16. Illustration depicting the effect of surface finishes on tensile and fatigue strength of steel.

The need for improved surface finish in manufacturing for increased fatigue limit is, therefore, not questionable. This becomes more important at a higher level of strength where notch sensitivity of steel increases.

In addition to these types of failure, there are other types of failure in service—namely creep, corrosion and corrosion fatigue—which have their own distinctive characteristics. Creep fracture takes place when a part experiences tensile load while operating at higher temperature (refer Chapter 5, Section 5.4, Fatigue and Creep Testing). Creep fracture is characteristically similar to tensile fracture, but corresponds to higher temperature tensile fracture. Because of higher operating temperature, creep fracture strength—generally called creep strength—is lower than the room temperature tensile strength of the steel. Depending on the temperature and applied load, rate of creep deformation could be

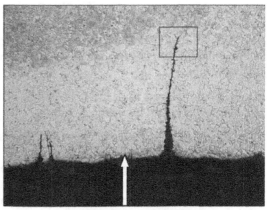

(Internal surface of the pipe)

Figure 9-17. Low magnification optical photomicrograph of the transverse cross-section of a stainless steel pipe, showing number of cracks along the internal pipe diameter, which came in contact with corrosive fluid.

slow or fast, which finally gives way to fracture when the applied load working on the deformed cross section exceeds the corresponding strength of the steel at that temperature. Such failure is very similar to tensile fracture, which takes place by coalescence of micro-voids. A preventive measure for creep failure is to select steel with complex alloying—e.g. alloyed with Cr and Mo—to form complex alloy carbide in the structure and resist high temperature softening of steel and deformation under tensile load.

Corrosion failure starts with pitting or oxidation of the steel, which ends up by eating out of metal or erosion of metal from the body of a component, leading to loss of mass and consequent failure. If the failure is under the combined action of pitting (corrosion) and dynamic load (alternating cyclical load), the failure is termed corrosion fatigue—where pitted sites provide sites for crack initiation and propagation. Figure 9-17 shows a typical corrosion fatigue crack in stainless steel pipe.

In corrosion fatigue, the fatigue crack growth rate is enhanced by corrosion, where the rate of growth of crack depends on both corrosion type and fatigue load levels. The other mechanism of corrosion failure is called the *stress corrosion cracking* (SCC), which is introduced in steels under combined tensile load and corrosion. The SCC is especially important in some alloy steels, including ferritic stainless steels. Further, SCC cracks are very fine till the rapid crack growth starts and causes catastrophic failure. Appearance of these cracks is very similar to corrosion fatigue cracks, but differs to some extent based on the corrosive media involved (e.g. sodium, chloride etc.).

Corrosion or corrosion fatigue failure can occur in both structural and engineering components working in corrosive and hazardous environment. Actions necessary for reducing or preventing types of failure include:

- Minimise amplitude of cyclic stresses,
- Reduce stress concentration or redistribute stress—balance strength and stress throughout the component,
- Select the correct shape of critical sections,

- Guard against rapid changes of loading, temperature or pressure,
- Avoid internal (i.e. residual) stress,
- Avoid fluttering and vibration producing or vibration transmitting design, and
- Limit corrosion factor in the corrosion-fatigue process—using more resistant material/less corrosive environment.

Thus, taking lessons from the field failure of steel parts, prevention of premature failure in service demands attention to following factors:

1. Right design of the component that avoids or minimises stress raisers and uneven distribution of acting stress.
2. Avoiding over design for fail-proof component life—instead focus on fail-safe practical design.
3. Setting right metallurgical specifications with reference to strength, toughness, impact strength, fatigue strength, corrosion resistance properties of the steel, etc. on the basis of end-use application requirements
4. Setting right and appropriate dimensional and metallurgical specifications after understanding the end-use application requirements and avoiding over/under specification for cost optimisation.
5. Ensuring quality of manufacturing—not only for dimensional correctness but also for heat treated properties, surface finish and freedom from any types of surface damage that can lead to failure initiation.
6. Avoiding or eliminating chances of surface residual stresses (tensile) in the manufacturing, including heat treatment and welding.
7. Monitoring and maintaining the functional performance by appropriate controls and preventions.

Selection of steel is a subject matter of optimisation of quality, cost and failure. Failure analyses, to reveal the causes and shortcomings of the steel behaviour, are important part of lessons for steel selection and application. Without the knowledge of foot-prints of field failure, efforts for right and cost effective selection and application of steels might not succeed.

SUMMARY

1. The chapter highlights the principles of controlling cost and quality for given applications, provides road map for analysing the failures and outlines preventive measures for improved performance. The areas of care, concern and prevention necessary for reliable performance at optimum cost and service level have been pointed out.
2. Cost and quality goals have been discussed with reference to Taguchi's 'Quality Loss Function' (QLF) and it has been shown that there has to be a realistic quality target for an application, beyond which either way of variation in quality will lead to loss i.e. the cost will not be justified. According to Taguchi, the means to addressing quality goals are not necessarily by specifying higher quality material, but by controlling the process variations by improved process capability.
3. Life cycle behaviour of engineering components has been illustrated, showing the various stages of failure in service. Probable reasons for fluctuating failure rates at different stages have been mentioned and the necessity for preventive measures in the design, selection and manufacturing has been emphasised. The purpose of such understanding is necessary for avoiding the tendency of

blindly going for higher grade steel as cure-all solution of performance related problems; such step only enhances cost without commensurate return on investment.

4. The combined responsibility of material selection, design, manufacturing and servicing for satisfactory component performance have been mentioned and the need for a synchronised studies to eliminate risks of failure, such as by component integrity study, have been emphasised. In the absence of detailed component integrity evaluation, this could turn out to be a complex process at times. Alternative steps of root-cause analysis have however been illustrated.

5. In this context, state of internal structure of materials, possible structural defects, effects of stress raisers and their influence on fracture initiation, effects of notches and other factors that can lead to failure and fracture have been examined and highlighted.

6. In order to identify preventive measures a number of failure and fracture modes have been examined and their causes discussed. Based on these observations, various preventive measures necessary for successful steel selection and application have been highlighted. It has been emphasised that all factors impairing the performance of engineering components, as discussed in this chapter, and their preventive measures should form a part of the process for right selection and application of steels.

FURTHER READINGS

Dieter, George E. *Mechanical Metallurgy.* New York: McGraw-Hill, 1961.

ASM Handbook, Vol. 19, *Fatigue and Fracture.* Ohio: ASM International, 1997.

Campbell, F.C., ed. *Fatigue and Fracture: Understanding the Basics.* Ohio: ASM International, 2012.

Reardon, Arthur C. *Metallurgy for the Non-Metallurgists,* 2nd edition. Ohio: ASM International, 2011.

ASM Handbook, Vol. 11, *Failure Analysis.* Ohio: ASM international, 2002.

CHAPTER 10

Selection and Application of Steels

Case Studies

The purpose of this chapter is to outline the methods of steel selection and applications by illustrative case studies and analyses. Principles and rules guiding steel selection, processing and applications have been already discussed in the previous chapters, and the aim of this chapter is to demonstrate how those principles and rules are applied in practice for actual selection and application of steels. The chapter illustrates the highpoints of steel application methodology by discussing some cases selected from industry practices.

10.1. Introduction

The ultimate purpose of steel technology is to make and process steels of appropriate grades and properties for different applications concerning the society and industries at large. Discussions about steel and steel technology, therefore, must have focus on *application of steels*. Application is the other face of steel technology that makes it meaningful, useful and complete. To illustrate further, be it for making 'safety pin' or 'rebar for high rise building' or 'chassis of a car' or 'aircraft landing gear' or 'the high-sea oil-rig platform', all require certain application specific properties and attributes, which must be fulfilled if steel has to be meaningful and useful material for the society and industries.

Application is a 'bottom-up process', requiring analyses of properties and attributes of steels starting from the end-user requirements (i.e. application requirements). Therefore, the process of application of steel starts with capturing the application requirements for end uses, and then goes on to identifying how those requirements can be fulfilled in a cost-effective manner by proper selection and processing of steel. The process chain of steel application, therefore, should examine:

a. End-use properties,
b. Quality requirements for manufacturing and processing of the steel,
c. Critical manufacturing and application parameters influencing the reliability in performance,
d. Identification of steel types and quality that can satisfactorily fulfil the manufacturing and end-use requirements, and
e. Identifying the 'steel grade' and final set of properties with reference to a 'standard' and 'specification', but not overlooking the cost and availability.

In practice, the process of application of steel translates to:

- Identification of physical, mechanical and chemical properties required for end-use;
- Identification of steel properties and attributes required for forming and manufacturing (e.g. forging, machining, casting, rolling, heat-treating, welding, cold forming etc.);
- Identification of appropriate steel grade by referring to available steel standards, cost and availability; and
- Finally, efficiently and effectively processing the steel through the manufacturing chains—for developing the target properties and finishing without introduction of defects or impairment.

Thus, the selection of steel could be a part of design—where action for manufacturing starts, but the application of steel is the part of larger chain of events and actions that include quality of manufacturing and conformance to what design had envisaged. The previous chapter (Chapter 9) has demonstrated the role of cost, quality and failure analysis for successful steel applications, and this chapter will attempt to illustrate how the steel is selected for specific applications in industries with reference to component characteristics and application properties. This chapter will not go further into the mechanisms of failure or its prevention as part of exercise for steel application, which has been already discussed in Chapter 9.

10.2. Road Map to Selection of Steel: A Review

Steel selection starts with understanding of end-use requirements, and thereafter goes on building-up the choices for required forming, manufacturing and developing mechanical and other service related properties. For the steel selection to be purposeful, suitability of forming, shaping, machining or heat treating the steel must be examined along with attainable properties of the steel. Steel is no good if it cannot be processed economically—various alternative grades of steels are to be considered, wherever necessary. For illustration, let's examine the selection of steel for the following applications:

A heavy duty commercial vehicle 'Long-member' frame part: Long member frame has to carry the working load of the vehicle, consisting of gross vehicle weight including consignment load. End-use requirement of the vehicle frame calls for high yield strength (YS of above 42 kgf/mm^2) and the forming process calls for high ductility i.e. elongation (e.g. over 22%), with good bendability and cold formability. The long member part also requires some rigidity to avoid whipping under bending moment of the frame while moving at speed with load on the road. Hence, a thicker rigid section in the range of 6–10 mm thick plate with above mentioned strength and ductility level is required for application. Such a combination of strength and ductility in plain carbon steel plates is not normal. Hence, there are two options:

1. To select C-Mn grade steel plate and subject this to additional heat treatment of normalising for grain refinement and improvement in strength and ductility. Or
2. To go for micro-alloyed HSLA steel plate where the strength and ductility are already built into the steel by controlled rolling.

The HSLA steel might be more expensive than the standard plain-carbon or C-Mn steel plate, but it saves the cost and cumbersomeness of normalising operation, an additional off-line operation. Such a choice may ultimately work out to be more productive, less in-process rejection, easier to handle in the shop-floor, and cost-effective in a mass production shop, and simultaneously providing a superior product.

Engineering applications of gears: selected steel would require high wear and fatigue resistance along with complicated shaping by extensive machining. Hence, the steel should have a number of combined properties—like good forgability, good machinability, strong and tough core and very high surface hardness for wear and fatigue resistance. These types of combined properties in the steel are difficult to develop by using straight through hardening grade steel. Straight through hardening grade steels can have high hardness, but may not have required toughness; limiting the application. This would necessitate considering two alternatives—(1) to select a case hardening grade steel and develop the properties by case carburising and hardening, or (2) select a through hardening grade steel, process it for tough core, and induction harden the surface for developing high surface hardness.

Since gear manufacturing is a mass production job, especially for engineering and automobile industries, the criterion for choice of option should be productivity—including cost and quality—flexibility and better reproducibility of results. In these respects, case carburised and hardening of steel offers better alternative for mass production of gears. Merits of case hardened properties compared to induction hardened properties can be assessed—as per specific application and choice of steel grade and the processing route to be adopted. In general, case hardening process is more flexible for developing closer range of properties along with some favourable compressive residual stress, which is beneficial for fatigue. Hence, for mass production of gear application, case carburising route of steel processing is often preferred.

There are many types of carburising grade steels, but selection should be done by taking the following into consideration:

- *Duty of the gear.* If very high duty requiring high impact loading during operation, then tougher nickel-bearing steel like SAE 4320 could be used e.g. landing gears of aircraft.
- *Cost to justify end-use.* If application involves standard wear and fatigue, as in most automobile or engine gears, steel like 20MnCr5—an economical grade of steel—compared to SAE 4320, can be used with certain amount of pre-emptive caution about cleanliness of the steel and effective control of carburising process.
- *Processing facility in the shop.* Steel like 20MnCr5, which is less in terms of toughness than Ni-Cr-Mo bearing steel, demands more careful processing and process control to take full advantage of the steel. German industries have pioneered such cost effective but efficient uses of steels by improved process design and control.

Steel is a versatile material and its composition and processing can be planned in very accurate manner for tailoring the application related properties of steels. Today's advanced steelmaking processes support closer control over composition, inclusion, grain size and other character of steels for developing right microstructure and properties. Such features of steels, steel grades and obtainable properties have been already identified and discussed in this book (refer Chapters 1 to 7) for facilitating right selection. A selector of steel has to judiciously choose the steel based on such knowledge and information.

Steels are to be selected for a given application and procured from steel suppliers by referring to some standards and specifications. The chain of steel selection and procurement, therefore, includes number of steps such as:

1. Identify the *nature of application* i.e. structural or engineering, necessitating uses of flat or long product steel (this step is needed to help focus on steel grades under the respective standards).
2. Know the end-use requirements of properties as per service load and service conditions, and *identify the critical ones which must get fulfilled.*

3. Identify the forming/manufacturing properties *required for cost-effective production or meeting the constraints of production shop*. Identify the critical path for convenient and economical forming/manufacturing.

4. Select the steel by referring to the appropriate *Steel Standards* with respect to fulfilling the critical properties as identified earlier.

5. Check production resources available in the shop and *plan out the production process* or route.

6. Work out steel availability, sources of supply and cost; finalise the grade as per cost, quality and availability.

7. Draw out a *supply specification* incorporating desired steelmaking route, composition control, rolled dimensions and internal soundness quality, properties to be guaranteed, testing methods and test schedule per 'heat' of steel, and method of certification for conformance to quality.

Identification of steel types—e.g. structural or engineering grade steel—at the very beginning is necessary for starting the process right, allowing focus on the appropriate grade of steels and set of properties. A structural steel part is generally subjected to static stress i.e. static loading with very little fluctuation of applied load, whereas engineering steel part is generally subjected to dynamic loading with fluctuation of load in cycles i.e. fatigue load. Steel grades—as discussed in Chapters 6 and 7—are different for these two broad areas of applications, both in composition and properties. They are unique in their own way, but a designer or selector of steel has the final choice of choosing the level of intrinsic quality—e.g. inclusion content, acceptable internal quality standard, killing methods, etc.—beside composition and attainable mechanical properties.

Application of steel calls for open approach to problem solving. To illustrate, high strength structural steel such as HSLA plate though offer higher yield strength, have limited weldability, because, localised heating involved in welding might disturb the microstructure of fine precipitates and result in localised weakening. In cases where extensive welding is involved, either very low-heat input electrode or alternate grade of normalised steel should be used—with necessary reinforcement of critical load-bearing area by bolting and fastening. Thus, the load bearing capacity of a structure can be increased by suitable reinforcement plates at loading points, and such an option gives a wider choice to choose from standard structural steel grades. Therefore, considerations that lead to final choice of structural steel in such cases include:

- Cost and availability of right quality steel, including HSLA grade steel, free from any lamination and internal defect;
- Design features for jointing and fastening the parts involved for the structure;
- Cost of raw-material and production, justifying the choice; and
- Forming press-tool capacity and flexibility, including availability of dies and tools in the shop floor (Higher the yield strength of the material, higher would be the required press-tool capacity for forming and chances of spring-back of shape, which might render the job defective).

Major considerations for structural steel are its tensile strength, yield strength and elongation as measures of ductility and formability. Structural steels have to invariably support some static load, but may also suffer from fluctuation of loads due to vibration, wind force and other natural phenomena. Hence, they require some ductility not only for forming and shaping, but also for absorbing shock load. Structural steel also requires consideration for welding, which is widely adopted for fabrication of structural parts and components.

The component of fluctuating load i.e. dynamic load in engineering parts is, however, higher than the structural parts, and plays a dominant role in the selection of steel. In addition to fluctuating load, engineering components in service may encounter bending moment, torsion and twists due to applied forces, high contact pressure on the surface area leading to wear and contact fatigue, creep like environment, sudden change in temperature of operation leading to environment induced brittleness in steel behaviour, and many such complex stress and environmental patterns. Hence, engineering steels require considerations for strength (bulk), toughness, surface hardness, impact strength and impact transition temperature, and stricter control over microstructural features like grain size, inclusions, micro-defects etc. Engineering applications also require freedom from any surface defects—such as dent, pits, cracks, laps, scabs, etc.—and stress raisers due to their adverse influence on fatigue strength of steel which has already been discussed in Chapter 9. In sum, engineering steels require such considerations as:

- Intrinsic quality of the steel, like the compositional homogeneity, internal soundness, low inclusions, finer grain size, etc.;
- Heat treatability—hardenability and ease of processing—and ability of the steel to develop high strength and toughness after heat treatment;
- Suitability of the steel for forming, forging or machining; and
- Low susceptibility from environmental fluctuation e.g. for corrosion, low-temperature toughness, etc.

Therefore, steel for engineering applications must metallurgically satisfy closer range of chemistry, sound and defect-free section, killed steel for control of grain size, low inclusion counts and adequate hardenability values. This does not imply that quality requirement of structural steel is any less; structural steels have their own end-use specific property requirements that must be considered for selection and application of steel. For example, structural steel for ship or high sea structure requires very good low-temperature toughness property for insurance against brittle fracture due to stress and temperature fluctuation. Hence, such steels should be of fine grain size with very low inclusion content, low in S and P, and free from any residuals that affect ductile-brittle transition temperature of the steel.

In general, while structural steels may pose problems about the methods to attain higher strength with improved bendability and weldability, engineering steels may pose problems about way to ensure good machinability of the steel for mass production along with the potential for attaining higher strength and toughness by appropriate heat treatment. Since steels with higher strength are not easily formable or machinable, considerations for manufacturing properties of the selected steel become important. The manufacturing property often dictates the term of steel selection due to the need for productivity, and then the weak areas of application-related properties are built into the components by additional measures, like heat treatment. Table 10-1 illustrates a generalised picture of steel properties required for different applications.

Fatigue strength, which is an important property for engineering steel applications, can be approximately equated to the tensile strength and ductility/toughness of the steel—generally, 40% of tensile strength is taken as fatigue strength in standard heat treated steels as ball-path figure in absence of specific data. In addition to these, there could be special requirements of corrosion/oxidation resistance, creep resistance, heat resistance etc. for some applications. Steels which fulfil such application requirements can be obtained from the appropriate 'standards' or list of steel standards and specifications provided in Chapters 6 and 7, and in the appendices section.

To conclude, these are the technical considerations and approaches involved in the selection and application of steels. The process of steel selection and application, however, demands a systematic approach and analysis of the 'task and solution'. Accordingly, few illustrative case studies of steel selection and application are presented in the following section.

Table 10-1. A guide to steel properties required for various applications.

Identify the Nature of Application—Structural or Engineering Steel				
	If 'Structural'		If 'Engineering'	
Starting Material	Cold-rolled	Hot-rolled	HR Billets/Bars	Bright-drawn Bars
*Check for	Hardness	Tensile strength	Tensile strength	Hardness
Properties	Cupping value	Elongation	Cleanliness	Surface quality
	r	Bendability	Soundness	Ductility
	Surface quality	Weldability	Ductility	Machinability
	Paintability	Surface quality	Machinability	Heat treatability
			Impact toughness	
			Heat treatability	

10.3. Case Studies for Steel Selection and Application: Structural and Flat Product Steels (Sheet Steel)

Case 1: Selection of Steels for Engineering Structures

Product information

Steel structures are widely used globally for construction of factory sheds, warehouses, high-rise buildings, ports and docks, bridges, conveyor chains, cranes and many such modern industrial requirements. The structure must be sturdy enough for bearing the load generated from its own weight and applied weight, flexible enough to withstand the external aerodynamic load, resistant to shock and accidental overloading, and also to natural and weather related decay due to oxidation, corrosion etc. These properties are required in addition to many other engineering features in accordance with civil engineering principles and practice. Figure 10-1 depicts a part of an industrial structure showing its members and cross-members.

In Figure 10-1, in order to build a stable and strong structure as per the design—the structure requires a combination of steel sections, sizes and thickness for the main support structures—trusses and beams, cross-beams and reinforcement plates. Thickness and section profile apart, mechanical and welding properties of such structural steels are governed by various steel standards, for example: IS-226, IS-961, IS-2062, IS-8500 etc. or their equivalent of other international standards, like ASTM-A36, BS-4360, DIN-17100 etc. (see Table 6A-3 in Appendix to Chapter 6).

Figure 10-1. Picture depicting a portion of typical industrial structure with its main support structures, beams, and cross-beams, etc.

Analysis of product features and determination of end-use requirements

- *Geometry and complexity of parts:* The structure consists of sections and angles of different thickness and heaviness with at least 3 sizes of material sections for: (1) Large beams, including vertical columns/trusses (2) Cross-beams (large and main members), (3) Cross-beams (lighter section), and (4) Reinforcement bars or sections within the lighter cross-beams.
- *Working environment:* Structure is an outdoor constructional application, requiring good oxidation resistance, especially of its main columns/trusses.
- *Working load/applied load include:* (1) Own weight, (2) Load from assumed operations of over-head 'cranes' for transporting goods and utilities, (3) External aero-dynamic forces of the area of operations, and (4) Safety load factor for unknown natural or man-made consequences.
- *Other functional/manufacturing requirements (if any):* Good weldability and good ductility for earthquake proofing.

Approach to the selection of steel

Technical considerations: (1) Main column (truss) should be of heavier section for strength as well as rigidity of the structure, (2) large cross-beams should be lighter but strong enough to offer strength and rigidity of the entire structure, (3) other cross-beams of lighter weight should also be of lighter section and higher strength. All steels should have good ductility for allowing absorption of fluctuating stress from wind and other natural phenomena, and with sufficient weldability for jointing. Reinforcing bars or angles could be of standard strength but with proper weldability. Thickness and strength of reinforcing bars and angles should be strong enough to ensure required stability of the frame structure. Weldability and welding care is of extreme importance, if the structure parts are to be welded to the frame for jointing and securing. Based on different technical considerations, Table 10-2 illustrates steel grades and properties that can be considered for this application.

Table 10-2. Different grades of steel popular for structural applications.

Description of Parts	Steel Grades	Tensile Strength (kgf/mm²)	Yield Strength (kgf/mm²)	Ductility (%Elongation) (%)	Weldability/ Paintability
Main column (truss) of heavy section steel	IS-226 or IS-961, if heavy welding is involved, using fine-grained killed steel	44 (min)	26 (min)	22 (min)	Good weldability (CE less than 0.40), and paintability for external protection
Large cross-beams	IS-961 or IS-8500 HSLA	44 (min)	IS-961: 26 (min) or IS-8500: 34 (min)	22 (min) / 26 (min)	Good weldability (CE less than 0.40) and good paintability
Cross-beam (lighter)	IS-961 for meeting extensive welding	42 (min)	26 (min)	22 (min)	Very good weldability; CE limited to 0.40 (max)
Reinforcing bars/sections	Conforming to IS-961, killed steel	42 (min)	26 (min)	22 (min)	Very good weldability; CE limited to 0.40 (max)

Analysis of critical parameters

Metallurgically, factors critical for the applications are: yield strength (YS), tensile strength (UTS) and %elongation as measure of ductility. In addition to these, weldability of selected steels is also important, because more often than not, such structure components are joined by fusion welding rather than bolting, which has certain disadvantage for such construction. To illustrate further: for bolting, holes are to be punched into the steel plates that might leave some sharp edges and burrs which can act as points for crack initiation. Further, bolted joints might become loose with vibration or swing or seasonal expansion of the structure and it is difficult to periodically check and tighten with right torque for continued functioning. For welding, fine-grained killed steel is preferred for ensuring ductility of the weld joints.

Decisions on steel quality

Main members should be of higher strength with sufficient ductility to resist brittle fracture and should have good weldability.

- Weldability and welding care are of extreme importance as structures are often welded to the frame for jointing and securing. *Recommended:* Steel grade as per IS-961.
- Cross-member sections used for reinforcing the structure are generally of lower thickness, but with higher strength and good weldability, because such members have to be extensively welded for fixing. *Recommended:* Steel grade as per IS-961.
- Reinforcement steels might not be of exotic quality, but must have strength and weldability to provide a strong and rigid frame structure. These structures are specially designed by enough reinforcement and tying, such that the structure becomes strong and stable, especially against the aerodynamic forces. Recommended: Steel grade as per IS-226 for economy.

Remarks:

a. Various forces working on such structures working outdoor demand that the structure should be capable of absorbing certain amount of stretch and strain at the junction points where the forces act most. This requires appropriate joint design and use of sufficiently ductile steel, as measured by uniform elongation.

b. Another point of interest in steel selection for structures is the surface quality—surface quality of structural steel should be free from laps, scabs, lamination etc. that can weaken the structure and can also locally weaken the welded junctions.

There are many similar structural applications of steel, such as for bridges, crane load-frames etc. where approach to steel selection is similar but with greater care and choice of fully killed fine grained steel (steel grade IS-961) or even better. Ductile to brittle fracture transition temperature of these steels is also equally important, if there are chances of exposure to temperature fluctuations due to local climate. In such cases, steel must have a 27 J impact transition temperature above the room-temperature.

Case 2: Selection of Steel for LPG Cylinder

Product information

LPG cylinder is an example of cylindrical forming, involving strength as well as deep drawing and welding properties. Figure 10-2 exhibits a typical LPG cylinder after forming and painting. The component has a large central cylinder, a base ring for seating, and a valve guard ring at the top over a punched hole where the gas regulating valve seats.

The forming, piercing (for valve fitting) and welding processes call for absolutely defect free joints that are free from any chance of hydraulic or gas leakage. After forming and welding, 100% cylinders are tested for gas leakage and hydraulic pressure test for ensuring that the weld is sound and leak free and the steel cylinder has enough ductility after forming to withstand sufficiently high hydraulic pressure. The pressure to be withstood comprises not only the internal pressure under which liquid petroleum is stored in the cylinder but also the necessity to guard against any accidental rise in pressure and consequent bursting.

An LPG cylinder is a safety critical component and demands very careful selection of steel quality and processing, including welding and meeting the strict specification standards like ISO-10464 of 2004, IS-3196 and BS-1442 for manufacturing an LPG cylinder. General steps in manufacturing this safety critical gas cylinder include:

Figure 10-2. A standard LPG cylinder used in India, having foot ring, valve guard ring, valve neck and central seam welding.

- Blanking and body forming by deep drawing
- Valve guard ring and Foot ring welding,
- Valve seat formation,
- Appropriate welding,
- Heat treatment for stress-relieving and restoring ductility affected by welding and forming,
- Testing: *Pneumatic* testing for leakage and *hydraulic* testing for ensuring safe burst pressure, and
- Finishing by painting.

Being an item of predominantly domestic and repetitive use due to extensive circulation, the quality of steel should be dent resistant and painting should be extremely adherent and peel proof. Thus, required properties of selected steel should be: adequate strength, deep drwability and good welding characteristics.

Analysis of the product features and end-use requirements

- *Geometry and complexity of parts:* This is a cylindrical component of large diameter. For absolute leak proof condition, the part has to be produced with as few seams and joints as possible.
- *Working environment:* The component contains highly inflammable gas inside and frequently exposed to heat and fire of the working place. Safety requirement from gas leakage and pressure proofing are very critical.
- *Working load/applied load include:* High compressed pressure exists inside the cylinder where petroleum gas is kept under liquefied condition.
- *Other functional/manufacturing requirements (if any):* Absolutely leak proof, capable of withstanding high pressure inside, ability to withstand pressure at sub-zero temperature due to liquid state of the gas and freedom from embrittlement effect.

Analysis of forming process

Since LPG cylinder involves critical forming and welding operations, consideration of factors controlling the forming process needs to be examined before selecting the steel.

1. Forming of main cylinder: To limit the extent of welding, cylinder portion should be made of two halves—having drawn out dished-end of each half—and welded at the middle seamlessly. This involves:

   • Deep drawing of half cylinder with dished-end, and punching out the valve seat area in the upper half; and
   • Drawing operation should not cause appreciable thinning of the sheet metal being used, and be limited to approximately 10% of the original thickness.

2. Foot rings and valve guard rings could be produced separately and welded.
3. To avoid any instances of defect, welding of the central portion of two halves should be under automatic gas-shield using TIG or MIG process.
4. Heat input of other welding area should be controlled by using low-hydrogen low-heating electrodes.
5. Residual stress generated from welding processes must be relieved by suitable heat treatment process. However, heat treating temperature should not be high to cause alteration of parent steel properties (generally, HT temperature is limited to about 700°C i.e. below the lower critical temperature).

Approach to the selection of steel

1. Steel for the main cylinder must be ductile enough to be formed by complex drawing operation without thinning of original thickness beyond a limit, generally set by the relevant standard.
2. Other steels for sub-parts—e.g. valve ring, foot ring, etc.—should be of suitable strength for long duration life with good formability and weldability.
3. Steel for the cylinder must have good strength and ductility to withstand applied holding pressure of LPG and safety against 'burst pressure' as specified in the LPG standards (e.g. IS-3196 of India).
4. Good weldability of all steels used.

Technically, for absolute safety from gas leakage, weld joints should be few and of very high quality, and free of possibility of joint stress. Any stress developed during forming and welding must be removed for restoring original ductility of the steel by suitable heat treatment. For longer durability of the cylinder in use, the part should also be well painted after appropriate surface preparation.

Decision on steel quality

To fulfil these application conditions and safety requirements, manufacturing of LPG cylinders is undertaken in India mainly by using a superior grade killed steel with limits of S and P control, under the specification of IS-6240. This type of steel is mandated by the safety certifying authority for LPG cylinder in the country.

Details of chemistry and mechanical properties of IS-6240 that are specified for LPG cylinder manufacturing are shown in Table 10-3. IS-6240 having higher strength and elongation is the recommended grade for the cylinder body, which is suitable for deep drawing for shaping the dish shape ends where the steel undergoes some stretching and might thin down. For higher and uniform elongation (i.e. ductility), the steel used is of fine grained killed steel quality that ensures not only ductility but also good weld joint.

Table 10-3. Steel manufacturing specifications for domestic LPG cylinders.

Steel Grades	%C (max)	%Mn (max)	%P (max)	%S (max)	%Si (max)	%Al* (max)	N ppm (max)	YS (min) (kgf/mm²)	TS (min) (kgf/mm²)	%El (min)
IS-6240 Gr.B (main cylinder)	0.20	0.90	0.035	0.035	0.25	0.02	100 ppm	25	36-46	25

* Fully Al-killed fine grained steel.
Note: Alternate Grade is: IS-10787 equivalent to ASTM A-621 grade steel.

Steel for production of other ring parts need not be of IS-6240; it can be chosen from other recommended softer and commonly available grades such as—IS-1079, ordinary. IS-1079 or similar grade of steel (Al-killed condition) with strength similar to mild steel but with higher elongation is recommended for ring materials. Since the final cylinder needs to be given special heat-treatment for relieving all types of welding and forming stress, the mechanical properties of steel must be stable and guaranteed after heat treatment.

Remarks: Other points on the steel characteristics for this application include:

1. Steel for cylinder body must be Al-killed, strong and tough, but simultaneously allowing deep drawing for forming into cylindrical halves. Hence, fine grain and uniform elongation is required in the steel.
2. Steel for cylinder body must be of weldable quality for producing sound joints, limiting the 'carbon equivalent' value within 0.40 CE.
3. Other ring materials should also have good weldability i.e. with low to medium carbon (e.g. IS-1079, IS-2062, etc.).

Case 3: Selection of Steel for Automotive Wheel-rims

Product information

Automotive wheel rim is another example of steel application requiring complex forming and welding operations. Though wheel rims are also manufactured from high strength aluminium alloys, steel wheel rim is still widely used. Steel wheel rim is produced from hot-rolled steel strips of suitable thickness and strength by using complex pressing and welding operations. Figure 10-3 shows few types of common wheel rims that are used in passenger cars.

Wheel rims of commercial vehicles are much bigger in size, and made of thicker sheets and plates with reinforcements. Wheel rims have to be extremely sturdy and resistant to shape change under high speed cyclical load, coming from vehicle load plus impact force and resistance on the road. Functionally, it works in conjunction with rubberised tyres of different descriptions.

Figure 10-3. Steel wheel-rims for passenger cars.

Manufacturing of steel wheel rim is a complex process involving precision engineering and automated production line. There are many patented designs and features of this item, but in simple form, the component will have two parts—the central wheel disk that supports the circular rim and the circular wheel rims with some profile to hold the tyre edges. These parts are manufactured separately and welded together and finally press-formed to give finish shape and profile for sturdiness and balance.

Analyses of product features and end-use requirements

Limiting factor for this application is that steel must have good initial strength but must also possess good weldability and high elongation characteristics for forming. Hence, the selection of steel is limited to lower carbon structural steel with CE of 0.40 or less and elongation of 23% minimum. The steel also has to be killed steel with fine grains for welding and forming.

The rigidity and shape stability of the part, therefore, has to be worked out by balancing the strength and thickness of the steel strip used. Thickness is generally chosen between 2 mm to 4 mm thick strip. Lower thickness might be easier to form, but may also cause service problems for heavier vehicles or cars on bad road due to tendency for buckling and deformation of shape.

Analysis of forming process

Some essential steps in the production process of wheel rims include:

- Blanking from coil stock for disc;
- Deep drawing of the disc by progressive forming under press for shaping and punching for central hole, vent hole and other design features;
- Blanking from coil stock for wheel rim (two halves), and pre-bending to round form;
- Two round forms are centrally welded into a circular blank;
- Profiling via roll stands to shape the rim as per design;
- Assembly of disc to the rim by press fitting and welding; and
- Stress relieving and finishing.

Two different steel thicknesses and quality are required for disc and rim. Critical steel properties are: strength, ductility for forming and complex shaping, and good weldability.

Approach to the selection of steel

Generally, acceptable steel for such applications is in the strength range of St.34–St.37 as per DIN 1623 or Indian standards equivalent of IS-5986 grade St.34/37, and thickness of steel sheet is between 2.0 mm to 4.0 mm—higher thickness can be used for lower strength. The steel property details of St.34/St.37 are shown in Table 10-4.

Table 10-4. Popular specification for deep drawing grade steels.

Steel Grade	%Carbon	% (min)	%Si	% S and P	N (max)	Tensile Strength (kgf/mm^2)	Yield Strength (kgf/mm^2)	%El	Supply Condition
St. 34	0.17 max	0.30–0.60	0.30	0.045	100 ppm	34–42	21 kgf/mm^2 (min)	26	Fully Al-killed
St. 37	0.20	0.30–0.60	0.30	0.045	100 ppm	37–45 kgf/mm^2	24 kgf/mm^2 (min)	25	Fully Al-killed

The steel should also be fine grained Al-killed with restricted nitrogen content, so that no post-forming ageing can take place causing lowering of ductility. Due to Al-killing and restricted chemistry, the steel has carbon equivalent below 0.4 CE and offers good weldability. Welding of this part is of critical importance as the welded rim undergoes further shaping and profile forming, which is a critical operation. Post operations, the wheel rim assembly should be heat-treated to relieve the residual stress caused by welding and forming operations—any residual stress may lead to fatigue crack initiation in service as the part undergoes repetitive cyclical load.

Decision on steel quality

- Steel grade St.37 as per DIN or IS standard is recommended for the central disc.
- Steel grade St.34 with higher ductility is recommended for the rim halves for intricate shaping and better welding.
- Disc steel thickness should be higher than the rim steel.

Remarks:

1. The steel should be clean and of fine grained quality for better ductility and strength.
2. Welding process should be automated under flux cover for preventing any hydrogen absorption.
3. Wheel rims have air vent holes punched out from the disc. Edges of air vent holes should be smooth so as not to add to any stress concentration.

Case 4: Steel for Automobile Body Parts

Product information

Car body is made up of number of intricate shaped parts of aerodynamic engineering design made from cold-rolled smooth and glittery steel sheets. Parts are assembled together mostly by spot welding and through screws and hinges. Figure 10-4 shows some assorted parts of car body—comprising of bumper, bonnet, hood, doors and other small parts.

The parts are formed under sheet-metal press with suitable male-female die-punch arrangement. Sheet thickness and quality differs with part being produced, ranging between 0.60 mm thick to 1.60 mm thick, and steel grade being simple drawing grade (D quality) to extra deep-drawing grade (EDD quality) or even better. The formed contours of parts have bows and edges, which are often very complex to form. The formed surface must be smooth and wrinkle free, and suitable for high quality painting for durability and aesthetics. The parts are so manufactured that they could be assembled for final body cowl on an automatic transfer line consisting of spot-welding guns and other jointing means.

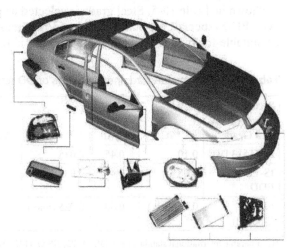

Figure 10-4. Assorted car body parts for assembly.

Analysis of product features and end-use requirements

The manufacturing process for individual car body parts starts with blanking from cold rolled sheets of appropriate size and grade, pressing to shape under male-female dies, inspection and trimming of edges, surfacing to remove any defect spots and smoothening, transferring to cowl assembly line for assembly by spot welding, reinforcing the identified locations by spot welding of plates and ribs, joining by screw wherever necessary.

Important features of car body parts manufacturing include:

- Forming is the over-riding parameter over strength. Strength is of less consequence for this application, except in the case of dent and bent resistance, which can be achieved by reinforcing vulnerable locations.
- Forming operation involves stretching and drawing in multiple directions.
- No thinning or wrinkling or pan-caking of surfaces is permissible.
- Parts must respond well to automatic spot welding with good spot strength.
- There should also be good paintability.

Discussion on steel selection

A special feature of the steel is that along with high uniform elongation the steel should have high $\bar{r}$ value for forming free of thinning and wrinkling. $\bar{r}$ value measures the anisotropy of the sheet property—high $\bar{r}$ values of the steel are beneficial for stretch forming. The steel should, therefore, be 100% Al-killed, cold-rolled, and annealed and temper-rolled for improved texturing and $\bar{r}$ value—$\bar{r}$ value represents 'vertical anisotropy' and a value of 1.8 and above is desirable for extra deep drawing steel sheet for body panel parts. Since the forming also involves extensive straining and stretching, the steel should have low strain-hardening co-efficient (called 'n'); n value of 0.25 is generally expected of this steel. In general, steel should have lowest possible yield strength, good cold formability and non-ageing characteristics.

Therefore, the selected steel should be of very low-carbon, fully Al-killed, with high elongation and good $\bar{r}$ value to permit extensive drawing and forming. Generally used steel specifications for auto body panel forming are IS-11513 Grade DD and EDD with following chemistry and properties as shown in Table 10-5. Steel grade is selected as per complexity of the body part, requiring EDD or even EDD (special) for bonnet and hood, whereas DD grade steel of a little higher thickness would be suitable for bumpers where some strength is required for end-use. Bonnet and hood steels might

Table 10-5. Indian standard specifications of DD and EDD grade steels suitable for auto-body parts.

Steel Specification and Grade	%C	%Mn	%P and S	%Si	%Al (min)	YS kgf/mm^2	UTS kgf/mm^2	%El (min.)	$\bar{r}$ Value
IS-11513 DD	0.10	0.45	0.035	—	0.02	23	27–37	23	1.3
IS-11513 EDD	0.08	0.40	0.03 (max)	—	0.02	22	27–35	25	1.6
EDD (Special)	0.06 (max)	0.40	0.025 (max)	—	0.02	22	27–35	32	1.8

*Comparable International standards are: DIN 1623, JIS G 3141, ASTM A366, and BS 1449.

be procured in surface coated condition—e.g. cold-rolled galvanised steel sheet—for better forming and painting.

Decision on steel quality and grade

- For upper body parts e.g. door, bonnet and hood, steel should be of EDD (S) grade with high $\bar{R}$ value—$\bar{R}$ of 1.60 and above;
- For bumper, steel grade could be EDD or DD type, depending on the manufacturer's choice, but with $\bar{r}$ value of about 1.25 and above, so that there is no wrinkling after forming.
- Other small parts can be made of DD steel as per IS-11513.

Remarks:
   a. Now-a-days, more exotic steels, such as extra low-carbon baked steel, having higher yield strength but similar or better forming properties than EDDS, and interstitial-free (IF) steel, can be used for superior forming and finishing of body parts of very complicated shapes.
   b. Because of large demand and consumption of such low-carbon DD/EDD type steel for automobile applications, large scale research and development work is going on in steel industries for making available superior materials. These steels are not only controlled with leaner chemistry and cleanliness, but also composed, rolled and processed for better forming textures and with higher r-bar value and lower n value; for example EDDQ grades of ASTM and DIN specifications.

In general, for good drawing and cupping operations, softest steels with good surface quality and very low oxide inclusion content should be used. Moreover, cupping involves biaxial stretching; hence the cold rolled steel sheet used for deep drawing and cupping should be so specified as to ensure removal of bi-axiality in the sheet by special temper rolling. *For many critical automobile body parts, requiring extra deep drawing and cupping, special grades of extra-low carbon cold rolled, vacuum-degassed clean steel with controlled grain size are used.* Testing and evaluation of these steels for the required cupping and stretching values are thus necessary for ensuring trouble-free processing.

10.4. Case Studies of Steel Selection and Application: Engineering and Long Product Steels

Case 5: Steel for Automotive Axle Shafts

Product information

Axle shafts are universal features in automobiles and engineering machine tools for transmission of power from one station of the system to other. One end of the shaft is flanged and bolted to one of the stations, generally the driven end e.g. wheel of a vehicle, and the other end of the shaft is fitted with gears to transmit the power for drive—e.g. gear box or differential gear box of a vehicle. Figure 10-5 illustrates few typical axle shafts used in for different engineering applications.

Analyses of product features and end-use requirements

Shafts function in wide ranging service conditions, experiencing a range of loading conditions which might comprise tension, compression, bending, torsion or even a combination of them. They can also experience lower temperature and corrosive environment. Axle shafts transmit high power under dynamic and rotational load experiencing high torque and bend. They may also encounter vibratory load at time.

Hence, the component is expected to carry high fluctuating load, requiring high level of toughness to withstand the different modes of fatigue failure, especially torsional fatigue. This necessitates the selection of steel which can develop high strength and toughness upon heat treatment i.e. hardening and tempering process. Since the resolved shear stress in torsion is highest on the surface area, shaft surface has to be flawless and free from any crack, pits, dents etc. that can act as point for fracture initiation. Similarly, the shaft should not have any soft spot on the surface.

Figure 10-5. Few typical axle shafts used in for different engineering applications.

Analysis of forming process

Production steps for shaft manufacturing are:

- Flanging by upset forging at one end;
- Machining the shaft as per design;
- Heat treating the shaft for attaining high strength and toughness;
- Crack testing by magnetic particle test for surface cracks;
- Shot blasting the surface to remove any scale or pits or soft spots arising from decarburised layer; and
- Finish grinding etc. as per drawing.

Shafts can also be manufactured by induction hardening the surface after through hardening for developing high core strength with good toughness.

Discussion on steel selection

The steel characteristics should, therefore, fulfil the following conditions:

1. The steel must have good 'forgability', required for up-set forging the flange. This calls for fine grained low-inclusion steel with controlled S and P for good forgability.
2. The steel must be capable of developing a high strength structure with adequate toughness for the high torsional load. This can be met either by using a higher hardenable grade steel by hardening and tempering or by 'case-hardening' the shaft by a process like 'induction surface hardening'. The choice of steel will be different as per the chosen route to attain the ultimate mechanical properties.
3. Since the steel undergoes high cyclical fatigue load and is susceptible to sudden impact load, steel microstructure should be toughened martensite to as much depth as possible, with low inclusion rating, and having high impact strength.

Based on these considerations, the following grades of steels (as indicated in Table 10-6) with controlled inclusion content and fine grain size can be considered for application.

Barring SAE 4320, other grades are through hardening grade steels, but with different properties at the end of heat treatment. In general, steel with Cr-Mo is superior in strength and toughness to Cr-V, and Cr-Ni-Mo is superior to Cr-Mo grade steels. However, considering the cost of steel, steel containing Cr-V would be cheaper than steel with Cr-Mo. Cr-Ni-Mo grade will be most expensive, but its toughness

Table 10-6. List of hardening grade steels commonly used for axle manufacturing.

Steel Grade	%C	%MN	%Si	%SandP	%Cr	%V	%Mo	%Ni	Inclusion Rating	Grain Size	Hardenability End-quench Value
SAE 4140	0.38–0.43	0.75–1.00	0.15–0.35	0.035 (max)	0.80–1.20	—	0.15–0.25	—	Type Thin/Thick A 2.0/1.0 B 2.0/1.0 C1.5/1.0 D1.5/1.0	ASTM size 5–8	As per 'Ruling Section', but should be specified
SAE 6150	0.48–0.53	0.70–0.90	0.15–0.35	0.035 (max)	0.80–1.10	0.15 (min)	—	—	"	"	"
SAE 8650	0.48–0.53	0.75–1.0	0.15–0.35	0.035 (max)	0.40–0.60	—	0.15–0.25	0.40–0.70	"	"	"
SAE 4340	0.38–0.43	0.60–0.80	0.15–0.35	0.035 (max)	0.70–0.90	—	0.20–0.30	1.65–2.00	"	"	"
*SAE 4320	0.17–0.22	0.45–0.65	0.15–0.35	0.035 (max)	0.40–0.60	—	0.20–0.30	1.65–2.00	"	"	Case Hardening steel

Table 10-7. Indicative properties of axle grade steels after heat treatment.

Grade of Steel	UTS (kgf/mm^2)	YS (kgf/mm^2)	%Elongation	Impact Izod (Joules)	Reduction in Area (%RA)	Ruling Section
SAE 4140	100–120	80 (min)	14 (min)	54 J at RT	45 (min)	30 mm
SAE 6150	120–140	90 (min)	12 (min)	47 J	35	45 mm
SAE 8650	130–150	100 (min)	10 (min)	43 J	30	50 mm
SAE 4340	130–150	100 (min)	14 (min)	47 J	45	63 mm
*SAE 4320	Surface Hard: 60 ±2 HRc	Case Depth: 2–3 mm	Core Strength 100–120 kgf/mm^2			

will also be proportionately high. Therefore, the selection has to be based on severity of the duty of the shaft; if the application demands very harsh operating conditions, one may prefer to go for SAE 4340 (En 24 equivalent grade), which would attain high strength as well as very high toughness due to presence of nickel and molybdenum. However, for most purposes of engineering applications steel SAE 6150 (DIN 50 CrV$_4$ equivalent) or SAE 4140—after induction hardening of the surface—would be adequate. Induction hardened steel will have stronger surface hardness and tougher core strength. But, the cost will be higher due to additional manufacturing operations and need not be used unless the application truly calls for. After heat treating, the shaft should have the following indicative properties for the end-uses as specified in Table 10-7.

For most common purposes, such as passenger car shafts where the ruling section is lower than 25 mm, even lower alloy steel—such as SAE 5140 or EN-18 (old BS specification) or a cleaner grade carbon steel of specification 45C8 (Old En-8 /En-9)—can also be considered with lower attainable mechanical properties.

Decision on steel quality

- For very heavy duty axle shaft, e.g. for dumper/tipper type of vehicles, steel should 4340 type or SAE 8650 type with higher strength and extreme toughness.
- For most automotive commercial vehicle shafts, SAE 4140 or SAE 6150 steel in hardened and tempered condition is adequate.
- For passenger cars, axle shaft could be made of clean plain carbon steel or SAE 5140 type steel, with surface induction hardening wherever necessary.

Remarks: The most common cause of shaft failure is fatigue. Fatigue failure commonly starts from points of stress raisers on the surface. Hence, adequate care should be taken to make the surface free from any harmful defects. In this regard, care is also necessary in manufacturing the shafts, ensuring that the steps, corners, key ways, grooves, neck radius, splines etc.—which are always present in shafts as part of design—do not cause or act as stress raisers.

Case 6: Steel for Pivot Pins

Product information

Pivot pins are a short rod or shaft about which a mounted part rotates or swings. Figure 10-6 shows an assortment of pivot pins that are used in the industry. The pivot pins along with the related parts form a mechanical system by which load is applied or a force is directed. Though the part is simple looking, it plays a critical role in the functioning of most engineering assemblies for transmission of force. In order to secure the assembly, the pins should have a locking arrangement, either in the form of key-way or a slot for anchor pins or a grooves for spring pins. The system is designed for transmitting varying degrees of load and rotational force acting between the connected parts—examples start with a simple door hinge pivot pin to automobile 'king-pin' that support steering function of vehicle wheels.

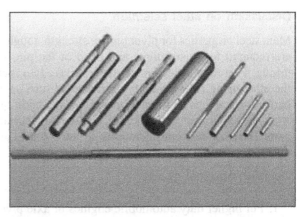

Figure 10-6. Assortment of pivot pins used in industries. The thick pin at the centre is called the 'king-pin' in automobile assembly.

There are many types of pivot pins for different applications. However, metallurgically, they should have some common properties for meeting the end-use requirements; for example:

- Wear and pitting resistance due to contact load—like the bearing surface, due to relative movement of the mounted parts with the pin;
- Shear stress due to applied load by the mounted parts; and
- Bending load (moment) due to non-axial load.

Common types of failure of pivot pins are caused by wear, pitting and fatigue. Fatigue generally starts from the key groove or slots made in the pin for fitting—due to sharp edges or points of stress raisers. Fatigue can also start if the fit between the parts is not right, causing localised dents and rubs.

Analyses of product features and end-use requirements

1. Pivot pins are generally round shaped long items with high surface hardness and very smooth surface quality. It has a groove on the surface for locking the pin in position.
2. In general, the pin experiences high shear stress in service; however, it can also experience bending load in application.
3. Since the part is in contact with another body under load, there are chances of wear and contact fatigue, in case of lose coupling.

Analysis of manufacturing process

The manufacturing process generally follows the following steps:

- Cutting (from rods or bars), machining and groove cutting;
- Heat treating to required strength and toughness. Alternately, the pin can be induction hardened or case carburised for developing high surface hardness;
- Threading, debarring etc., if necessary (For induction hardening parts, thread should be left soft.);
- Grinding to required surface finish, which should be very fine and smooth; and
- The finished parts must be of accurate dimensions and have very smooth and crack free surfaces.

Discussion on steel selection

Main steel properties for pivot pin are strength, toughness and surface hardness—strength and toughness are required to resist fatigue and bend of the pin, while high surface hardness is to resist wear and pitting. Therefore, selection of steel can take two routes: one using *through hardening grade steel* and the other using *case-hardening grade steel* such as by carburising or induction hardening. Since case hardening requires additional operations and additional process facility, this route is to be resorted to when the application calls for high load-bearing surface with chances of relative movement between the parts. Based on thse considerations, following steel grades (shown in Table 10-8) can be considered for different pivot pin applications.

Decision on steel quality

1. For higher duty automobile engines or axle pivot pins (e.g. king pins), steel should be SAE 4140 type and used after through hardening and light tempering. For further increasing hardness, where necessary, the pin can be either induction hardened or even chrome plated for high surface hardness.
2. For standard pivot pins, steel could be SAE 1050 or C 1045 with boron for developing high hardness after hardening. This type of steel provides adequate strength and toughness for many unknown areas of application.
3. For very heavy duty applications, as in earth moving equipment and vehicles, SAE 4320 steel after carburising and hardening to high surface hardness with tough core could be used.

Table 10-8. Popular grade steels for different pivot pin applications.

Steel Grade	Condition of Application	Recommended Properties*
SAE 1050	General and low duty shaft/ pivot pin	Hardened and tempered. Tensile Strg. 70 to 100 kgf/mm^2 and elongation of 8%–12%. Al-killed, fine grained steel
SAE 1045 H with 0.0005–0.003% Boron	Pin shaft requiring high hardness with higher toughness	H and T. TS: 80–100 kgf/mm^2 with elongation of 12%–14%, fine grained Al-killed steel
†SAE 5140	Higher strength pins and shafts for higher endurance	H&T. TS: 100–120 kgf/mm^2 with elongation of 8% min. Al-killed fine grained
†SAE 4140	Higher duty with superior toughness	H&T. TS: 120–140 kgf/mm^2 with elongation of 12% min. Recommended for higher section and heavier load. Al-killed Fine grained
‡SAE 8620— case carburised	For higher wear and abrasion	Case depth: 0.6–1.0 mm with surface hardness 58–60 HRc and core strength of 100–120 kgf/mm^2. Al-killed, fine grained
‡SAE 4320 for— case carburising	For even higher duty and tougher application than SAE 8620 steel	Case depth 0.80–1.20 mm with surface hardness of 58–62 HRc and core strength of 120–135 kgf/mm^2. Al-killed, fine grained

* Will depend on the section diameter.

† SAE 4140 and 5140 can also be induction hardened (after hardening and tempering) with case dpth of 1.0–2.0 mm (depending on the pin diameter) with core strength ranging between 100–140 kgf/mm^2 for higher duty pins ad shafts.

‡ Carburising of these steels should be done after normalising for producing tougher core.

Remarks: For better service life, pins/shafts should be fine ground and all burrs removed from corners and key-holes, leaving no scope for stress raisers to facilitate fatigue crack initiation from those points. At times, pins can be also made by hardening and tempering route and thereafter giving a layer of hard–chrome plating, which has very high surface hardness to counter any pitting. In general, the application requires high surface hardness, tough core and flawless surface.

Case 7: Steel for Fastening Bolts

Product information

A bolt is a threaded fastener used to secure two or more parts together in an engineering system for carrying or transmitting the required load/torque. There are as many varieties of fasteners as their wide spread uses in industries. They are available with different thread, shank and head configurations. Common examples of fasteners are hex-head bolts, socket head bolts, studs, screws etc. For screws, softer steel for ease of screw machining is used. But, studs and bolts are meant for higher duty industrial applications and require extra strength and toughness. Bolts and nuts are marked by 'grade', based on their strength properties; the grade mark is generally stamped on the head of the bolt or face of the nut. Grading system of fasteners follows a standard rule. Table 10-9 below indicates the grading system and the required strength criteria.

Table 10-9 is indicative of the way steel bolts are graded as per strength requirements and associated yield strength or permanent set strength. The latter is necessary for ensuring that the bolt can take required tightening torque with adequate pre-stress. Figure 10-7 shows an assortment of threaded fasteners of hexagonal head configuration. Part of the bolt length is threaded with a standard pitch followed by a plain shank, collar and hex-head. There is a matching nut with the bolt that is used for tightening—generally, the nut is one grade lower than the bolt for the reason of proper tightening with enough pre-stress.

The nut-bolt assemblies are tightened by applying a measured torque as per grade and diameter of the bolt with the purpose of ensuring proper tightening between the parts with adequate pre-tension. At times, additional locking mechanism is used, like lock washers, locking nuts, cotter-pins, etc., in order to prevent loosening of the fastened joint over time under the service conditions. Special precautions are necessary to prevent loosening of fastened joints under vibratory environment. This is because, slight loosening of a joint would cause sudden enhancement of shear force on the joint, leading to shear failure under the applied or tensile load.

Table 10-9. Grading system and required strength level for different fastener grades.

Designation ➡️ Strength (kgf/mm²)	4.6	4.8	5.6	5.8	6.6	6.8*	8.8 <16 mm	8.8 >16 mm	9.8	10.9	12.9
Tensile Strength	40	40	50	50	60	60	80	80	90	100	120
Yield Strength (min)	24 (4 × 6 = 24)	32 (4 ×8 = 32)	30	40	36	48	64	64	72	90	108
Stress at permanent set	—	—	—	—	—	—	64	64/66	72	90	108

* Bolt designation up to 6.8 can be without heat treatment (H&T), but 8.8 and above grades are always heat treated (H&T) bolts.
Note: In accordance with BS 3692:2001 and BS EN IOS 898-1:1999.

Major causes of bolt failure include:

- Failure under higher tensile load—caused by loose joint and over load or inadequate mechanical properties;
- Failure under reverse bending fatigue—starting from a stress raiser or thread root; and
- Failure due to improper tightening torque and locking mechanism.

Tightening torque is very critical for securing the fastened joint; it is ensured by pre-stressing the joint within the elastic limit. Over-torquing might elongate the bolt by plastic deformation-and under-torquing will leave the joint insecured. Hence, all fastened joints should be given a-measured tightening torque for securing the joint against working and vibratory load. Torque charts as per bolt garde and diameter are available for all common applications.

Figure 10-7. Assorted types of industrial fasteners with threads.

Analyses of product features and end-use requirements

Bolts have a threaded portion followed by shank and head. Since, many bolts are of high tensile varieties, the steel should be tough enough not to cause excessive notch effect at the root of the threads in service. Therefore, very often threads are rolled—into the shank for reducing stress concentration at the threaded areas.

In fastened condition, bolt experiences tensile force along the length in the shank areas, and any plastic deformation of the bolt will cause loosening of the joint. Hence, end-use properties of bolts in grade 8.8 and above should possess the following qualities:

- High strength in combination with good elongation (i.e. toughness) are necessary.
- Steel should be free from harmful inclusions so that they don't interfere with the thread rolling process and also act against fatigue failure.
- Steel should be of fine grain size for better elongation and ductility for improved cold formability.

Analyses of manufacturing processes

Common manufacturing processes of bolts include:

- Selection of steel grade as per grade of the bolt;
- Softening the steel by special heat-treatment, like annealing, spheridised annealing, etc., for ensuring formability or using 'cold-heading' grade clean and ductile steel rod;
- Coating the wire rod for easy drawing (e.g. phosphate coat);
- Forming the bolt blank by upset forging the head (generally by cold heading);
- Trimming and preparing the bolt blank for threading;
- Thread rolling to dimensions—this can be done after the HT, if necessary;
- Heat-treating by hardening and tempering under control atmosphere for avoiding any soft layer or thread decarburisation—only for quenching and tempering (Q&T) grades; and
- Phosphating, plating, finishing, etc.

There could be more steps or different types of order of processing, depnding on the type of bolt or the facility in the shop. Critical parametres for successful fastening operations are as follows:

- Strength (both tensile and yield strength) and toughness of the steel;
- Accuracy of the thread profile;
- Dimensional accuracy of shank and the collar where the bolt seats; and
- Accuracy of tightening torque.

Discussions on steel selection

From the preceding discussion, it is evident that quality and steel grade will be different for different grades of bolt. Steels can be selected from regular low-carbon to medium carbon alloy steel, depending on the bolt grade and diameter, but the critical parameter is the cold formability of the steel selected *vis-à-vis* its ability to develop the strength and toughness. This is particularly so for bolts of higher grade, requiring Q&T process for properties.

For low duty fastening of static joints, annealed or normalised steel i.e. without Q&T can be used in grades 4.4–6.6. But, for many other applications, steel must be richer in carbon or alloying elements where cold formability could be a problem. Hence, steels for such higher grade bolts should be procured as 'cold-heading' grade steel, which is fine grained and lower in inclusion content. Steels for high tensile fasteners should be specially made through secondary steelmaking technology with low inclusion content and given a special softening treatment to make the steel 'cold-heading' grade. Since bolt forming is done in highly automated high speed machines, the quality consistency of steel is of critical importance. Therefore, the following parameters are to be given due consideration for selecting steels meant for fasteners:

- Cold formability of the steel—this can be improved by minimising inclusions in the steel and giving some special types prior heat treatment such as like annealing, normalising, spheridising etc.
- Achievable strength and toughness level of the steel—this will depend on the composition of the selected steel.
- Heat treatibility (for Q&T) of the steel—this is important for larger diameter bolts of higher grade.

Steels should also be less 'notch-sensitive', because all bolts have threads and the roots of threads are prone to act as 'notches' on the bolt (this is the reason why thread-rolling of high tensile bolts are preferred). Considering these conditions and strength criteria as per grade table for bolts, the following grades of steels, as shown in Table 10-10, can be recommended for different applications. These are only indicative; actual steel garde will depend on the section thickness or bolt diameter, especially for Q&T gardes.

Decisions on steel quality

1. For screws and nuts of lower grade i.e. < 4.4, plain low carbon steel can be used for better forming and machning behaviour.
2. For bolt grades 4.4–6.6, medium carbon clean steel with normalising or spheriodising can be used, in order to meet the strength properties as well for better formability.
3. For bolt grade 8.8, recommended steel is SAE 4140 or similar medium carbon low-alloy steel, to be used after hardening and tempering.
4. For higher grade bolts, calling for higher than 100 kgf/mm^2 tensile strength and 8%–10% elongation, higher carbon alloy steels like SAE 4145, 4150 and richer grade should be used.

Table 10-10. Recommended steel grades for diffrent grades of bolts.

Bolt Grade	Steel Grades	Conditions/Remarks
4.6	Mild steel	Machined screws and bolts of general description
4.8	Low to medim carbon steel—preferably light cold drawn for good surface	Fine grained normalised steel in tensile strength range of 45–60 kgf/mm^2 and yield stress or proof stress of 32 kgf/mm^2 min.
5.6	Low to medium carbon—annealed and cold rolled steel	Fine grained medium carbon steel, like AISI 1030 /1035 grades or equivalents with TS of 50–65 kgf/mm^2 and YS (PS) of 30 kgf/mm^2 min. (Cold drawing is required to increase the YS or PS, but should not drastically bring down the ductility of the steel)
5.8	Medium carbon steel; light cold drawn	Fine grained medium carbon cold drawn steel in grades AISI 1045 or equivalent with TS of 50–65 kgf/mm^2 and PS of 40 kgf/mm^2 min. (Cold drwaing is to increase the YS or PS in the rod, but without much sacrifice to elongation.)
6.6	Medium carbon light cold drawn steel	Fine grained, annealed or normalised in grade AISI 1045/1048 or equivalent and with light cold drawn for improved surface—TS 60–75 Kgf/mm^2 and YS (PS) of 36 kgf/mm^2 min.
6.8	Medium carbon steel, light cold drawn	Fine grained normalised steel, with controlled cold drawn, in grade AISI 1045/1048 or equivalent in TS of 60–75 kgf/mm^2 and YS (PS) of 48 kgf/mm^2.
8.8: ≤ 16 mm	Medium carbon alloy steel grade AISI 5140, 4140 or Boron containing AISI 10B45 steel or equivalents	Quenched and tempered (Q&T) steel with low inclusion in TS range of 80–100 kgf/mm^2 and PS of 64 kgf/mm^2 min, and elongation of 12% min. (Cold-heading grade steel) For 8.8 grade, often 'Boron steel' containing very low boron—in the range of 0.0005–0.003—is used.
8.8 ≥ 16 mm	AISI 4140, 50B44H, 5147H or equivalent grades	Q&T steel with low inclusions in TS range of 80–100 kgf/mm^2 and PS of 64 kgf/mm^2 and elongation of 12% min. (Steel composition should be a bit more rich than the standard 8.8 grade steel above)
9.8	AISI 4140, 4142, 4145 or equivalent grades	Q&T to 90–105 kgf/mm^2 tensile strength and PS of 72 kgf/mm^2 with elongation of about 10%
10.9	AISI 4145, 8640 or similar grade of steel with low inclusion content	Q&T to 100–120 kgf/mm^2 tensile strength and PS of 90 kgf/mm^2 min and elongation of about 8%–10%
12.9	AISI 4150, 4340 or similar grade with low inclusion.	Q&T to 120 kgf/mm^2 and PS of 108 kgf/mm^2 with elongation of about 8% (Choice of steel will also depend on the bolt diameter)

Note: Steel grades following the AISI standard mentioned above are only indicative of the types of steel that can be used— other alternative grades as per different national and international standards can be also selected and used as per forming requirements and properties of the bolt.

Remarks: Steel bolts are also used in a wide variety of environments, such as corrosive environment of a chemical plant, marine environment, heat and dust, etc. Steel quality requied for such special bolts would be different as per their end-use. There are a number of stainless steel grades and Cr-Mo-V grade alloy steels that are used for special fasteners in such corrosive or heated environment, respectively. To list a few examples: AISI 304, 316, 347, etc. which are essential for tensile strength range of 80 kgf/mm^2 and above as per severity of environment. CR-Mo-V alloy steel grades—e.g. 1Cr-½Mo or 2Cr-½Mo etc.—are used for applications involving creep type environment.

Case 8: Steel for Ball Bearings

Product information

Bearing is an anti-frictional device that carries load for transmission from one component to the other. Without bearing, there will be considerable friction and wear of mating parts and loss of transmitted load. It is an engineering essential for almost all machine parts and appliances. Its applications range from domestic fan to big (rotation) turbines. Figure 10-8 shows the typical ball bearing arrangement in work.

There are many types of bearings: for example, ball bearing, roller bearing, taper-roller bearing, needle bearing, etc. The basic nature of anti-friction behaviour is same for all, but they differ as per their load bearing and load distribution capacity. Of these bearing types, ball bearing is very common in industrial applications e.g. in domestic fans and other domestic appliances.

Analysis of product features and end-use requirements

Ball bearing facilitates effective transmition of high dynamic loads. As such, the component is subjected to high fatigue and stress condition. The bearing is mounted on a shaft or seat with interference fit for firm anchoring, ensuring that there is no relative movement between the shaft and the bearing. While functioning, the bearing transmits the dynamic load through its inner reces via the balls to the outer reces and to the moving parts in the system. Functionally, the bearing has to rotate under high speed carrying the load. Hence, the balls inside the track of races expeience high contact load and need low coefficient of friction for free ball movement between the races or the cage. The transmitted load for ball bearing is generally 'radial load'— perpendicular to the shaft or axle—with minimum axial load, but there are other types of bearings that can take high axial load as well. Hence, the ball grooves have to be adequately deep to anchor the balls or rollers under such complex loading pattern.

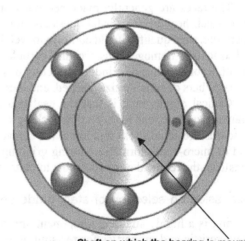

Shaft on which the bearing is mounted

Figure 10-8. A typical 'ball bearing' working arrangement with outer races, inner races and rotating balls.

Engineering properties required for ball-bearing applications include: strength, toughness, surface hardness, corrosion/pitting resistance, and above all fatigue resistance for rolling contact fatigue and fatigue under axial loading. For uses and applications, bearing steel has to fulfill the tasks of manufacturing by forging, machining, heat-treating, grinding, and fine finishing, demanding good hardenability and very low inclusion level in the steel. Focus of selection of ball bearing steel is reliability in service and high life cycle, and focus in processing of the steel is to make the components absolutely defect-free with high surface hardness. Microstructually, ball bearing calls for uniformly fine structure of tempered martensite with hardness range of 60 ±2 HR_C. Ball bearing also calls for very fine surface finish for minimum friction.

Failure of bearings could be mainly due to metal fatigue caused by high dynamic load, on the one hand, and pits and wear due to high contact load, on the other. Metal to metal contact inside the bearing is protected by effective lubrication. To avoid wear and pitting under contact load, all bearing applications require good lubricating system and high quality lubricants for long service life. The metal to metal contact and lubrication to prevent friction is common to all bearing applications, whether the bearing is ball bearing or other types.

Because of very special nature of bearing applications—involving very high rotary speed and carrying high dynamic load, the rolling track and balls experience high contact load. Hence, bearing steels require special grade of steels, termed as 'Bearing Steel' under different steel standards.

Analysis of forming/manufacturing process

Ball bearing has atleast three parts in its construction; the outer and inner races, the balls, and the cage—which is not shown in the figure, but necessary for holding the balls in position within the grooves of two races. The cage is formed by drawing into shapes and then punching holes for balls. Normally deep drawing quality (DD) steel with high ductility would be good enough.

The races are generally machined out from annealed seamless steel tubes of appropriate bearing grade steel, heat treated for hardness and microstructure and then finshed and ground to dimensions. Therefore, steel quality should conform to well known bearing grade steel containing very low inclusions and structually homogeneous. Original carbide size in the steel must be very fine so that the carbide spots do not become points of grinding difficulty or sources of micro-cracks. The balls are machined out from annealed bars of suitable size, shaped, and heat treated for hardening, and finished ground.

Critical requirement is that the heat treated parts should not have high residual stress, and also retained austenite, due to thermal quenching while hardening for high hardness specified by the bearing standards. Residual stress will lower the contact fatigue strength of the steel and retained austenite may lead to micro-crack formation during grinding which generates some heat, which could be enough to transform the retained austenite spots to matrensite (untempered).

Discussion on selection of steel grade and quality

Bearing is a highly stressed component and a bearing's life expectancy is mainly based on material characteristic and properties for resisting metal fatigue and wear. Other factor that can cause bearing failure is the 'environmental factors'—like dust and dirt, temperature fluctuation, water ingress, etc. that occur in service. Based on these considerations, selection of bearing steel can take three routes: (1) use of high hardness and low-inclusion content clean steel for fatigure and wear resistance; (2) use of carburised grade surface hardened steel for improved fatigue, wear and pitting resistance—this is often used for high duty railway axle bearings; and (3) Use of stainless grade steel for resistance against

Table 10-11. Details of through hardening ball bearing grade steels.

Material Specification	%C	%Si	%Mn	% S&P	%Cr	Others	Grain Size	Inclusion level/ Oxygen level
AISI 52100 (other equivalent: EN-31, DIN 100Cr6)	0.95–1.10	0.15–0.35	0.25–0.45	0.025 max	1.30–1.60		ASTM-8	*Inclusion examples: see below. O₂: 10 ppm max preferred

*Max recommended inclusion levels for ball bearing applications				
	A-type	**B-type**	**C-type**	**D-type**
Thin	2.0	2.0	1.5	1.0
Thick	1.0	1.0	1.0	0.5

Note: Steel must be of ladle metallurgy treated or vacuum degassed with as low oxide inclusion as possible.

severe environmental factors. However, ball bearing grade chrome-bearing high hardenable steel is the most common amongst these alternatives.

Hence, there is a separate specification of ball bearing steel under all national standards, e.g. AISI 52100 steel, EN-31 or DIN 100Cr6 grade steel, detailing the composition, inclusion level, permissible limits of gas and other impurities, residual elements etc., which can otherwise affect a bearing's performance. The details of bearing grade steels are given in Table 10-11.

For the special bearing application i.e. for bearing races and balls, there could be other types of steels e.g. carburised grade and stainless steel grade steels. Popular carburising grade steel is the SAE 4320, Cr-Ni-Mo bearing steel known for its toughness, which is used for high duty railway axle-bearings by some reputed bearing manufacturers. Whatever be the steel grade, the steel has to be used after hardening the steel for surface hardness of 60 ± 2 HRc for countering the rolling-contact fatigue and core strength of 130–140 Kgf/mm^2 for sustaining high loads. For improving the wear characteristics, chromium carbide, which is the hall-mark of this type of steel, must be in fine and uniform dispersion. Bearing steel should also have high bending strength of about 2400 MPa (i.e. about 240 kgf/mm^2) for withstanding high stress and centrifugal forces.

Stainless steel varieties can be used for special environmental conditions, such as severe corrosive environment and where temperature fluctuation can cause brittle fracture. Popular stainless steel grades are 400 series martensitic stainless steels which can be hardened to high levels of hardness e.g. AISI 410, 416, 440, etc.

Decision on steel quality

For ordinary ball bearing applications, SAE 52100 steel with standard specified limits of composition and inclusion would be adequate. But, for high duty ball bearing applications, the same steel has to be procured based on the following conditions:

- Steel composition should be around the mean values for higher hardenability.
- Steel should be procured with guaranteed carbide size and uniform distribution of carbides for uniform properties.

- Hydrogen and oxygen levels in the steel should be as low as possible.
- Steelmaking route should specify vacuum degassing of liquid steel for control of gases, inclusions and uniformity of chemical composition.

Remarks: Success of the application of bearing steel also depends on the care taken in manufacturing. Manufacturing process, including heat treatment and grinding, must not cause micro-cracking and burning of the ground surface, and leave any residual stress on the contact surfaces.

In sum, from the analyses of different types of applications, it is apparent that applications of steel for different purposes require technical analyses of (a) end-use properties, (b) steps in manufacturing and processing, and (c) identification of *critical performance as well as manufacturing parameters.* The grade of steel should then be selected from various alternatives based on *fitness for the purpose,* cost and availability—though the latter two parameters have been kept outside the scope of discussions in this chapter. Such an approach may, however, seem simple, but it has huge implications for understanding of—(a) What is required?, (b) What are the critical controlling factors?, (c) How different requirements can be matched?, and (d) What is the optimum choice with regard to cost and availability factors for a realistic performance? These call for knowledge based decisions, which this book has described, discussed and illustrated through Chapters from 1 to 10. While Chapters 1 to 8 have focused on the properties and grades of steels, Chapter 9 has discussed the role of design quality, cost, manufacturing quality and steps for prevention of instances of premature failure based on filed failure analyses. Chapter 10 has described the methods of going about selecting steels for different applications. Importantly, what emerged from the discussions and illustrations is that the successful application of steels not only demands knowledge about the steel grades and their attainable properties, but also the factors limiting the performance in service due to inadequate design or manufacturing lapses. Therefore, steel selection and application are also the functions of quality of design and manufacturing for prevention of premature failure. It is the sum of [design quality + steel quality + manufacturing accuracy + maintenance quality] × environment and service conditions. It is this complete cycle of quality along with the service conditions and environment, which governs the probability of success of steel selection for a given application. All applications of steel should have to be approached from this holistic angle of total quality; *specifications and properties of steel are the hub of this wheel of quality but not the whole wheel.* For successful application of steels, attention must be paid to the design and manufacturing quality of components as well—any deficiency in these areas would frustrate the efforts of steel selection and application.

SUMMARY

1. The chapter emphasises that application is a 'bottom-up process', requiring analyses of required properties and characteristics starting from the bottom end of actual application and progressing upward for:

   - Identification of physical (mechanical) and chemical properties required for end-use;
   - Identification of steel properties and characteristics required for forming and manufacturing— e.g. forging, machining, casting, rolling, heat-treating, welding etc.—the steel parts;

- Identification of appropriate steel grade by referring to available steel standards, cost and availability; and
- Finally, efficiently and effectively processing the steel through the manufacturing chains—for developing the target properties and finishing without introduction of defects.

2. Steel is a versatile material and its composition and process can be planned in very accurate manner for tailoring the application related properties of steel. Advanced steelmaking processes support closer control over composition, inclusion, grain size and other characteristics of steels for developing right microstructure and properties. A selector of steel has to judiciously choose steel based on such knowledge and information.

3. The chapter outlines the chain of steel selection and procurement steps following the above mentioned approach, which include:

- Identification of the *nature of application* i.e. structural or engineering, to help focus on steel grades under the respective standards.
- Knowing the end-use requirements of properties as per service load and service conditions, and *identifying the critical ones which must get fulfilled*.
- Identifying the forming/manufacturing properties *required for cost-effective production or meeting the constraints of production shop*.
- Selecting the steel by referring to the *appropriate Steel Standards* with respect to fulfilment of the critical properties as identified earlier.
- *Planning out the production process* or route to achieve defect free production of the part.
- Working out steel availability, *sources of supply and cost*; and finalising the grade as per cost, quality and availability.
- Drawing out a *Supply Specification* incorporating desired steelmaking route, composition control, rolled dimensions and internal soundness quality, properties to be guaranteed, testing methods and test schedule per 'heat' of steel, and methods of certification for conformance to quality.

4. Based on this approach, a road map for steel selection has been outlined emphasising that choice of right steels for different applications is a subject of: Judicious assessment of service conditions, knowledge of mechanical, chemical and physical properties of steels, relating to failure in service by fatigue and fracture, corrosion and oxidation, wear and tear, etc., knowledge of steel grades available for the purpose (including cost), information about resources and facilities available for steel processing—e.g. forging, forming, machining, heat-treating, etc.—and quality capability to manufacture and process the steel for application.

5. The chapter also illustrates the process of steel selection by discussing eight different applications covering the uses of structural, sheet metal and engineering types of steel. In each case, user background of the component and selection methodology has been narrated.

6. Finally, it has been emphasised that successful application of steels is not only the function of good steel selection but also the function of quality of design and manufacturing for prevention of pre-mature failure. It is the sum of [design quality + quality of selected steel + manufacturing accuracy + maintenance quality] × Environment and service conditions. All challenges to steel application should have to be approached, from this holistic angle of steel metallurgy dealing with properties, specifications and applications.

FURTHER READINGS

1. Llewellyn, D.T. and Hudd, B.C. *Steels: Metallurgy and Applications*, 3rd edition. Butterworth-Heinemann, 1998.
2. ASM Handbook Vol. 11, *Failure Analysis and Prevention*. 2002.
3. ASM Handbook, Vol. 20, *Materials Selection and Design*. 1997.
4. Rollason E.C. *Metallurgy for Engineers*. The English Language and Edward Arnold, 1973.
5. Dieter, George E. *Mechanical Metallurgy*. New York: McGraw-Hill, 1961.

APPENDICES

Appendix A: List of SAE/AISI Steel Grades and Composition

Table A-A1. Plain carbon steels.

SAE No.	%C	%Mn	%S (max)	%P (max)
1005	0.06 max	0.35 max	0.040	0.050
1006	0.08 max	0.25/0.40	0.040	0.050
1008	0.10 max	0.30/0.50	0.040	0.050
1010	0.08/0.13	0.30/0.60	0.040	0.050
1011	0.08/0.14	0.60/0.90	0.040	0.050
1012	0.10/0.15	0.30/0.60	0.040	0.050
1013	0.11/0.16	0.30/0.60	0.040	0.050
1015	0.13/0.18	0.30/0.60	0.040	0.050
1016	0.13/0.18	0.60/0.90	0.040	0.050
1017	0.15/0.20	0.30/0.60	0.040	0.050
1018	0.15/0.20	0.60/0.90	0.040	0.050
1020	0.18/0.23	0.30/0.60	0.040	0.050
1021	0.18/0.23	0.60/0.90	0.040	0.050
1022	0.18/0.23	0.70/10.00	0.040	0.050
1023	0.20/0.25	0.30/0.60	0.040	0.050
1025	0.22/0.28	0.30/0.60	0.040	0.050
1026	0.22/0.28	0.60/0.90	0.040	0.050
1029	0.25/0.31	0.60/0.90	0.040	0.050
1030	0.28/0.34	0.60/0.90	0.040	0.050
1035	0.32/0.38	0.60/0.90	0.040	0.050
1038	0.35/0.42	0.60/0.90	0.040	0.050
1039	0.37/0.44	0.70/10.00	0.040	0.050
1040	0.37/0.44	0.60/0.90	0.040	0.050
1042	0.40/0.47	0.60/0.90	0.040	0.050
1043	0.40/0.47	0.70/1.00	0.040	0.050
1044	0.43/0.50	0.30/0.60	0.040	0.050
1045	0.43/0.50	0.60/0.90	0.040	0.050
1046	0.43/0.50	0.70/1.00	0.040	0.050

SAE No.	%C	%Mn	%S (max)	%P (max)
1049	0.46/0.53	0.60/0.90	0.040	0.050
1050	0.48/0.55	0.60/0.90	0.040	0.050
1053	0.48/0.55	0.70/1.00	0.040	0.050
1055	0.50/0.60	0.60/0.90	0.040	0.050
1060	0.55/0.65	0.60/0.90	0.040	0.050
1065	0.60/0.70	0.60/0.90	0.040	0.050
1070	0.65/0.75	0.60/0.90	0.040	0.050
1074	0.70/0.80	0.50/0.80	0.040	0.050
1078	0.72/0.85	0.30/0.60	0.040	0.050
1080	0.75/0.88	0.60/0.90	0.040	0.050
1086	0.80/0.93	0.30/0.50	0.040	0.050
1090	0.85/0.98	0.60/0.90	0.040	0.050
1095	0.90/1.03	0.30/0.50	0.040	0.050

Notes:

1. Steels requiring strength for applications are generally chosen with higher C and Mn levels, such as for forged and heat-treated parts. But, for sheet metal uses where forming is critical, steels with lower level of C and Mn grades are preferred.
2. At high carbon level, grades with controlled Mn are generally used for cold rolled strips for springs and allied applications.

Table A-A2. Standard SAE grade alloy steels.*

SAE No.	%C	%Mn	%Cr	%Ni	%Mo	Others
1330	0.28/0.33	1.60/1.80	—	—	—	—
1335	0.33/0.38	1.60/1.90	—	—	—	—
1340	0.38/0.43	1.60/1.90	—	—	—	—
4023	0.20/0.25	0.70/0.90	—	—	0.20/0.30	—
4027	0.25/0.30	0.70/0.90	—	—	0.20/0.30	—
4037	0.35/0.40	0.70/0.90	—	—	0.20/0.30	—
4047	0.45/0.50	0.70/0.90	—	—	20/0.30	—
4118	0.18/0.23	0.70/0.90	0.40/0.60	—	0.08/0.15	—
4120	0.18/0.23	0.90/1.20	0.40/0.60	—	0.13/0.20	—
4130	0.28/0.33	0.40/0.60	0.80/1.10	—	0.15/0.25	—
4137	0.35/0.40	0.70/0.90	0.80/1.10	—	0.15/0.25	—
4140	0.38/0.43	0.75/1.00	0.80/1.10	—	0.15/0.25	—
4142	0.40/0.45	0.75/1.00	0.80/1.10	—	0.15/0.25	—
4145	0.43/0.48	0.75/1.00	0.80/1.10	—	0.15/0.25	—

SAE No.	%C	%Mn	%Cr	%Ni	%Mo	Others
4150	0.48/0.53	0.75/1.00	0.80/1.10	—	0.15/0.25	—
4320	0.17/0.22	0.45/0.65	0.40/0.60	1.65/2.00	0.20/0.30	—
4340	0.38/0.43	0.60/0.80	0.70/0.90	1.65/2.00	0.20/0.30	—
4620	0.17/0.22	0.45/0.65	—	1.65/2.00	0.20/0.30	—
4820	0.18/0.23	0.50/0.70	—	3.25/3.75	0.20/0.30	—
50B46	0.44/0.49	0.75/1.00	0.20/0.35	—	—	Boron
5120	0.17/0.22	0.70/0.90	0.70/0.90	—	—	—
5130	0.28/0.33	0.70/0.90	0.80/1.10	—	—	—
5132	0.30/0.35	0.60/0.80	0.75/1.00	—	—	—
5140	0.38/0.43	0.70/0.90	0.70/0.90	—	—	—
5150	0.48/0.53	0.70/0.90	0.70/0.90	—	—	—
5160	0.56/0.64	0.75/1.00	0.70/0.90	—	—	—
51B60	0.56/0.64	0.75/1.00	0.70/0.90	—	—	Boron
51100	0.98/1.10	0.25/0.45	0.90/1.15	—	—	—
E52100	0.98/1.10	0.25/0.45	1.30/1.60	—	—	P/S 0.025
52100	0.93/1.05	0.25/0.45	1.35/1.60	—	—	P/S 0.025
6150	0.48/0.53	0.70/0.90	0.80/1.10	—	—	—
8615	0.13/0.18	0.70/0.90	0.40/0.60	0.40/0.70	0.15/0.25	—
8617	0.15/0.23	0.70/0.90	0.40/0.60	0.40/0.70	0.15/0.25	—
8620	0.18/0.23	0.70/0.90	0.40/0.60	0.40/0.70	0.15/0.25	—
8622	0.20/0.25	0.70/0.90	0.40/0.60	0.40/0.70	0.15/0.25	—
8630	0.28/0.33	0.70/0.90	0.40/0.60	0.40/0.70	0.15/0.25	—
8640	0.38/0.43	0.75/1.00	0.40/0.60	0.40/0.70	0.15/0.25	—
8645	0.43/0.48	0.75/1.00	0.40/0.60	0.40/0.70	0.15/0.25	—
8720	0.18/0.23	0.70/0.90	0.40/0.60	0.40/0.70	0.20/0.30	—
8822	0.20/0.25	0.75/1.00	0.40/0.60	0.40/0.70	0.30/0.40	—
9259	0.56/0.64	0.75/1.00	0.45/0.65	—	—	Si 0.70/1.10
9260	0.56/0.64	0.75/1.00	—	—	—	Si 1.8/2.2

* Chemical composition ranges of alloys and limits.

Notes:
1. Unless specified Si: .15/.35, P: .035 max, S: .040 max. Trace elements: Ni: .25 max, Cr: .25 max, Mo: .06 max.
2. Boron level is .0005 to .003% in 'Boron' grade steels.
3. These standard grades can have modification of chemistry by agreement between supplier and user.
4. Steels with sufficient carbon and alloy can be obtained under 'Hardenability' guaranteed condition (see AISI/SAE Steel Specification chart for 'H' steels).

Table A-A3. SAE grade free cutting resulphurised steels: chemical composition.

SAE No.	%C	%Mn	%P	%S
1117	0.14/0.20	1.0/1.30	0.03 max	0.08/0.13
1118	0.14/0.23	1.30/1.60	0.03 max	0.08/0.13
1126	0.23/0.29	0.70/1.00	0.03 max	0.08/0.13
1132	0.27/0.34	1.35/1.65	0.03 max	0.08/0.13
1137	0.32/0.39	1.35/1.65	0.03 max	0.08/0.13
1138	0.34/0.40	0.70/1.00	0.03 max	0.08/0.13
1140	0.37/0.44	0.70/1.00	0.03 max	0.08/0.13
1141	0.37/0.45	1.35/1.65	0.03 max	0.08/0.13
1144	0.40/0.48	1.35/1.65	0.03 max	0.08/0.13
1146	0.42/0.49	0.70/1.00	0.03 max	0.08/0.13
1151	0.48/0.55	0.70/1.00	0.03 max	0.08/0.13
*1212	0.13 max	0.70/1.00	0.07/0.12	0.16/0.23
*1213	0.13 max	0.70/1.00	0.07/0.12	0.24/0.33
*1215	0.09 max	0.75/1.05	0.04/0.09	0.26/0.35

* 12XX grades are rephosphorised and resulphurised steels.

Note: Resulphurised steels are customarily furnished without specified silicon content because of adverse effect silicon on machinability.

Table A-A4. Stainless steel designations and compositions.

SAE Designation	UNS Designation	% Cr	% Ni	% C	% Mn	% Si	% P	% S	% N	Other
					Austenitic					
201	S20100	16–18	3.5–5.5	0.15	5.5–7.5	0.75	0.06	0.03	0.25	—
202	S20200	17–19	4–6	0.15	7.5–10.0	0.75	0.06	0.03	0.25	—
205	S20500	16.5–18	1–1.75	0.12–0.25	14–15.5	0.75	0.06	0.03	0.32–0.40	—
254	S31254	20	18	0.02 max	—	—	—	—	0.20	6 Mo; 0.75 Cu; "Super austenitic"; All values nominal
301	S30100	16–18	6–8	0.15	2	0.75	0.045	0.03	—	—
302	S30200	17–19	8–10	0.15	2	0.75	0.045	0.03	0.1	—
302B	S30215	17–19	8–10	0.15	2	2.0–3.0	0.045	0.03	—	—
303	S30300	17–19	8–10	0.15	2	1	0.2	0.15 min	—	Mo 0.60 (optional)
303Se	S30323	17–19	8–10	0.15	2	1	0.2	0.06	—	0.15 Se min
304	S30400	18–20	8–10.50	0.08	2	0.75	0.045	0.03	0.1	—
304L	S30403	18–20	8–12	0.03	2	0.75	0.045	0.03	0.1	—
304Cu	S30430	17–19	8–10	0.08	2	0.75	0.045	0.03	—	3–4 Cu
304N	S30451	18–20	8–10.50	0.08	2	0.75	0.045	0.03	0.10–0.16	—
305	S30500	17–19	10.50–13	0.12	2	0.75	0.045	0.03	—	—
308	S30800	19–21	10–12	0.08	2	1	0.045	0.03	—	—
309	S30900	22–24	12–15	0.2	2	1	0.045	0.03	—	—
309S	S30908	22–24	12–15	0.08	2	1	0.045	0.03	—	—
310	S31000	24–26	19–22	0.25	2	1.5	0.045	0.03	—	—
310S	S31008	24–26	19–22	0.08	2	1.5	0.045	0.03	—	—
314	S31400	23–26	19–22	0.25	2	1.5–3.0	0.045	0.03	—	—
316	S31600	16–18	10–14	0.08	2	0.75	0.045	0.03	0.10	2.0–3.0 Mo

Type	UNS	Cr	Ni	C	Mn	Si	P	S	N	Other
316L	S31603	16–18	10–14	0.03	2	0.75	0.045	0.03	0.10	2.0–3.0 Mo
316F	S31620	16–18	10–14	0.08	2	1	0.2	0.10 min	—	1.75–2.50 Mo
316N	S31651	16–18	10–14	0.08	2	0.75	0.045	0.03	0.10–0.16	2.0–3.0 Mo
317	S31700	18–20	11–15	0.08	2	0.75	0.045	0.03	0.10 max	3.0–4.0 Mo
317L	S31703	18–20	11–15	0.03	2	0.75	0.045	0.03	0.10 max	3.0–4.0 Mo
321	S32100	17–19	9–12	0.08	2	0.75	0.045	0.03	0.10 max	Ti 5(C + N) min, 0.70 max
329	S32900	23–28	2.5–5	0.08	2	0.75	0.04	0.03	—	1–2 Mo
330	N08330	17–20	34–37	0.08	2	0.75–1.50	0.04	0.03	—	—
347	S34700	17–19	9–13	0.08	2	0.75	0.045	0.030	—	Nb + Ta, 10 × C min, 1 max
348	S34800	17–19	9–13	0.08	2	0.75	0.045	0.030	—	Nb + Ta, 10 × C min, 1 max, but 0.10 Ta max; 0.20 Ca
384	S38400	15–17	17–19	0.08	2	1	0.045	0.03	—	—
Ferritic										
405	S40500	11.5–14.5	—	0.08	1	1	0.04	0.03	—	0.1–0.3 Al, 0.60 max
409	S40900	10.5–11.75	0.05	0.08	1	1	0.045	0.03	—	Ti 6 × C, but 0.75 max
429	S42900	14–16	0.75	0.12	1	1	0.04	0.03	—	—
430	S43000	16–18	0.75	0.12	1	1	0.04	0.03	—	—
430F	S43020	16–18	—	0.12	1.25	1	0.06	0.15 min	—	0.60 Mo (optional)
430FSe	S43023	16–18	—	0.12	1.25	1	0.06	0.06	—	0.15 Se min
434	S43400	16–18	—	0.12	1	1	0.04	0.03	—	0.75–1.25 Mo
436	S43600	16–18	—	0.12	1	1	0.04	0.03	—	0.75–1.25 Mo; Nb + Ta 5 × C min, 0.70 max
442	S44200	18–23	—	0.2	1	1	0.04	0.03	—	—
446	S44600	23–27	0.25	0.2	1.5	1	0.04	0.03	—	—

Martensitic

403	S40300	11.5–13.0	0.60	0.15	1	0.5	0.04	0.03	—	—
410	S41000	11.5–13.5	0.75	0.15	1	1	0.04	0.03	—	—
414	S41400	11.5–13.5	1.25–2.50	0.15	1	1	0.04	0.03	—	—
416	S41600	12–14	—	0.15	1.25	1	0.06	0.15 min	—	0.060 Mo (optional)
416Se	S41623	12–14	—	0.15	1.25	1	0.06	0.06	—	0.15 Se min
420	S42000	12–14	—	0.15 min	1	1	0.04	0.03	—	—
420F	S42020	12–14	—	0.15 min	1.25	1	0.06	0.15 min	—	0.60 Mo max (optional)
422	S42200	11.0–12.5	0.50–1.0	0.20–0.25	0.5–1.0	0.5	0.025	0.025	—	0.90–1.25 Mo; 0.20–0.30 V; 0.90–1.25 W
431	S41623	15–17	1.25–2.50	0.2	1	1	0.04	0.03	—	—
440A	S44002	16–18	—	0.60–0.75	1	1	0.04	0.03	—	0.75 Mo
440B	S44003	16–18	—	0.75–0.95	1	1	0.04	0.03	—	0.75 Mo
440C	S44004	16–18	—	0.95–1.20	1	1	0.04	0.03	—	0.75 Mo

Heat Resisting

501	S50100	4–6	—	0.10 min	1	1	0.04	0.03	—	0.40–0.65 Mo
502	S50200	4–6	—	0.1	1	1	0.04	0.03	—	0.40–0.65 Mo

Appendix B. Equivalent Grades of Steel

Table A-B1. Comparison of SAE/AISI/ASTM steel grades with other international standards.

Comparison of USA Standards to International Standards for Chemistry						
USA	**EUROPE**	**GERMANY**		**ENGLAND**	**ITALY**	**JAPAN**
A.S.T.M. S.A.E. A.I.S.I.	Euronorm	Werkstoff W.-Nr	Kurzname DIN	BS 970	UNI	JIS
CARBON						
1018	C15D	1.1141	CK15	040A15	C15	S15
1018	C18D	1.0401	C15	080M15	C16	S15CK
1018		1.0453	C16.8	080A15	1C15	S15C
1018				EN3B		
1045	C45	1.0503	C45	060A47	C45	S45C
1045		1.1191	CK45	080A46	1C45	S48C
1045		1.1193	CF45	080M46	C46	
1045		1.1194	CQ45	EN8D	C43	
1140/1146	35S20	1.0726	35S20	212M40		
1140/1146	45S20	1.0727	45S20	En8M		
1215	11SMn37	1.0715	9SMn28	230M07	CF9SMn28	SUM 25
1215		1.0736	9SMn36	En1A	CF9SMn36	SUM 22
12L14	11SMnPb30	1.0718	9SMnPb28	230M07Leaded	CF9SMnPb28	SUM 22L
12L14	11SMnPb37	1.0737	9SMnPb36	En1A Leaded	CF9SMnPb36	SUM 23L
12L14						SUM 24L
ALLOY						
4130		1.7218	25CrMo4	708A30	25CrMo4 (KB)	SCM 420
4130			GS-25CrMo4	CDS110	30CrMo4	SCM 430
4130					30CrMo4	SCCrM1
4140/4142	42CrMo4	1.7223	41CrMo4	708M40	41CrMo4	SCM 440
4140/4142		1.7225	42CrMo4	708A42	38CrMo4 KB	SCM 440H
4140/4142		1.7227	42CrMoS4	709M40	G40 CrMo4	SNB 7
4140/4142		1.3563	43CrMo4	En19	42CrMo4	SCM 4M
4140/4142				En19C		SCM 4
4340	34CrNiMo6	1.6582	34CrNiMo6	817M40	35NiCrMo6 KB	SNCM 447
4340		1.6562	40 NiCrMo8 4	En24	40NiCrMo7 KB	SNB24-1-5
8620	20NiCrMo2-2	1.6543	21NiCrMo22	805A20	20NiCrMo2	SNCM 220 (H)
8620		1.6523	21NiCrMo2	805M20		
STAINLESS						
303	X8CrNiS18-9	1.4305	X10CrNiS18-9	303S 21	X10CrNiS 18 09	SUS 303
303				En58M		
304		1.4301	X5CrNi 18 9	304S 15	X5CrNi 18 10	SUS 304
304	X2CrNi19-11		X5CrNi 18 10	304S 16		SUS 304-CSP
304	X2CrNi18-10		XCrNi 19 9	304S 18		
304				304S 25		
304				En58E		
304L	X2CrNi19 11	1.4306		304S 11		SUS304L
316	X5CrNiMo17-12-2	1.4401	X5CrNiMo17 12 2	316S 29	X5CrNiMo17 12	SUS 316
316	X5CrNiMo18-14-3	1.4436	X5CrNiMo17 13 3	316S 31	X5CrNiMo17 13	SUS 316TP
316			X5CrNiMo 19 11	316S 33	X8CrNiMo17 13	
316			X5CrNiMo 18 11	En58J		
316L	X2CrNiMo17 12 2	1.4404		316S 11		SUS316L
316Ti		1.4571	X6CrNiMoTi17 12	320S 33		
321		1.4541	X6CrNiTi18 10	321S 31		SUS321
430		1.4016	X6Cr17	430S 17		SUS430
430F		1.4104	X14CrMoS17			SUS430F
TOOL STEEL						
A-2	X100CrMoV5	1.2363	X100CrMoV51	BA 2	X100CrMoV5 1 KU	SKD 12
D-2	X153CrMoV12	1.2379	X155CrVMo12 1	BD 2	X155CrVMo12 1	SKD 11
O-1		1.2510	100MnCrW4	BO 1	95MnWCr 5 KU	

Substitution is only possible after a complete examination of the individual specifications.

Table A-B2. Common case-carburising grade steels and their near equivalents.

IS Grade: IS-1570: 1961	Nominal Composition (Average)					BS 970: Old/(New) Designation	SAE	DIN
	%C	Mn	Cr	Ni	Mo			
C-10	0.15	0.60	—	—	—	EN 32A (045M10)	1012/1015	—
17Mn1Cr95	0.17	1.1	0.95	—	—	—	—	16MnCr5
20MnCr1	0.20	1.2	1.1	—	—	—	—	20MnCr5
20Ni2Mo25	0.17	0.60	—	1.8	0.25	EN 34 (665M17)	4620	—
15Cr65	0.15	0.60	0.75	—	—	EN 207(527A19)	5115	15Cr3
16Ni80Cr60	0.17	0.90	0.70	0.80	—	En 351 (635M15)	—	—
20Ni55CR50Mo20	0.18	0.80	0.50	0.60 (op)	0.20	En362 (805M20)	8620, 8720	—
13Ni3Cr80	0.14	0.60	0.80	3.4	—	En 36A, B (655M13)	3310	—
15NiCr1Mo12	0.16	0.90	1.0	1.22	0.12	EN 353 (822M17)	—	—
15Ni2Cr1Mo15	0.17	0.80	1.0	1.7	0.15	En 354 (815M17)	4320	—
16NiCr2Mo20	0.18	0.60	1.5	2.0	0.20	En 355 (822M17)	—	(15CrNi6)

Note: S and P limited to 0.050 and 0.035 max; Si is generally, 0.35 max. Comparative grades are near equivalents.'

Table A-B3. Miscellaneous grades of popular steels (special).

Types	IS Grade	Nominal Composition (Average)						BS Grade	SAE	DIN
		%C	Mn	Cr	V	Mo	Others			
Alloy spring steels	50Cr1	0.50	0.70	1.1	—	—		EN 48	5147/5150	—
	50Cr1V23	0.50	0.75	1.1	0.20	—		EN 47	6150	50Crv4
Nitriding steel	40Cr2A11Mo18	0.40	0.60	1.6	—	0.20	Al, 10	En 41B	—	34CrAlMo5
	40Cr3Mo1V20	0.40	0.60	0.32	0.20	1.0	—	En 40C	—	—
Bearing steels	1003Cr1	1.0	0.60	1.0			S&P: 0.025 max	—	51100	—
	103Cr2	1.0	0.60	1.40			S&P: 0.25 max	EN 31	52100	100Cr6
Carbon tool steels	T70	0.70	0.30	—			S&P: 0.03 max	—	W-1	—
	T80	0.80	0.30	—			S&P: 0.03 max	—	W-1	—
	T90	0.90	0.30	—			S&P: 0.03 max	—	W-1	—

Appendix C. Properties of Some Common UNS and SAE/AISI Grade Steels

Table A-C1. Properties of steel.

UNS Number	Processing Method	Yield Strength (kpsi)	Tensile Strength (kpsi)	Yield Strength (MPa)	Tensile Strength (MPa)	Elongation in 2 in. (%	Reduction in Area (%)	Brinell Hardness (H_b)
G10100	Hot Rolled	26	47	179	324	28	50	95
G10100	Cold Drawn	44	53	303	365	20	40	105
G10150	Hot Rolled	27	50	186	345	28	50	101
G10150	Cold Drawn	47	56	324	386	18	40	111
G10180	Hot Rolled	32	58	220	400	25	50	116
G10180	Cold Drawn	54	64	372	441	15	40	126
G10350	Hot Rolled	39	72	269	496	18	40	143
G10350	Cold Drawn	67	80	462	551	12	35	163
G10350	Drawn 800 °F	81	110	558	758	18	51	220
G10350	Drawn 1000 °F	72	103	496	710	23	59	201
G10350	Drawn 1200 °F	62	91	427	627	27	66	180
G10400	Hot Rolled	42	76	289	524	18	40	149
G10400	Cold Drawn	71	85	489	586	12	35	170
G10400	Drawn 1000 °F	86	113	593	779	23	62	235
G10500	Hot Rolled	49	90	338	620	15	35	179
G10500	Cold Drawn	84	100	579	689	10	30	197
G10500	Drawn 600 °F	180	220	1240	1516	10	30	450
G10500	Drawn 900 °F	130	155	896	1068	18	55	310
G10500	Drawn 1200 F	80	105	551	723	28	65	210
G15216	Hot Rolled, Annealed	81	100	558	689	25	57	192
G41300	Hot Rolled, Annealed	60	90	413	620	30	45	183
G41300	Cold Drawn, Annealed	87	98	599	675	21	52	201
G41300	Drawn 1000 °F	133	146	916	1006	17	60	293
G41400	Hot Rolled, Annealed	63	90	434	620	27	58	187
G41400	Cold Drawn, Annealed	90	102	620	703	18	50	223
G41400	Drawn 1000 °F	131	153	903	1054	16	45	302
G43400	Hot Rolled, Annealed	69	101	475	696	21	45	207
G43400	Cold Drawn, Annealed	99	111	682	765	16	42	223

G43400	Drawn 600 °F	234	260	1612	1791	12	43	498
G43400	Drawn 1000 °F	162	182	1116	1254	15	40	363
G46200	Case Hardened	89	120	613	827	22	55	248
G46200	Drawn 800 °F	94	130	648	896	23	66	256
G61500	Hot Rolled, Annealed	58	91	400	627	22	53	183
G61500	Drawn 1000 °F	132	155	909	1068	15	44	302
G87400	Hot Rolled, Annealed	64	95	441	655	25	55	190
G87400	Cold Drawn, Annealed	96	107	661	737	17	48	223
G87400	Drawn 1000 °F	129	152	889	1047	15	44	302
G92550	Hot Rolled, Annealed	78	115	537	792	22	45	223
G92550	Drawn 1000 °F	160	180	1102	1240	15	32	352

Note: G indicates UNS system for SAE/AISI steels and First 4 digits after G indicate the SAE/AISI grades.

Appendix D. Physical Data and Conversion Tables

Table A-D1. Handy physical constants.

HANDY PHYSICAL CONSTANTS

Acceleration of gravity, g	32.17 ft/s^2 = 9.807 m/s^2
Density of water	62.4 lbm/ft^3 = 1 g/cm^3
	1 gal H$_2$O = 8.345 lbm
Gas Constant, R	1545 ft-lbf/pmole-R = 8.314 J/gmole-K
Gas volume (STP: 68°F, 1 atm)	359 ft^3/pmole = .02241 m^3/gmole
Joule's Constant, J	778 ft-lbf/BTU
Poisson's ratio, μ	.3 (for steel)

STEEL CONSTANTS

Fe-Fe$_3$C eutectoid composition	0.77 w/o carbon
Fe-Fe$_3$C eutectoid temperature	1340°F (727°C)
Modulus of Elasticity (steel)	30 × 10^6 psi

Densities:

Carbon & Low-Alloy Steels	0.283 lbm/in^3 = 7.84 g/cm^3
304 SS	0.29 lbm/in^3 = 7.88 g/cm^3
Tool Steels	Carbon Steels × 1.000
Moly High Speed	Carbon Steels × 1.035
Multiphase Alloys	Carbon Steels × 1.074
Steel Tensile Strength (psi)	~ 500 × Brinell Number

COMPARISON MATERIALS

Material	Density (g/cm^3)	Modulus of Elasticity (psi)	Poisson's Ratio
Aluminum Alloys	2.6-2.9	10.0 × 10^6	0.33
Nickel-base Superalloys	8.0-8.9	28.5 -- 31.0 × 10^6	0.31
Titanium Alloys	4.4-5.0	15.0 × 16.8 × 10^6	0.34

SI PREFIXES

giga	G	10^9
mega	M	10^6
kilo	k	10^3
hecto	h	10^2
deka	da	10^1
deci	d	10^{-1}
centi	c	10^{-2}
milli	m	10^{-3}
micro	μ	10^{-6}
nano	n	10^{-9}

Table A-D2. Conversion factors.

CONVERSION FACTORS

EQUATION: A × B = C		
A	**B**	**C**
Area [L]²		
ft²	0.092903	m²
in²	645.16	mm²
in²	6.45160	cm²
Energy, Work or Heat [M] [L]² [t]⁻²		
Btu	1.05435	kJ
Btu	0.251996	kcal
Calories (cal)	4.184*	Joules (J)
ft-lbf	1.355818	J
ft-lbf	0.138255	kgf-m
hp-hr	2.6845	MJ
KWH	3.600	MJ
m-kgf	9.80665*	J
N-m	1.	J
Flow Rate [L]³ [t]⁻¹		
ft³/min	7.4805	gal/min
ft³/min	0.471934	l/s
gal/min	0.063090	l/s
Force or Weight [M] [L] [t]⁻²		
kgf	9.80665*	Newton (N)
lbf	4.44822	N
lbf	0.453592	Kgf
Fracture Toughness		
ksi√in	1.098800	MPa√m
Heat Content		
Btu/lbm	0.555556	cal/g
Btu/lbm	2.324444	J/g
Btu/ft³	0.037234	MJ/m³
Heat Flux		
Btu/hr-ft²	7.5346 E-5	cal/s-cm²
Btu/hr-ft²	3.1525	W/m²
cal/s-cm²	4.184*	W/cm²
Length [L]		
Foot (ft)	0.304800	Meter (m)
Inch (in)	25.4000	Millimeter (mm)
Mile (mi)	1.609344	Kilometer (km)

* Indicates exact conversion(s)

Table A-D2. Conversion factors *(continued)*.

EQUATION: A × B = C		
A	**B**	**C**
Mass Density [M] [L]$^{-3}$		
lbm/in^3	27.68	g/cm^3
lbm/ft^3	16.0184	kg/m^3
Power [M] [L]^2 [t]$^{-3}$		
Btu/hr	0.292875	Watt (W)
ft-lbf/s	1.355818	W
Horsepower (hp)	745.6999	W
Horsepower	550.*	ft-lbf/s
Pressure (fluid) [M] [L]$^{-1}$ [t]$^{-2}$		
Atmosphere (atm)	14.696	lbf/in^2
atm	1.01325 E5*	Pascal (Pa)
lbf/ft^2	47.88026	Pa
lbf/in^2	27.6807	in. H$_2$O at 39.2°F
Stress [M] [L]$^{-1}$ [t]$^{-2}$		
kgf/cm^2	9.80665 E-2*	MPa
ksi	6.89476	MPa
N/mm^2	1.	MPa
kgf/mm^2	1.42231	ksi
Volume [L]^3 & Capacity		
in^3	16.3871	cm^3
ft^3	0.028317	m^3
ft^3	7.4805	Gallon
ft^3	28.3168	Liter (l)
Gallon	3.785412	Liter
Specific Heat		
Btu/lbm-°F	1.	cal/g-°C
Temperature*		
Fahrenheit	(°F−32)/1.8	Celsius
Fahrenheit	°F+459.67	Rankine
Celsius	°C+273.16	Kelvin
Rankine	R/1.8	Kelvin
Thermal Conductivity		
Btu-ft/hr-ft^2-°F	14.8816	cal-cm/hr-cm^2-°C

* Indicates exact conversion(s)

Table A-D3. Metric-English stress conversion.

METRIC-ENGLISH STRESS CONVERSION TABLE
Kg Per Sq Mm to Psi to M Pa

Kg per sq mm	Psi	M Pa	Kg per sq mm	Psi	M Pa	Kg per sq mm	Psi	M Pa	Kg per sq mm	Psi	M Pa
10	14,223	98.1	50	71,117	490.3	90	128,011	882.6	130	184,904	1274.9
11	15,646	107.9	51	72,539	500.1	91	129,433	892.4	131	186,327	1284.7
12	17,068	117.7	52	73,962	510.0	92	130,855	902.2	132	187,749	1294.5
13	18,490	127.5	53	75,384	519.8	93	132,278	912.0	133	189,171	1304.3
14	19,913	137.3	54	76,806	529.6	94	133,700	921.8	134	190,594	1314.1
15	21,335	147.1	55	78,229	539.4	95	135,122	931.6	135	192,016	1323.9
16	22,757	156.9	56	79,651	549.2	96	136,545	941.4	136	193,438	1333.7
17	24,180	166.7	57	81,073	559.0	97	137,967	951.2	137	194,861	1343.5
18	25,602	176.5	58	82,496	568.8	98	139,389	961.0	138	196,283	1353.3
19	27,024	186.3	59	83,918	578.6	99	140,812	970.9	139	197,705	1363.1
20	28,447	196.1	60	85,340	588.4	100	142,234	980.7	140	199,128	1372.9
21	29,869	205.9	61	86,763	598.2	101	143,656	990.5	141	200,550	1382.7
22	31,291	215.7	62	88,185	608.0	102	145,079	1000.3	142	201,972	1392.5
23	32,714	225.6	63	89,607	617.8	103	146,501	1010.1	143	203,395	1402.4
24	34,136	235.4	64	91,030	622.6	104	147,923	1020.0	144	204,817	1412.2
25	35,558	245.2	65	92,452	637.4	105	149,346	1029.7	145	206,239	1422.0
26	36,981	255.0	66	93,874	647.2	106	150,768	1039.5	146	207,662	1431.8
27	38,403	264.8	67	95,297	657.0	107	152,190	1049.3	147	209,084	1441.6
28	39,826	274.6	68	96,719	666.9	108	153,613	1059.1	148	210,506	1451.4
29	41,248	284.4	69	98,141	676.7	109	155,035	1068.9	149	211,929	1461.2
30	42,670	294.2	70	99,564	686.5	110	156,457	1078.7	150	213,351	1471.0
31	44,093	304.0	71	100,986	696.3	111	157,880	1088.5	151	214,773	1480.8
32	45,515	313.8	72	102,408	706.1	112	159,302	1098.3	152	216,196	1490.6
33	46,937	323.6	73	103,831	715.9	113	160,724	1108.2	153	217,618	1500.4
34	48,360	333.4	74	105,253	725.7	114	162,147	1118.0	154	219,040	1510.2
35	49,782	343.2	75	106,675	735.5	115	163,569	1127.8	155	220,463	1520.0
36	51,204	353.0	76	108,098	745.3	116	164,991	1137.6	156	221,885	1529.8
37	52,627	362.8	77	109,520	755.1	117	166,414	1147.4	157	223,307	1539.6
38	54,049	372.7	78	110,943	764.9	118	167,836	1157.2	158	224,730	1549.5
39	55,471	382.5	79	112,365	774.7	119	169,258	1167.0	159	226,152	1559.3
40	56,894	393.3	80	113,787	784.5	120	170,681	1176.8			
41	58,316	402.1	81	115,210	794.3	121	172,103	1186.6			
42	59,738	411.9	82	116,632	804.1	122	173,525	1196.4			
43	61,161	421.7	83	118,054	814.0	123	174,948	1206.2			
44	62,583	431.5	84	119,477	823.8	124	176,370	1216.0			
45	64,005	441.3	85	120,899	833.6	125	177,792	1225.8			
46	65,428	451.1	86	122,321	843.4	126	179,215	1235.6			
47	66,850	460.9	87	123,744	853.2	127	180,637	1245.4			
48	68,272	470.7	88	125,166	863.0	128	182,059	1255.3			
49	69,695	480.5	89	126,588	872.8	129	183,482	1265.1			

Table A-D4. Temperature conversion.

TEMPERATURE CONVERSION TABLES
Albert Sauveur type of table. Values revised.

−459.4 to 0			0 to 100						100 to 1000					
C	F/C	F	C	F/C	F	C	F/C	F	C	F/C	F	C	F/C	F
-273	-459.4		-17.8	0	32	10.0	50	122.0	38	100	212	260	500	932
-268	-450		-17.2	1	33.8	10.6	51	123.8	43	110	230	266	510	950
-262	-440		-16.7	2	35.6	11.1	52	125.6	49	120	248	271	520	968
-257	-430		-16.1	3	37.4	11.7	53	127.4	54	130	266	277	530	986
-251	-420		-15.6	4	39.2	12.2	54	129.2	60	140	284	282	540	1004
-246	-410		-15.0	5	41.0	12.8	55	131.0	66	150	302	288	550	1022
-240	-400		-14.4	6	42.8	13.3	56	132.8	71	160	320	293	560	1040
-234	-390		-13.9	7	44.6	13.9	57	134.6	77	170	338	299	570	1058
-229	-380		-13.3	8	46.4	14.4	58	136.4	82	180	356	304	580	1076
-223	-370		-12.8	9	48.2	15.0	59	138.2	88	190	374	310	590	1094
-218	-360		-12.2	10	50.0	15.6	60	140.0	93	200	392	316	600	1112
-212	-350		-11.7	11	51.8	16.1	61	141.8	99	210	410	321	610	1130
-207	-340		-11.1	12	53.6	16.7	62	143.6	100	212	413.6	327	620	1148
-201	-330		-10.6	13	55.4	17.2	63	145.4	104	220	428	332	630	1166
-196	-320		-10.0	14	57.2	17.8	64	147.2	110	230	446	338	640	1184
-190	-310		-9.4	15	59.0	18.3	65	149.0	116	240	464	343	650	1202
-184	-300		-8.9	16	60.8	18.9	66	150.8	121	250	482	349	660	1220
-179	-290		-8.3	17	62.6	19.4	67	152.6	127	260	500	354	670	1238
-173	-280		-7.8	18	64.4	20.0	68	154.4	132	270	518	360	680	1256
-169	-273	-459.4	-7.2	19	66.2	20.6	69	156.2	138	280	536	366	690	1274
-168	-270	-454	-6.7	20	68.0	21.1	70	158.0	143	290	554	371	700	1292
-162	-260	-436	-6.1	21	69.8	21.7	71	159.8	149	300	572	377	710	1310
-157	-250	-418	-5.6	22	71.6	22.2	72	161.6	154	310	590	382	720	1328
-151	-240	-400	-5.0	23	73.4	22.8	73	163.4	160	320	608	388	730	1346
-146	-230	-382	-4.4	24	75.2	23.3	74	165.2	166	330	626	393	740	1364
-140	-220	-364	-3.9	25	77.0	23.9	75	167.0	171	340	644	399	750	1382
-134	-210	-346	-3.3	26	78.8	24.4	76	168.8	177	350	662	404	760	1400
-129	-200	-328	-2.8	27	80.6	25.0	77	170.6	182	360	680	410	770	1418
-123	-190	-310	-2.2	28	82.4	25.6	78	172.4	188	370	698	416	780	1436
-118	-180	-292	-1.7	29	84.2	26.1	79	174.2	193	380	716	421	790	1454
-112	-170	-274	-1.1	30	86.0	26.7	80	176.0	199	390	734	427	800	1472
-107	-160	-256	-.6	31	87.8	27.2	81	177.8	204	400	752	432	810	1490
-101	-150	-238	0	32	89.6	27.8	82	179.6	210	410	770	438	820	1508
-96	-140	-220	.6	33	91.4	28.3	83	181.4	216	420	788	443	830	1526
-90	-130	-202	1.1	34	93.2	28.9	84	183.2	221	430	806	449	840	1544
-84	-120	-184	1.7	35	95.0	29.4	85	185.0	227	440	824	454	850	1562
-79	-110	-166	2.2	36	96.8	30.0	86	186.8	232	450	842	460	860	1580
-73	-100	-148	2.8	37	98.6	30.6	87	188.6	238	460	860	466	870	1598
-68	-90	-130	3.3	38	100.4	31.1	88	190.4	243	470	878	471	880	1616
-62	-80	-112	3.9	39	102.2	31.7	89	192.2	249	480	896	477	890	1634
-57	-70	-94	4.4	40	104.0	32.2	90	194.0	254	490	914	482	900	1652
-51	-60	-76	5.0	41	105.8	32.8	91	195.8				488	910	1670
-46	-50	-58	5.6	42	107.6	33.3	92	197.6				493	920	1688
-40	-40	-40	6.1	43	109.4	33.9	93	199.4				499	930	1706
-34	-30	-22	6.7	44	111.2	34.4	94	201.2				504	940	1724
-29	-20	-4	7.2	45	113.0	35.0	95	203.0				510	950	1742
-23	-10	14	7.8	46	114.8	35.6	96	204.8				516	960	1760
-17.8	0	32	8.3	47	116.6	36.1	97	206.6				521	970	1778
			8.9	48	118.4	36.7	98	208.4				527	980	1796
			9.4	49	120.2	37.2	99	210.2				532	990	1814
						37.8	100	212.0				538	1000	1832

Look up reading in middle column. If in degrees Celsius, read Fahrenheit equivalent in right hand column; if in Fahrenheit degrees, read Celsius equivalent in left hand column.

Table A-D4. Temperature conversion *(continued)*.

TEMPERATURE CONVERSION TABLES – continued
Albert Sauveur type of table. Values revised.

1000 to 2000						2000 to 3000					
C	F/C	F	C	F/C	F	C	F/C	F	C	F/C	F
538	1000	1832	816	1500	2732	1093	2000	363?	1371	2500	4532
543	1010	1850	821	1510	2750	1099	2010	3650	1377	2510	4650
549	1020	1868	827	1520	2768	1104	2020	3668	1382	2520	4568
554	1030	1886	832	1530	2786	1110	2030	3686	1388	2530	4586
560	1040	1904	838	1540	2804	1116	2040	3704	1393	2540	4604
566	1050	1922	843	1550	2822	1121	2050	3722	1399	2550	4622
571	1060	1940	849	1560	2840	1127	2060	3740	1404	2560	4640
577	1070	1958	854	1570	2858	1132	2070	3758	1410	2570	4658
582	1080	1976	860	1580	2876	1138	2080	3776	1416	2580	4676
588	1090	1994	866	1590	2894	1143	2090	3794	1421	2590	4694
593	1100	2012	871	1600	2912	1149	2100	3812	1427	2600	4712
599	1110	2030	877	1610	2930	1154	2110	3830	1432	2610	4730
604	1120	2048	882	1620	2948	1160	2120	3848	1438	2620	4748
610	1130	2066	888	1630	2966	;166	2130	3866	1443	2630	4766
616	1140	2084	893	1640	2984	1171	2140	3884	1449	2640	4784
621	1150	2102	899	1650	3002	1177	2150	3902	1454	2650	4802
627	1160	2120	904	1660	3020	1182	2160	3920	1460	2660	4820
632	1170	2138	910	1670	3038	1188	2170	3938	1466	2670	4838
638	1180	2156	916	1680	3056	1193	2180	3956	1471	2680	4856
643	1190	2174	921	1690	3074	1199	2190	3974	1477	2690	4874
649	1200	2192	927	1700	3092	1204	2200	3992	1482	2700	4892
654	1210	2210	932	1710	3110	1210	2210	4010	1488	2710	4910
660	1220	2228	938	1720	3128	1216	2220	402N	1493	2720	4928
666	1230	2246	943	1730	3146	1221	2230	4046	1499	2730	4946
671	1240	2264	949	1740	3164	1227	2240	4064	1504	2740	4964
677	1250	2282	954	1750	3182	1232	2250	4082	1510	2750	4982
682	1260	2300	960	1760	3200	1238	2260	4100	1516	2760	5000
688	1270	2318	966	1770	3218	i243	2270	4118	1521	2770	5018
693	1280	2336	971	1780	3236	1249	2280	4136	1527	2780	5036
699	1290	2354	977	1790	3254	1254	2290	4154	1532	2790	5054
704	1300	2372	982	1800	3272	1260	2300	4172	1538	2800	5072
710	1310	2390	988	1810	3290	1266	2310	4190	1543	2810	5090
716	1320	2408	993	1820	3308	1271	2320	4208	1549	2820	5108
721	1330	2426	999	1830	3326	1277	2330	4226	1554	2830	5126
727	1340	2444	1004	1840	3344	1282	2340	4244	1560	2840	5144
732	1350	2462	1010	1850	3362	1288	2350	4262	1566	2850	5162
738	1360	2480	1016	1860	3380	1293	2360	4280	1571	2860	5180
743	1370	2498	1021	1870	3398	1299	2370	4298	1577	2870	5198
749	1380	2516	1027	1880	3416	1304	2380	4316	1582	2880	5216
754	1390	2534	1032	1890	3434	1310	2390	4334	1588	2890	5234
760	1400	2552	1038	1900	3452	1316	2400	4352	1593	2900	5252
766	1410	2570	1043	1910	3470	1321	2410	4370	1599	2910	5270
771	1420	2588	1049	1920	3488	1327	2420	4388	1604	2920	5288
777	1430	2606	1054	1930	3506	1332	2430	4406	1610	2930	5306
782	1440	2624	1060	1940	3524	1338	2440	4424	1616	2940	5324
788	1450	2642	1066	1950	3542	1343	2450	4442	1621	2950	5342
793	1460	2660	1071	1960	3560	1349	2460	4460	1627	2960	5360
799	1470	2678	1077	1970	3578	1354	2470	4478	1632	2970	5378
804	1480	2696	1082	1980	3596	1360	2480	4496	1638	2980	5396
810	1490	2714	1088	1990	3614	i366	2490	4514	1643	2990	5414
			1093	2000	3632				1649	3000	5432

Look up reading in middle column. If in degrees Celsius, read Fahrenheit equivalent in right hand column; if in Fahrenheit degrees, read Celsius equivalent in left hand column.

Table A-D4. Hardness conversion.

HARDNESS CONVERSION TABLES
BASED ON BRINELL
(APPROXIMATE)

BRINELL HARDNESS		ROCKWELL HARDNESS				Diamond Pyramid Hardness Number (Vickers)	Approx. Tensile Strength 1000 psi
Diameter mm 3000 Kg	Tungsten Carbide 10 mm Ball	A-Scale 60 Kg Brale	B-Scale 100 Kg 1/16" Ball	C-Scale 150 Kg Brale	Superficial 30 N		
....	....	86.5	....	70.0	86.0	1076	....
....	....	86.0	....	69.0	85.0	1004	....
....	....	85.6	....	68.0	84.4	940	....
....	....	85.0	....	67.0	83.6	900	....
....	757	84.4	....	65.9	82.7	860	....
2.25	745	84.1	....	65.3	82.2	840	....
....	722	83.4	....	64.0	81.1	800	....
....	710	83.0	....	63.3	80.4	780	....
2.35	682	82.2	....	61.7	79.0	737	....
2.40	653	81.2	....	60.0	77.5	697	....
2.45	627	80.5	....	58.7	76.3	667	323
2.50	601	79.8	...	57.3	75.1	640	309
2.55	578	79.1	....	56.0	73.9	615	297
2.60	555	78.4	....	54.7	72.7	591	285
2.65	534	77.8	....	53.5	71.6	569	274
2.70	514	76.9	....	52.1	70.3	547	263
2.75	495	76.3	....	51.0	69.4	528	253
2.80	477	75.6	....	49.6	68.2	508	243
2.85	461	74.9	....	48.5	67.2	491	235
2.90	444	74.2	....	47.1	65.8	472	225
2.95	429	73.4	....	45.7	64.6	455	217
3.00	415	72.8	....	44.5	63.5	440	210
3.05	401	72.0	....	43.1	62.3	425	202
3.10	388	71.4	....	41.8	61.1	410	195
3.15	375	70.6	....	40.4	59.9	396	188
3.20	363	70.0	....	39.1	58.7	383	182
3.25	352	69.3	(110.0)	37.9	57.6	372	176
3.30	341	68.7	(109.0)	36.6	56.4	360	170
3.35	331	68.1	(108.5)	35.5	55.4	350	166
3.40	321	67.5	(108.0)	34.3	54.3	339	160
3.45	311	66.9	(107.5)	33.1	53.3	328	155
3.50	302	66.3	(107.0)	32.1	52.2	319	150
3.55	293	65.7	(106.0)	30.9	51.2	309	145
3.60	285	65.3	(105.5)	29.9	50.3	301	141
3.65	277	64.6	(104.5)	28.8	49.3	292	137

Table A-D5. Hardness conversion *(continued)*

BRINELL HARDNESS		ROCKWELL HARDNESS				Diamond Pyramid Hardness Number (Vickers)	Approx. Tensile Strength 1000 psi
Diameter mm 3000 Kg	Tungsten Carbide 10 mm Ball	A-Scale 60 Kg Brale	B-Scale 100 Kg 1/16" Ball	C-Scale 150 Kg Brale	Superficial 30 N		
3.70	269	64.1	(104.0)	27.6	48.3	284	133
3.75	262	63.6	(103.0)	26.6	47.3	276	129
3.80	255	63.0	(102.0)	25.4	46.2	269	126
3.85	248	62.5	(101.0)	24.2	45.1	261	122
3.90	241	61.8	100.0	22.8	43.9	253	118
3.95	235	61.4	99.0	21.7	42.9	247	115
4.00	229	60.8	98.2	20.5	41.9	241	111
4.05	223	59.7	97.3	(18.8)	....	234	....
4.10	217	59.2	96.4	(17.5)	....	228	105
4.15	212	58.5	95.5	(16.0)	....	222	102
4.20	207	57.8	94.6	(15.2)	....	218	100
4.25	201	57.4	93.8	(13.8)	....	212	98
4.30	197	56.9	92.8	(12.7)	....	207	95
4.35	192	56.5	91.9	(11.5)	....	202	93
4.40	187	55.9	90.7	(10.0)	....	196	90
4.45	183	55.5	90.0	(9.0)	....	192	89
4.50	179	55.0	89.0	(8.0)	....	188	87
4.55	174	53.9	87.8	(6.4)	....	182	85
4.60	170	53.4	86.8	(5.4)	....	178	83
4.65	167	53.0	86.0	(4.4)	....	175	81
4.70	163	52.5	85.0	(3.3)	....	171	79
4.80	156	51.0	82.9	(.9)	....	163	76
4.90	149	49.9	80.8	....	....	156	73
5.00	143	48.9	78.7	....	....	150	71
5.10	137	47.4	76.4	....	....	143	67
5.20	131	46.0	74.0	....	....	137	65
5.30	126	45.0	72.0	....	....	132	63
5.40	121	43.9	69.8	....	....	127	60
5.50	116	42.8	67.6	....	....	122	58
5.60	111	41.9	65.7	....	....	117	56

Values in ( ) are beyond normal range and are given for information only.

The Brinell values in this table are based on the use of a 10mm tungsten carbide ball; at hardness levels of 429 Brinell and below, the values obtained with the tungsten carbide ball, the Hultgren ball, and the standard ball are the same.

The Hardness Conversion Tables are based on SAE J417 and ASTM E140.

Table A-D6. Work-energy conversion.

ft.-lb$_f$		joules	ft.-lb$_f$		joules
0.7376	1	1.356	37.6157	51	69.147
1.4751	2	2.712	38.3532	52	70.503
2.2127	3	4.067	39.0908	53	71.858
2.9502	4	5.423	39.8284	54	73.214
3.6878	5	6.779	40.5659	55	74.570
4.4254	6	8.135	41.3035	56	75.926
5.1629	7	9.491	42.0410	57	77.282
5.9005	8	10.847	42.7786	58	78.637
6.6381	9	12.202	43.5162	59	79.993
7.3756	10	13.558	44.2537	60	81.349
8.1132	11	14.914	44.9913	61	82.705
8.8507	12	16.270	45.7289	62	84.061
9.5883	13	17.626	46.4664	63	85.417
10.3259	14	18.981	47.2040	64	86.772
11.0634	15	20.337	47.9415	65	88.128
11.8010	16	21.693	48.6791	66	89.484
12.5386	17	23.049	49.4167	67	90.840
13.2761	18	24.405	50.1542	68	92.196
14.0137	19	25.761	50.8918	69	93.551
14.7512	20	27.116	51.6294	70	94.907
15.4888	21	28.472	52.3669	71	96.263
16.2264	22	29.828	53.1045	72	97.619
16.9639	23	31.184	53.8420	73	98.975
17.7015	24	32.540	54.5796	74	100.331
18.4391	25	33.895	55.3172	75	101.686
19.1766	26	35.251	56.0547	76	103.042
19.9142	27	36.607	56.7923	77	104.398
20.6517	28	37.963	57.5298	78	105.754
21.3893	29	39.319	58.2674	79	107.110
22.1269	30	40.675	59.0050	80	108.465
22.8644	31	42.030	59.7425	81	109.821
23.6020	32	43.386	60.4801	82	111.177
24.3396	33	44.742	61.2177	83	112.533
25.0771	34	46.098	61.9552	84	113.889
25.8147	35	47.454	62.6928	85	115.245
26.5522	36	48.809	63.4303	86	116.600
27.2898	37	50.165	64.1679	87	117.956
28.0274	38	51.521	64.9055	88	119.312
28.7649	39	52.877	65.6430	89	120.668
29.5025	40	54.233	66.3806	90	122.024
30.2400	41	55.589	67.1182	91	123.379
30.9776	42	56.944	67.8557	92	124.735
31.7152	43	58.300	68.5933	93	126.091
32.4527	44	59.656	69.3308	94	127.447
33.1903	45	61.012	70.0684	95	128.803
33.9279	46	62.368	70.8060	96	130.159
34.6654	47	63.723	71.5435	97	131.514
35.4030	48	65.079	72.2811	98	132.870
36.1405	49	66.435	73.0186	99	134.226
36.8781	50	67.791	73.7562	100	135.582

Examples: 1 ft-lb$_f$ = 1.356 joules
1 joule = 0.7376 ft-lb$_f$

GLOSSARY

Important Metallurgical Terms

Ageing: It is a time-temperature dependent process where changes in properties take place in certain steel types, such as mild and low-carbon steels. It is generally a result of precipitation from the matrix of a solid solution—e.g. austenite or ferrite in steel—whose solubility decreases with temperature. Steel with smaller solute atoms containing nitrogen has the marked tendency for ageing.

Allotropic Transformation: Allotropy refers to changes in atomic structure which occur at a definite transformation temperature. In steel four allotropic changes occur known as α, β, γ, and δ. Of these, α, β and δ are different allotropic forms of 'ferrite' with body centred cubic structure (BCC) and γ has face centred cubic (FCC) structure.

Annealing: Annealing is a thermal treatment consisting of heating uniformly to a temperature, within or above the upper critical range, and cooling at a slow but controlled rate to a temperature below the critical range, generally to room temperature. This treatment is used to produce a softer microstructure, usually designed for improved machinability, homogenisation of structure, and to reduce or nullify stress.

Austenite: A high temperature crystallographic phase of steel having 'face centred cubic' (FCC) arrangement of atoms in the lattice structure. Austenite is not stable at temperature below the 'lower critical temperature' of steel (around 723°C), unless the steel contains higher percentage of alloying elements like Cr, Ni, and Mn. It is non-magnetic in nature.

Austenitising: The process of producing 100% austenite in the steel structure by heating above the transformation temperature A_3 and holding it for the required time to transform to austenite.

Austempering: It is a process of quenching the steel from austenitising temperature to a temperature just above the martensite transformation temperature (M_S) in a suitable media having cooling rate higher than required critical cooling rate, and then holding at that temperature until transformation is complete. Generally, lower bainitic structure is obtained by this treatment to alloy steels and alloy cast irons.

Burning: It is a phenomenon of overheating of steel during rolling or heat-treating causing incipient melting at the grain boundaries or inter-granular oxidation. Burning causes permanent damage to the steel.

Carbon Potential: A term used to define the ability of an environment containing active carbon to contribute, alter and maintain a defined carbon content of the steel surface exposed to it under a given condition. In any particular environment, the carbon level attained by the surface will depend on such factors as temperature, time and steel composition.

Cementite: It is a chemical compound of iron (Fe) and carbon (C), expressed by chemical equation Fe_3C, containing carbon percent of 6.67%. It is a hard compound and can be present in steel in globular, lamellar or dendritic form.

Controlled Cooling: A term used to describe a process by which a steel object is cooled from an elevated temperature, usually from the final hot-forming or hot rolling operation in a predetermined manner of cooling to avoid hardening, cracking, internal damage or to produce a desired microstructural combination.

Corrosion: It is an electrochemical phenomenon by which metal surface gets attacked by atmospheric or acidic environment causing decay or damage to the surface. Presence of moisture and oxygen is a vital factor in the corrosion process. With salt solution and presence of electric current, corrosion process gets greatly accelerated.

Critical Cooling Rate: The cooling rate necessary for a steel to transform to 100%martensite upon quenching (see CCT-diagrams in the text). Critical cooling rate is dependent on the composition of the steel; higher the carbon and alloying (i.e. hardenability of the steel) lower is the critical cooling rate required for 100%martensite formation.

Critical Temperatures: Steel undergoes different allotropic changes during cooling or heating, which are associated with evolution of heat at the points where allotropic (i.e. phase change) changes take place. This causes a momentary arrest of fall of temperature at those points—called critical points or critical temperatures, denoted by A. Since steel undergoes number of allotropic changes during cooling or heating, there would be number of 'critical points', which are denoted by A. Of these critical points, A_3, marking the change of austenite (γ) to ferrite (β variety), is referred as 'upper critical temperature' (910°C on cooling) and A_1, marking the change from austenite to pearlite which occurs at lower temperature, is called the 'lower critical temperature' (695°C on cooling). Though critical points occur both on heating and cooling, there is minor difference in their practical values; values on heating are a bit more.

Decarburisation: The loss of carbon from the steel surface as the result of heating in a medium that reacts with the carbon in the steel. Higher the temperature, higher is the chance of decarburisation. Decarburisation results to loss of strength of the steel in the surface areas.

Ductility: It may be defined as the property which enables the steel to be drawn into wire by use of tensile force. Ductility is used as a measure of how much steel can deform plastically before fracture.

Eutectic: It is a composition in an alloy system, e.g. Fe-C system, where the alloy system solidifies at a lower temperature than any other combination of composition in the same system. The composition is called *eutectic composition* and the temperature is called the *eutectic temperature*. In Fe-C system, eutectic composition is about 4.3% carbon.

Eutectoid: It is a composition when similar analogous transformation occurs in solid state, rather than liquid, where one solid solution e.g. austenite in steel, transforms to other phases e.g. ferrite and carbide—forming pearlite, at a temperature lowest in the system. In steel, eutectoid composition in plain carbon steel is about 0.85% carbon. This might change with alloy content in steel.

Ideal Critical Diameter (di): The diameter of a round steel bar that will harden at the centre to a given percentage of martensite or a definite hardness value when subjected to an ideal quench—i.e. Grossman quench severity H = infinity.

Isothermal Annealing: A process involving austenitising and then cooling to and holding at a temperature where austenite transforms to a relatively softer ferrite-carbide structure. The isothermal holding temperature is generally below the lower critical temperature of the steel.

Isothermal Transformation: A change in phase by transformation of austenite at constant temperature.

Elongation: In tensile testing, the increase in gauge length, measured after the fracture of a specimen within the gauge length; usually expressed as a percentage of the original gauge length. This is taken as measure of ductility in the steel.

Fatigue: Fatigue is the progressive and localised structural damage leading to fracture which occurs when a material is subjected to cyclic loading. Fatigue damages occur at stress level less than the tensile stress limit of the steel and can be even below the yield stress level.

Fracture Toughness: It is a property which describes the ability of the steel containing a pre-existing crack to resist fracture. Since all materials will have some kind of pre-existing flaws, which are not exactly avoidable—some fracture toughness is necessary to avoid catastrophic failure, especially in high strength steels. A parameter called the stress-intensity factor (K_t) is used to determine the fracture toughness of most materials, and the value is expressed in the unit: $MPa\sqrt{m}$ ($psi\sqrt{in}$)

Free Energy: A thermodynamic quantity that defines the change in energy between the internal energy of the system and the product that forms from the system at standard temperature and pressure (STP). If free energy change is negative i.e. energy is lowered due to the product formation, the process is thermodynamically favourable. It is also called 'Gibbs Free Energy'.

Hardness: Resistance of a metal to plastic deformation, usually by indentation. However, this may also refer to stiffness or temper or to resistance to scratching, abrasion or cutting.

Hardenablity: It a measure of the property of steel that determines the depth and distribution of hardness obtained by quenching. Hardenability of steel depends on the carbon and alloy content in the steel; higher carbon and alloy contributes to higher hardenability of the steel. It is not a measure of hardness, but the depth and distribution of a pre-fixed minimum hardness value measured by Jominy Test.

Impact Test: It is a test to determine the behaviour of materials when subjected to high rates of loading, usually in bending, tension or torsion. The parameter measured is the energy absorbed in breaking the specimen by a single blow, as in the Charpy or Izod tests.

Killed Steel: Steel treated with a strong deoxidizer—like Al,Si, etc.—to reduce oxygen to a level where no reaction occurs between carbon and oxygen during solidification.

Lever Rule: It is a tool used to determine weight % of each phase in a binary (two component system) equilibrium phase diagram. In an alloy with two phases, α and β, which themselves contain two elements, A and B, the lever rule states that the weight percentage of the α phase is:

$$X_a = \frac{c - b}{a - b}$$

where at some fixed temperature:

a = the weight percentage of element B in the α phase,
b = the weight percentage of element B in the β phase, and
c = the weight percentage of element B in the entire alloy.

Machinability: This is a generic term used for describing the ability of a material to be machined. To relate to practice, machinability is qualified in terms of tool wear, tool life, chip control, and/or surface finish, and a machinability index value is assigned. Overall machining performance is affected by a series of variables relating to the machining operation and the work piece microstructure.

Malleability: It is the property of being permanently extended in all directions when subjected to compressive force.

Martempering: A hardening method in which the austenitised steel piece is quenched into an appropriate medium maintained at around M_S temperature of the steel and held in the medium until the temperature in the piece is uniform throughout the body but not long enough to form bainite, and then cooled in air. Aim of the process is to form 100% martensite in high alloy steels without cracking or distortion due to residual stresses. Martempering should follow by tempering.

Normalizing: It is a thermal treatment consisting of heating uniformly to temperature at least $30°C$ above the critical temperature range and cooling in still air at room temperature. The treatment produces recrystallisation and refinement of the grain structure and gives uniformity in hardness and structure to the product. Strength of the steel also increases due to refinement of grains.

Pickling: An operation by which oxide scale of the surface which gets formed during heating or holding to higher temperature is removed by chemical action. Sulphuric acid is typically used for carbon and low-alloy steels; after the acid bath, the steel is rinsed in water.

Quenching: A treatment consisting of heating uniformly to a predetermined temperature and cooling rapidly in air or a suitable fluid medium to produce one or a set of desired microstructure structure. This is a critical step for successful hardening of steels.

Recrystallisation: It is a thermal process by which deformed grains are replaced by a new set of un-deformed grains that nucleate and grow from the original grains until the original grains have been entirely consumed. Recrystallisation temperature can vary from below the upper critical temperature to the lower critical temperature, depending on the degree of prior working and composition of the steel.

Reduction of Area (Ra): The difference, expressed as a percentage of original area, between the original cross-sectional area of a tensile test specimen and the minimum cross-sectional area measured after complete separation of test piece by fracture. This is a measure of ductility / toughness in steel.

Residual Stress: Macroscopic stresses that are set up within a metal as a result of non-uniform plastic deformation. The deformation can be caused by hot-working, cold working or by drastic temperature gradients created by accelerated cooling operations such as quenching. Residual stresses so formed are tensile in nature and add up with the applied load in service.

Rimmed Steel: It is a low carbon steel having enough oxygen in the melt to form iron oxide, causing continuous evolution of carbon monoxide gas from the reaction of iron oxide with carbon at higher temperature. This gas evolution causes deep seated blow-holes to form below the surface, counteracts the shrinkage and gives rise to 'rim' of pure metal free from voids in the surface area. Rimmed steel has low carbon rim, is free from impurities near the surface, but uniformly dispersed blow-holes and impurities in the core.

Scale: Oxidation of metal due to heat, resulting in relatively heavy surface layers of oxide.

Seam: A defect on the surface of steel which appears as a crack. Seams appear to form during casting due to burst of the surface and form seam during rolling. Seams are located on the surface of the steel that has been covered during rolling but does not get welded into the original metal. They are generally formed by a pre-existing defect produced during casting or cold/hot working (i.e. laps). Seams act like cracks on the surface of the steel and bad for upsetting and forging.

Secondary Hardening: An increase in hardness observed in the tempering of steels that occurs during the tempering of certain alloy steels (containing Mo and W) at temperatures ranging from 500°C to 650°C.

Semi-Killed Steel: Incompletely deoxidized steel which contains enough dissolved oxygen to react with the carbon to form carbon monoxide to offset solidification shrinkage.

Shear Stress: It is defined as the component of stress acting coplanar with the material; it is expressed as τ (Tau). It is the force tending to cause deformation of a material by slippage along a plane or planes parallel to the imposed stress.

Soaking: Prolonged heating of steel at a selected temperature in order to ensure it has achieved a uniform temperature throughout the entire cross section.

Spheroidise Anneal: A special type of annealing carried out below the lower critical temperature of the steel for spheroidising the carbides in the microstructure, thereby softening the steel. This treatment generally requires long cycle, but produces globular carbides and maximum softness required for good machinability or for improving cold formability.

Strain: The elastic or plastic deformation of steel caused by stress. Within the elastic limit in steel, strain is proportional to stress.

Strain Hardening: An increase in hardness and strength in steel caused by plastic deformation at temperatures which could be anywhere between room-temperature to lower than the recrystallisation range of the material.

Stress Relieving: A thermal treatment to relieve the residual stress produced during heating and cooling of steels, thereby restoring elastic-plastic properties in the steel. The operation minimises distortion on subsequent machining or hardening operations. This treatment is usually applied to material that has been heat treated—quenched and tempered. Normal practice is to heat and hold the steel below the conventional tempering temperature. Ordinarily, no straightening is performed after the stress relieving.

Tempering: A thermal treatment process consisting of heating uniformly to some predetermined temperature under the lower critical temperature range, holding at that temperature for a designated period of time and then cooling in air or liquid. This treatment is used after hardening of steel in order to take out internal stress associated with martensite formation and to induce formation of fine carbides out of the carbon that comes out of martensite due to mobility of carbon atoms at the tempering temperature. The operation is intended to produce toughness in the hardened structure of steels.

Temper Brittleness: A phenomenon observed in certain steels when cooled slowly through the temperature range of $400°C–550°C$ after tempering from above $>600°C$ or soaked at that range of temperature for long. Temper brittleness is believed to be due to grain boundary segregation of embrittling elements like lead, tin, antimony, arsenic, etc.

Tensile Strength: It is the level of maximum stress (maximum load in tensile testing) which the steel can withstand without giving into fracture, expressed as the ratio of maximum load (stress) to original cross sectional area in unit Kgf/mm^2.

Texture: It is the distribution of crystallographic orientations of a polycrystalline material. Material where crystal orientations are fully random, it is said to have no texture. And where orientations are not random and have some preferred orientation, it is said to have weak, moderate or strong texture—the degree is depending on the percentage of crystals having preferred orientation. Texture can have considerable influence on forming behaviour of steels.

Toughness: It is the ability of steel to deform plastically and to absorb energy in the process before fracture. The key to toughness is a good combination of strength and ductility; just because a material is ductile does not make it tough. There are several variables that have a profound influence on the toughness of a material. These variables are: strain rate (rate of loading), temperature, and notch effect.

Under Cooling: A phenomenon observed during solidification of metals when a difference between so-called 'freezing temperature' and actual 'nucleation temperature'—necessary for starting the solidification process—is observed; the latter being lower than the former. The degree of difference—which may range from fraction of a centigrade to few degrees of centigrade—is called the under-cooling effect. Under-cooling occurs due to difficulty of momentary nucleation on reaching the freezing temperature during rapid cooling in the absence of any foreign particle or initial nucleating points in the steel. Faster the cooling more is the under-cooling.

Widmanstatten Ferrite: A plate or lath like ferrite microstructure that forms when steels are cooled at a faster rate from the higher temperature, showing a cross-hatched appearance due to ferrite formation along certain crystallographic planes. Widmanstatten ferrite can form close to upper critical temperature with very low under-cooling effect, growing directly from austenite grain boundaries or it can also form at relatively lower temperature from any allotropic ferrite present in the structure. Both types may have similar appearance, but the latter is called secondary widmanstatten ferrite.

Work Hardening: An increased resistance to deformation due to increase in hardness produced by cold working.

Wrought Steel: Steel that has undergone some hot deformation after casting e.g. by thermo-mechanical working like rolling, forging, etc.

Yield Point: A point in the stress–strain curve of steel where the first inflexion in the rising stress strain curve occurs, usually at less than the maximum attainable stress, at which an increase in strain occurs without an increase in stress. This point is generally associated in steel with decrease in stress after yielding, coming to lower value before starting to increase again. The former is called upper yield point and the latter is called lower yield point.

Yield Strength: The stress at which a material exhibits a specified deviation from proportionality of stress and strain. An offset of 0.2% is commonly used to mark the 'yield strength', where there is no sharp yield point.

BIBLIOGRAPHY

1. United States Steel, USA, *Making, Shaping and Treating of Steels*, 10th edition. AISE,
2. Dieter, George E. *Mechanical Metallurgy*. McGraw-Hill, 1961.
3. Rollason E.C. *Metallurgy for Engineers*. The English Language Book Society & Edward Arnold (Publishers) Ltd., 1973.
4. Reed-Hill, Robert E. *Physical Metallurgy Principles*. Van Nostrand Reinhold Co., 1973.
5. Llewellyn, D.T. and Hudd B.C. *Steels: Metallurgy and Applications*. 3rd edition. Butterworth-Heinemann, 1998.
6. Smallman, R.E. *Physical Metallurgy and Advance Materials*. Butterworth-Heinemann, 2007.
7. Bhadeshia, H. and Honeycombe, R. *Steels: Microstructures and Properties*. Oxford: Butterworth-Heinemann, 2006.
8. Raghavan, V. *Physical Metallurgy: Principles and Practice*. New Delhi: PHI Learning, 2006.
9. Leslie, W.C. *The Physical Metallurgy of Steels*. Hemisphere Press, McGraw-Hill
10. Stephens, Ralph I. and Henry, O. *Metal Fatigue in Engineering*. New York: John Wiley & Sons, 2001.
11. Kiessling, Ronald. *Non-Metallic Inclusions in Steels*. Iron & Steel Institute: UK, 1965.
12. Nayar, Alok. *Testing of Metals.*. New Delhi: Tata McGraw-Hill, 2005.
12. Price, David, ed., *Ironmaking and Steelmaking: Processes, Products and Applications*. Institute of Materials, Minerals and Mining, Vol. 39 (2012).
13. Ghosh, Ahindra and Chatterjee, Amit. *Ironmaking and Steelmaking: Theory and Practice*. New Delhi: PHI Learning, 2011.
14. Ginzburg, Vladimir B. and Dekker M. *Steel-Rolling Technology: Theory and Practice*. 1989
15. UK Steel, *Steel Specification Handbook*, 13th edition. London:
16. British Standard Institution. *Summary of British and American Standard Specifications*, London:
17. Timken Steel Ltd. *Timken Practical Data for Metallurgists*, 17th edition, Canton, USA.
18. British Steel Corporation. *Iron & Steel Specifications*, London:
19. Bringas, John E. *Handbook of Comparative World Steel Standards*, 3rd edition. ASTM,
20. Totten, George E., ed., *Steel Heat-Treatment Handbook*, 2nd edition. CRC Press, 2006.
21. Brooks, Charlie R. *Principles of Heat-Treatment of Plain Carbon and Low-Alloy Steels*. ASM International, USA, 1996.
22. *'Stahlschlussel' (Key to Steel)*. Springer Verlag, 2006.
23. Nadkarni, S.V. *Modern Arc Welding Technology*. Oxford & IBH Publishing Co. Pvt. Ltd., 1988.
24. Sharma, Ramesh C. *Principles of Heat-Treatment of Steels*. Delhi: New Age International, 2007.
25. Campbell, F.C., ed. *Fatigue and Fracture: Understanding the Basics*. ASM International,
26. Reardon, Arthur C. *Metallurgy for the Non-Metallurgists*. ASM International,
27. Nayar, Alok. *The Steel Handbook*. New Delhi: McGraw-Hill, 2001.
28. ASM Handbook, Vol. 1, *Properties and Selection: Irons, Steels, and High Performance Alloys*. ASM International, USA, 1990.
29. ASM Handbook, Vol. 4, *Heat Treating*. ASM International, USA, 1991.
30. ASM Handbook, Vol. 5, *Surface Engineering*. ASM International, USA, 1994.
31. ASM Handbook, Vol. 6, *Welding, Brazing and Soldering*. ASM International, USA, 1994.
32. ASM Handbook, Vol. 8, *Mechanical Testing and Evaluation*. ASM International, USA, 2000.
33. ASM Handbook, Vol. 11, *Failure Analysis*. ASM international, USA, 2002.

34. ASM Metals Handbook, Vol. 14, *Forming and Forging,* ASM International, USA, 2000.
35. ASM Handbook, Vol. 17, *Non-Destructive Evaluation and Quality Control.* 1989.
36. ASM Handbook, Vol. 19, *Fatigue and Fracture, ASM International.* ASM International, USA, 1997.
37. ASM Handbook, Vol. 20, *Materials Selection and Design.* ASM International, 1997.
38. ASM International, USA. *Worldwide Guide to Equivalent Irons and Steels,* 5th edition. 2006.

INDEX

DATE DUE